Biochemical Aspects of
Plant–Parasite Relationships

Annual Proceedings of the Phytochemical Society

Biochemical Aspects of Plant-Parasite Relationships

PROCEEDINGS OF THE
PHYTOCHEMICAL SOCIETY SYMPOSIUM
UNIVERSITY OF HULL, ENGLAND
APRIL, 1975

Edited by

J. FRIEND and D. R. THRELFALL

*Department of Plant Biology,
University of Hull, England*

1976

ACADEMIC PRESS
LONDON   NEW YORK   SAN FRANCISCO

A Subsidiary of Harcourt Brace Jovanovich, Publishers

ACADEMIC PRESS INC. (LONDON) LTD.
24/28 Oval Road,
London NW1

United States Edition published by
ACADEMIC PRESS INC.
111 Fifth Avenue
New York, New York 10003

Library of Congress Catalog Card Number: 76-27189
ISBN: 0-12-267950-4

PRINTED IN GREAT BRITAIN BY
WILLIAM CLOWES & SONS LIMITED
LONDON, BECCLES AND COLCHESTER

Contributors

D. F. Bateman, *Department of Plant Pathology, Cornell University, Ithaca, New York 14853, U.S.A.* (p. 79).

J. A. Callow, *Department of Plant Sciences, University of Leeds, Leeds, England* (p. 305).

J. R. Coley-Smith, *Department of Plant Biology, The University, Hull, England* (p. 11).

W. W. Currier, *Department of Plant Pathology, Montana State University, Bozeman, Montana, U.S.A.* (p. 225).

J. M. Daly, *Laboratory of Agricultural Biochemistry, University of Nebraska, Lincoln, Nebraska 68503, U.S.A.* (p. 117).

B. J. Deverall, *Department of Plant Pathology and Agricultural Entomology, University of Sydney, N.S.W. 2006, Australia* (p. 207).

J. Friend, *Department of Plant Biology, The University, Hull, England* (p. 291).

D. S. Ingram, *Botany School, Downing Street, Cambridge, England* (p. 43).

R. Johnson, *Plant Breeding Institute, Cambridge, England* (p. 25).

H. W. Knoche, *Laboratory of Agricultural Biochemistry, University of Nebraska, Lincoln, Nebraska 68503, U.S.A.* (p. 117).

J. Kuć, *Department of Plant Pathology, University of Kentucky, Lexington, Kentucky, U.S.A.* (p. 225).

J. C. Overeem, *Institute of Organic Chemistry, T.N.O., Utrecht, Netherlands* (p. 195).

T. F. Preece, *Agriculture Building, University of Leeds, Leeds, England* (p. 1).

S. G. Pueppke, *Department of Plant Pathology, Cornell University, Ithaca, New York 14853, U.S.A.* (p. 239).

D. J. Royle, *Department of Hop Research, Wye College (University of London), near Ashford, Kent, England* (p. 161).

J. A. Sargent, *A.R.C. Unit of Developmental Botany, 181A Huntingdon Road, Cambridge, England* (p. 43).

M. J. Shih, *Department of Biological Sciences, Simon Fraser University, Burnaby 2, B.C., Canada* (p. 225).

G. A. Strobel, *Department of Plant Pathology, Montana State University, Bozeman, Montana 59715, U.S.A.* (p. 135).

I. C. Tommerup, *A.R.C. Unit of Developmental Botany, 181A Huntingdon Road, Cambridge, England* (p. 43).

H. D. VanEtten, *Department of Plant Pathology, Cornell University, Ithaca, New York 14853, U.S.A.* (p. 239).

R. K. S. Wood, *Department of Botany, Imperial College of Science and Technology, London, England* (p. 105).

Preface

In recent years there has been increasingly more emphasis on the use of biochemical methods and techniques for research in plant pathology and it has now become possible to offer biochemical explanations for several phytopathological phenomena.

It was the intention of the organizers of the Phytochemical Society Symposium held at Hull in April, 1975, on Biochemical Aspects of Plant Parasite Relationships to display some of the more biochemical of the recent research, particularly on the mechanisms involved in the invasion of plants by pathogens, the production of disease symptoms, and the mechanisms involved in the resistance of plants to the invading microorganisms.

Papers on the genetics of fungal–plant interactions and on structural features both of infection and of resistance are included in the volume. The reason for this inclusion is that there has been a tendency for plant bio-chemists to neglect both structure and genetics as aspects of their investigations and yet it is often through an understanding of the structural and genetical basis of the plant–parasite interaction that a sensible biochemical explanation can be given.

For various reasons it was not possible to produce this volume as soon after the Symposium as has been common for previous volumes of proceedings of the Phytochemical Society. However the authors have revised their manuscripts during the period of enforced delay.

October, 1976   J. FRIEND
                                                             D. R. THRELFALL

Contents

CHAPTER 1

Some Observations on Leaf Surfaces During the Early Stages of Infection by Fungi

T. F. Preece

CHAPTER 2

Some Interactions in Soil Between Plants, Sclerotium-forming Fungi and other Microorganisms

J. R. Coley-Smith

CHAPTER 3

Development and Use of some Genetically Controlled Lines for Studies of Host–Parasite Interactions

R. Johnson

CHAPTER 4

Structural Aspects of Infection by Biotrophic Fungi

D. S. Ingram, J. A. Sargent and I. C. Tommerup

CHAPTER 5

Plant Cell Wall Hydrolysis by Pathogens

D. F. Bateman

CHAPTER 6

Killing of Protoplasts

R. K. S. Wood

CHAPTER 7

Hormonal Involvement in Metabolism of Host–Parasite Interactions

J. M. Daly and H. W. Knoche

CHAPTER 8

Toxins of Plant Pathogenic Bacteria and Fungi

Gary A. Strobel

CHAPTER 9

Structural Features of Resistance to Plant Diseases

D. J. Royle

CHAPTER 10

Pre-existing Antimicrobiol Substances in Plants and their Role in Disease Resistance

J. C. Overeem

CHAPTER 11

Current Perspectives in Research on Phytoalexins

B. J. Deverall

CHAPTER 12

Terpenoid Phytoalexins

J. Kuć, W. W. Currier and M. J. Shih

CHAPTER 13

Isoflavonoid Phytoalexins

H. D. VanEtten and S. G. Pueppke

CHAPTER 14

Lignification in Infected Tissue

J. Friend

CHAPTER 15

Nucleic Acid Metabolism in Biotrophic Infections

J. A. Callow

CHAPTER 1

Some Observations on Leaf Surfaces During the Early Stages of Infection by Fungi

T. F. PREECE

Agriculture Building, The University of Leeds, Leeds, England

I. INTRODUCTION

It is sometimes difficult for plant pathologists to see relevance to controlling plant disease in elegant and detailed basic studies of particular host–parasite interactions, perhaps concerned with membrane damage or changes in the nucleic acids of host plants. I personally accept without reservation the definition of plant pathology given by Moore (1949), which is that it is the job of plant pathologists to "influence the practices of crop husbandry". Crop husbandry goes on in fields, orchards, forests and market gardens. Although more protected kinds of crop husbandry are found in glasshouses and special structures such as mushroom houses, it does not take place in laboratories. One way of reminding ourselves of the essentials of our subject is to look at diseased plants in the field, and in particular to examine, microscopically, the pre-penetration stages of disease, under field conditions. We would then be assured that we know more of the disease before trying to understand it and before attempting control measures.

However, that is not to say that current effort in physiological and bio-chemical plant pathology is not essential for further progress. Of course it is; what is more, some of our models of the host–parasite interaction are in an exciting stage of development at the present time, as the other contributions in this volume will testify.

II. THE PRE-PENETRATION STAGES OF INFECTION OF LEAVES

Almost all of the current basic work in plant pathology is concerned with events which occur *after* parasites have penetrated their hosts. *Before* these later stages of disease can occur, the early pre-penetration stages must be successful. Earlier plant pathologists, such as Marshall Ward (Large, 1940) clearly grasped, and exploited, the fact that during the arrival, adhesion and external growth stages fungal spores on leaves are the most vulnerable to changes in the environment and in particular to the action of fungicidal sprays. The spore of a pathogenic fungus arriving at a leaf surface is a living organism in a most delicate phase of its life history, struggling for existence. It eventually has to make contact with host cell membranes if it is to achieve the possibility (at least) of a more compatible, safe environment; alternatively, for the expression of the host genotype–fungus genotype interaction.

What factors interplay in microevents prior to penetration? Do the events on leaf surfaces differ from the early stages of growth from fungal spores on artificial surfaces? We, as good experimental plant pathologists, use apparently clean fungal spore suspensions and leaves grown in closely controlled environments. But the leaves of crop plants in the field exposed to the weather are much more "messy" than the simple models we may have in our minds (Preece, 1963). The natural history of field infection, and in particular of the pre-penetration stages of infection, is complex, and awkward to handle experimentally, but are we "missing the wood whilst looking at the trees"? This is an extreme suggestion, but, if it generates observation followed by experiments, then it is justified. Our minds are conditioned. Two things especially have helped to induce the idea that the early microscopic stages of infection in the field do not need modern work. These are (1) the comforting generalized notion of a "typical" infecting spore, and (2) the ease with which many spores germinate in water on glass slides. It is a fact that what we know of the earliest stages of infection in particular diseases—the pre-penetration stages—often depends on a single drawing or photomicrograph in a paper concerned with the earliest attempts to control a disease, usually by a field plant pathologist working against time. It is time for a reexamination (with the light microscope) of the details of leaf surface phenomena in infections under field conditions, prior to more sophisticated work. Modern analysis of the various leaf surface environmental factors is needed, as is a more detailed study of the microbial components which we now know are present on every leaf (Last, 1971). With further work we might discern patterns in the "Achilles heel" pre-penetration biology of fungal diseases. These might well be general patterns or be generalizations associated with particular groups of diseases, or hosts, or parasites, or environmental situations. It seems likely that much is about to be discovered about the leaf surfaces during the early stages of infection. In the most recent issue of the "Annals of Applied Biology", for example,

Russell (1976) reports on the significance of mere position on the leaf surface of wheat in the germination of *Puccinia striiformis* uredospores. (The percentage germination was higher on the adaxial surface, particularly on the distal parts, than on the abaxial surface of leaves of adult wheat plants.)

I have been asked to present here some of the particular contributions my research students have made to our picture of leaf pre-penetration biology. In doing so I would like to emphasize that not only is our ignorance immense, but also that whole areas of questioning are neglected here. Interactions between the phyllophane microflora and pathogens are discussed elsewhere (Preece and Dickinson, 1971; Dickinson and Preece, 1976). Light effects on spore germination on leaves need a separate review; present indications are that light may be much more important than hitherto suspected. We are now studying the effects of chemical additives to the leaf surface environment, whether by accident (e.g. pollutants, dusts) or by design (e.g. pesticides, fertilizers). The more obvious (but little studied) animals and their products at the leaf surface (e.g. the microbiological effects of the movements of insects) need study. I omit in this account too, considerations of viruses, bacterial infections and actinomycetes.

For the development of a fungal lesion on a leaf we need a source of spores, and an available leaf. Then follows the external pre-penetration stage. Penetration must occur. There must be internal development of the fungus, followed by release of spores from the lesion. Ultimately we might consider the fate of materials in the lesion (the death of the fungus in the leaf included). Our knowledge of these phases varies. I am concerned in this paper with some aspects of the external pre-penetration stage which includes (1) the arrival of the spore, (2) adhesion to the leaf surface and often (3) external growth prior to penetration. This external growth may (or may not) show each of the common morphologically definable phases of swelling, germ tube production and appressorium formation. We need to focus on where these stages occur on the leaf surface, how long each stage takes to occur, and what environment conditions prevail during each stage. The (apparently) saprophytic microflora (Preece and Dickinson, 1971; Dickinson and Preece, 1976) is part of the microenvironment of the arriving spores. Together with this microflora there may be unexpectedly significant objects —also part of the microenvironment of the pathogen, such as pollen grains (Chou and Preece, 1968).

III. The Arrival of Spores on Leaves

The numbers of airborne spores of particular fungi near a leaf out-of-doors is astoundingly variable with time, as the quantitative measurement of air spora using the Hirst (1952) spore trap reveals. Meredith (1966) noted that airborne conidia of *Helminthosporium* did not exceed three per cubic metre above affected plants, whereas Shanamuganathan and Arulpragasam (1966) working in tea fields, found concentrations of 10 000 basidiospores of *Exobassidium* per cubic metre above bushes affected by blister blight. Hirst (1953) reported the first quantitative records of diurnal patterns of spores in the air, in this finding very marked differences between fungal species. There may be, for example, distinct "wet" and "dry" period situations. Spores of *Erysiphe*, *Alternaria* and *Cladosporium* are dry air spores; Ascospores, such as those of *Venturia* and *Ophiobolus* are constituents of the damp air spora. The ascospores of *Mycosphaerella melonis* are found in highest concentration inside glasshouses when it rains outside (Fletcher and Preece, 1966).

The processes of change in concentration of spores near leaves are complex and little understood. Gregory (1961) gives much information and considers problems of movement in the air and deposition on to leaves by sedimentation and impaction. Turbulence is very important—it is however possible for sedimentation to occur in a moving air stream if it is non-turbulent (Chamberlain, 1967). As a non-turbulent air mass moves over the crop, particles such as rust spores sediment down to the laminar boundary layer of still air surrounding all objects such as leaves. Some spores will then penetrate the boundary layer—others float away from the leaf. Impaction is more complex. Efficiency of arrival at the leaf surface falls off with decreasing spore size, and increases with reduction in width of leaf. Efficiency of impaction is low on dry leaves. "Collection" of spores on natural surfaces is better, for example, than on sticky slides or tape. Rishbeth (1959) found maximum arrival rates of 20 spores of *Fomes* per 100 cm^2 of tree stump surface per hour, and much more commonly recorded 1–5 spores/100 cm^2 per hour deposited quite near sporophores. Barnes (1969) at Leeds compared the number of conidia of *Erysiphe polygoni* arriving on the surface of red clover leaves after exposure near a source of spores in 24 h periods with the atmospheric concentration of spores near the leaves recorded by a Hirst spore trap. These "airborne" and "deposited on leaves" counts are not related directly and there is much to be investigated. Numbers of *Erysiphe* conidia, deposited on clover leaves were often low when trap catches were high. The maximum daily count recorded during a twelve-month study was of 56 per leaf; many daily values being between 0 and 10 per leaf. Relatively high counts of powdery mildew conidia on leaves coinciding with low trap catches were a notable feature of the observations by Barnes. In the case of already infected plants, the number of spores trapped by leaves is directly related to the number of mature sporulating lesions, as was shown in *Exobasidium vexans* infections of tea (Kerr and Rodrigo, 1967). These authors also reported on unexplained greater deposition on more susceptible cultivars. Bock (1962) showed that the final distribution of infecting uredospores of *Hemiliea vastatrix* on coffee leaves can be related to daily rainfall amounts. It becomes clear that each host–parasite–environment situation needs separate study inasfar as arrival of spores on leaves is concerned. It is important to realize, too, that under field conditions it may be essential for adequate numbers of viable spores to arrive in close proximity to each other on the leaf surface to allow infection to occur at all. A single *Botrytis cinerea* conidium will not infect a bean leaf; the normal "inoculum threshold" is 160 conidia in a droplet of water (Chou and Preece, 1968). In some cases the inoculum threshold has not been determined satisfactorily in particular interactions. We do not yet understand it, but we cannot ignore it. Is it a nutritional phenomenon? Old *B. cinerea* conidia (i.e. more than four weeks old) cannot normally be germinated on bean leaf surfaces (Chou and Preece, 1968). Recording the arrival of the appropriate number of spores is not enough, difficult though this is; they must be viable and be able to infect. More detailed work is essential on many pathogens.

IV. ADHESION OF SPORES ON LEAVES

The question of particle retention on natural surfaces has received little study. Not only must spores arrive, they must stick in position. Kerling (1958) has shown that a decrease in spore numbers on leaves of sugar beet between two sample dates was the result of heavy rainfall just before the second sample was taken. On the other hand, many of the various components of the microflora of apple leaves (Preece, 1963) seemed to be firmly attached to the leaf, as they were not removed during many washes in a complex periodic-acid-Schiff staining and washing procedure (Preece, 1959). Barnes (1969) showed that rainfall, air turbulence and leaf movement brought about removal of conidia of clover powdery mildew from leaf surfaces after arrival. "Mucilaginous substances" produced by spores are concerned with their attachment to leaves, and when the spore germinates the mucilage also can be often seen surrounding the germ tube. A good example is in *B. cinerea* conidia germination on bean leaves (Blackman and Welsford, 1916; McKeen, 1974). It is sometimes difficult to work at all with particular spores because they adhere so readily to surfaces. Ascospores of *Venturia inaequalis*, suspended in water for inoculation experiments, sediment on to glass surfaces inside bottles on the bench very rapidly, and are difficult to keep in satisfactory suspension. More information is needed about spore adhesion.

V. EXTERNAL GROWTH PRIOR TO PENETRATION

Deverall (1969) provides a bird's-eye view of the events which may influence spores after adhesion. Spores of obligately parasitic organisms such as rusts germinate without assistance from their hosts. Facultative parasites germinate much more vigorously if nutrients such as sugars and amino-acids are available, either from intact or damaged plant surfaces. Infection structures or appressoria (Emmett and Parbery, 1975) may or may not form on leaf surfaces, and the site of penetration may be very precise. Germination may be stimulated or inhibited in the "infection-drop" (of water) on the leaf surface. The physical nature of the surface may be important (Dickinson, 1974). It seems to me that only direct microscopy can eliminate serious errors of interpretation of events. For example, ascospores of *V. inaequalis* produce long germ tubes in distilled water. But it is not possible to find any germ tube formation in successful ascospore infections of leaves using enzymatically detached cuticles stained with phenol-acetic-acid-aniline blue (Preece, 1962). A small conical protrusion develops on one side of the spore, and direct penetration of the cuticle occurs. On the other hand other fungi such as *Erysiphe polgyoni* conidia on red clover (*Trifolium pratense*) leaves produce both germ tubes and appressoria. In this latter case the site of penetration is only at the junctions between adjacent epidermal cells (Preece *et al.*, 1967). However, when *Erysiphe cruciferarium* germ tube growth and penetration were studied on swede (*Brassica napus*) leaves it became clear that they penetrated the outer, periclinal walls of the epidermal cells (Purnell, 1971a). A variety of direct microscopical procedures is available (Preece, 1971) including, of course, transmission and scanning electron microscopy. Following direct light microscopy, C. K. Chou (1970) was able to confirm that *Peronospora parasitica* forms appressoria over and penetrates only between adjacent epidermal cells of cabbage (*Brassica oleracea*). The transmission electron microscope (T.E.M.) pictures showed that after penetrating the cuticle between these cells the fungus actually enters the plant by penetrating the anticlinal wall of an epidermal cell. The clean fracture of the cuticle, Chou suggests, makes it likely that penetration is a physical process. The question of whether penetration is mechanical or chemical in nature is still unresolved. McKeen (1974), however, after showing esterase activity by the tips of hyphae of germinating *B. cinerea* on bean leaves, suggested that the clean passage of the germ tube through the cuticle may favour an enzymatic interpretation. Following the first use on leaf surfaces of the scanning electron microscope (S.E.M.) for investigating the growth of *Erysiphe* on clover leaves at Leeds (Barnes and Neve, 1968) the use of the critical point drying method has given much improved pictures of the early stages of leaf infection. The pictures of flagellate zoospores settling on and penetrating the stomates of hop leaves (*Humulus lupulus*) by Royle and Thomas (1971) are possibly the best S.E.M. pictures yet produced of infection in any disease. Leaf and root surfaces are not only stimulatory to the general growth of certain fungi such as *Rhizoctonia*, but also the precise form of the infection hyphae which develop on the surface of the host clearly depends in some instances on the presence of the living host surface or some exudate from it. The morphology of hyphal growth developing from spores of *Helminthosporium sativum* on wheat coleoptiles is dependent on some chemical property of the exudate; it can be reproduced on cellophane on exudate, but not on cellophane over distilled water or on killed epidermis over distilled water (Ullah and Preece, 1966).

VI. RAIN-WATER WASHING LEAF SURFACES

Wet films at the leaf surface can be shown to be necessary for the early stages of infection to occur. Most work by plant pathologists has concentrated on this positive effect of rain and dew. The length of time wetness persists on leaf surfaces can be recorded in field situations and simultaneous direct staining and microscopy used to define the hours during which infection takes place (Preece, 1964), but as well as providing an essential environmental component, rain falling on leaves has a leaching action. This negative effect—the removal of substances from the leaf into water films which then drip off, has long been known to occur. We have examined this effect, of negative removal of materials from leaves, on subsequent germination and growth of powdery mildew conidia on swede leaves (Purnell, 1971a). The quantities of wax, carbohydrates, amino-acids, organic acids and other materials removed from the leaf surfaces by rain can be measured, and (contrasting with the essential role of water films in diseases other than powdery mildews) the effect of this loss of materials on the early stages of infection on dry leaves examined by a variety of microscopical techniques including S.E.M. (Purnell, 1971b). Washing leaves with simulated rain for short periods caused a marked reduction in the number of subsequently deposited conidia which produced primary and secondary hyphae. However, when inoculation of leaves was delayed for five days after the "rain", germinating conidia showed a more enhanced production of hyphae than occurs on leaves which have not been rain-washed. Inoculation of washed leaves later than five days showed a much reduced production of hyphae. There was no regeneration of wax on the surface of washed fully expanded leaves, but the rain-removed carbohydrates were mostly replaced within seven days. It was convenient to use *B. cinerea* spores to assay rain-washing of leaves for biological activity. One hypothesis which explains these at first negative effects of rain-wash on spore germination (after a wetness period) is that two main interacting substances are washed off mature swede leaves when it rains. The first is likely to be carbohydrate in nature. Within five days after rain-washing, this is replaced from within the leaves, and subsequent increase in the number of conidia which produce primary and secondary growth results. The second substance is, it seems, an inhibitor of fungal growth, replaced more gradually on the leaf surface after washing.

Rain water itself contains calcium, sodium, magnesium and nitrogen in appreciable amounts; but not carbohydrate (Carlisle *et al.*, 1966). The natural history of spores germinating in water on leaves at any particular point in time, such as in *Botrytis* infections, will be affected by the final concentration of the materials in the infection droplet. This resultant concentration of interacting materials in water round spores will be made up principally from (1) those arriving on the spore surface itself, (2) those arriving in rainwater, (3) those which can diffuse out of the leaf and (4) those inhibitory or stimulatory materials which diffuse from nearby constituents of the microflora of the leaf. Of these, we have shown (3) to be more dependent on previous rain-water washing and its timing and duration, in relation to spore-arrival than was expected; (4) will depend on the make-up of the living and non-living components of the leaf surface microflora at that particular time and place on the leaf.

VII. POLLEN ON LEAVES AND INFECTION

Pollen grains are one of the commonest components of the plant surface microflora. Counts of 1 000 per cm^2 on apple leaves (Preece, 1963) and 3 000 per cm^2 on red clover leaves (Barnes, 1969) have been noted. Barnes also noted that whole anthers were quite common on the surfaces of clover leaves in the field. Jenkins (1974) has provided an illustration of such anthers on leaf surfaces of barley, associated with *Botrytis* infections. We have been able to demonstrate (Chou and Preece, 1968; M. Chou, 1970) that pollen grains have very marked effects on the early stages of infection of *Botrytis* species on broad bean leaves and strawberry petals and fruits. The effects can be summarized as:

1. increasing the speed and rate of spore germination;
2. restoring the germinability and infectivity of old spores;
3. reducing the infection threshold;
4. enhancing the speed and severity of infection;
5. increasing the virulence of *Botrytis* spp.

The active principle in pollen is water soluble, dialysable, heat stable and belongs to the cationic fraction. The chemical identity of the substance or substances has not been determined, but certain sugars, amino-acids, vitamins and organic acids tested have been shown not to be effective in affecting pathogen virulence. Orange juice has an effect on the early stages of infection of *Botrytis* which is indistinguishable from that caused by pollen grains. Frozen pollen grains are more effective than freshly collected pollen. Pollen diffusate in water is as effective as pollen grains, and this aqueous extract is still effective at a dilution of 1:10 000. Pollen grains from different species of plants have similar effects. Barnes (1969) has shown that the sites where pollen grains have been situated on leaves may be made out by S.E.M. after the grains have been removed. He reports that when examining the leaf surface microflora of red clover leaves in the field 80% of the germinated saprophytic spores on leaves were seen to have produced their hyphae round pollen grains or pollen grain sites.

The present position would appear to be as follows: *Botrytis* (Chou and Preece, 1968); *Alternaria* (Channon, 1970); *Fusarium* (Strange and Smith, 1971); *Phoma* (Warren, 1972) and *Helminthosporium* (Fokkema, 1971) infections have been clearly shown to be affected by pollen grains or entire anthers. There is no evidence of effects on rusts or powdery mildews, nor on either apple scab or potato blight. If we regard fungicides as "inoculum-reducing" it would seem useful to regard pollen grains as "inoculum increasing" in infections caused by facultatively parasitic fungi. It has been suggested that the chemical growth stimulants explaining the extreme susceptibility of wheat anthers to *Fusarium* are betaine and choline (Strange *et al.*, 1972). It is not yet clear whether these substances are active in pollen grains. Pollen enables *B. cinerea* to overcome the inhibitory action of Wyerone acid, an antifungal product of infected bean leaves (Mansfield and Deverall, 1971).

Much of fungal mycelium present on lemmas and paleas of normal healthy barley grains and within the pericarp of the barley caryopsis has been shown to be derived from the anthers which (Warnock, 1972) are infected very easily by saprophytic fungi such as *Cladosporium*. Much more work on the part played by pollen grains in infection and in the microbiology of the aerial surfaces of plants is needed, and there may be similar particles (as yet undiscovered) on leaves in specific crop situations.

REFERENCES

Barnes, G. (1969). "A micro-ecological study of fungi on the leaves of red clover". Ph.D. Thesis, University of Leeds.

Barnes, G. and Neve, N. F. B. (1968). *Trans. Br. mycol. Soc.* **51**, 811–812.

Blackman, V. H. and Welsford, E. J. (1916). *Ann. Bot.* **30**, 389–398.

Bock, K. R. (1962). *Trans. Br. mycol. Soc.* **45**, 63–74.

Carlisle, A., Brown, A. H. F. and White, L. J. (1966). *J. Ecol.* 54, 87–98.

Chamberlain, A. O. (1967). *In* "Airborne Microbes" (P. H. Gregory and J. L. Montheith, eds). Cambridge University Press, Cambridge and London.

Channon, A. G. (1970). *Ann. appl. Biol.* **65**, 481–487.

Chou, C. K. (1970). *Ann. Bot.* N.S. **34**, 189–204.

Chou, M. (1970). "Biological interactions on the host surface influencing infection by *Botrytis cinerea* Fr. and other fungi: Pollen Grains". Ph.D. Thesis, University of Leeds.

Chou, M. and Preece, T. F. (1968). *Ann. appl. Biol.* **62**, 11–22.

Deverall, B. J. (1969). "Fungal Parasitism". Edward Arnold, London.

Dickinson, C. H. and Preece, T. F. (1976). "Microbiology of Aerial Plant Surfaces". Academic Press, London and New York.

Dickinson, S. (1974). *Physiol. Pl. Path.* **4**, 373–377.

Emmett, R. W. and Parbery, D. G. (1975). *A. Rev. Phytopath.* **13**, 147.

Fletcher, J. T. and Preece, T. F. (1966). *Ann. appl. Biol.* **58**, 423–430.

Fokkema, N. J. (1971). *In* "Ecology of Leaf Surface Micro-organisms" (T. F. Preece and C. H. Dickinson, eds), p. 278. Academic Press, London and New York.

Gregory, P. H. (1961). "The Microbiology of the Atmosphere". Leonard Hill, London.

Hirst, J. M. (1952). *Ann. appl. Biol.* **39**, 257–265.

Hirst, J. M. (1953). *Trans. Br. mycol. Soc.* **36**, 375.

Jenkins, J. E. E. (1974). *Pl. Path.* **23**, 83–84.

Kerling, L. C. P. (1958). *Tijdsch PlZiekt.* **64**, 402–410.

Kerr, A. and Rodrigo, W. R. F. (1967). *Trans. Br. mycol. Soc.* **50**, 49–55.

Large, E. C. (1940). "The Advance of the Fungi". Jonathan Cape, London.

Last, F. T. (1971). *In* "Ecology of Leaf Surface Micro-organisms" (T. F. Preece and C. H. Dickinson, eds). Academic Press, London and New York.

Mansfield, J. W. and Deverall, B. J. (1971). *Nature, Lond.* **232**, 339.

McKeen, W. E. (1974). *Phytopathology* **64**, 461–467.

Meredith, D. S. (1966). *Phytopathology* **56**, 949–952.

Moore, W. C. (1949). *Ann. appl. Biol.* **36**, 295–306.

Preece, T. F. (1959). *Pl. Path.* **8**, 127–129.

Preece, T. F. (1962). *Nature, Lond.* **193**, 902–903.

Preece, T. F. (1963). *Trans. Br. mycol. Soc.* **46**, 523–529.

Preece, T. F. (1964). *Pl. Path.* **13**, 6–9.

Preece, T. F. (1971). *In* "Ecology of Leaf Surface Micro-organisms" (T. F. Preece and C. H. Dickinson, eds). Academic Press, London and New York.

Preece, T. F. and Dickinson, C. H. (1971). "Ecology of Leaf Surface Micro-organisms". Academic Press, London and New York.

Preece, T. F., Barnes, G. and Jill M. Bayley (1967). *Pl. Path.* **16**, 117–118.

Purnell, T. J. (1971a). "Environmental and Physiological studies of leaf infection of swede (*Brassica napus*) by *Erysiphe cruciferarium*. Opiz ex Junell". Ph.D. Thesis, University of Leeds.

Purnell, T. J. (1971b). *In* "Ecology of Leaf Surface Micro-organisms" (T. F. Preece and C. H. Dickinson, eds), pp. 269–275. Academic Press, London and New York.

Rishbeth, J. (1959). *Trans. Br. mycol. Soc.* **42**, 243–260.

Russell, G. E. (1976). *Ann. appl. Biol.* **82**, 71–78.

Royle, D. J. and Thomas, G. G. (1971). *Physiol. Pl. Path.* **1**, 345–349.

Shanamuganathan, N. and Arulpragasam, P. V. (1966). *Trans. Br. mycol. Soc.* **49**, 219–229.

Strange, R. N. and Smith, H. (1971). *Physiol. Pl. Path.* **1**, 141–150.

Strange, R. N., Smith, H. and Majer, J. R. (1972). *Nature, Lond.* **238**, 103–104.

Ullah, A. K. M. O. and Preece, T. F. (1966). *Nature, Lond.* **210**, 1369–1370.

Warnock, D. W. (1972). "A study of the distribution and development of fungal mycelium in grains of barley, *Hordeum sativum*". Ph.D. Thesis, University of Leeds.

Warren, R. C. (1972). *Neth. J. Pl. Path.* **78**, 89–98.

CHAPTER 2

Some Interactions in Soil Between Plants, Sclerotium-forming Fungi and other Microorganisms

J. R. COLEY-SMITH

Department of Plant Biology, The University, Hull, England

I. INTRODUCTION

The soil is an extremely complex medium which contains large and fluctuating populations of different microorganisms. These interact with each other and with the chemical and physical components of the soil. In the case of plant pathogens there is a wealth of information on such inter-actions (Baker and Snyder, 1965; Toussoun *et al.*, 1970; Baker and Cook, 1974) but despite this it has recently been claimed (Baker and Cook, 1974) that:

"Man does not know the total roster of microorganisms in a single tiny lump of soil, let alone their complex interactions with each other and with their physical and chemical environment".

In an attempt to clarify some of the confusion which existed about the biology of plant pathogens in soil, Park (1963) produced a simple scheme (Fig. 1) in which three components, the host plant (H), the pathogen (P) and the soil microbial population (S), interacted in a series of six pathways. The inorganic environment could affect each component directly or each pathway. In Park's scheme these pathways could, in theory, carry a positive (+), a neutral (0) or a negative (−) sign, although in some cases it was recognized that one of the pathways might generally be dominant. Many examples showing the operation of these pathways are given in Park's review. In addition the complexity of the situation in soil was contrasted with the simple two component host–pathogen interaction existing for most airborne diseases. The number of more complex interactions discovered for airborne diseases has increased dramatically since 1963 (see Chapter 1) but despite this the situation in soil is undoubtedly of a much higher order of complexity.

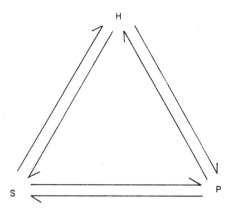

FIG. 1. Interactions involving soil pathogens (P), soil microflora (S) and host plant (H) (Park, 1963).

Many of the interactions described are still applicable today and there is little point in extending the list. However, in my work with sclerotium-forming fungi I have encountered two areas in which the situation has changed to some extent and this change has general implications as far as plant–pathogen interrelationships in soil are concerned. It is these two areas of research which I shall describe and discuss in this short review.

II. THE NATURE OF FUNGAL SCLEROTIA

The term "sclerotium" has been used to describe fungal bodies which may have diverse methods of development (Butler, 1966; Coley-Smith and Cooke, 1971; Willetts, 1971, 1972). In this review the term "sclerotium" will refer to a "multicellular resting body usually with a distinct outer layer, the rind, and an inner region, the medulla". The sclerotia of some fungi have a considerably more complex structure than this basic type however (Coley-Smith and Cooke, 1971). The rind consists of highly pigmented hyphae and has a protective function whereas the medulla is usually formed of colourless hyphae containing considerable quantities of food reserve materials. Sclerotia are the important perennating structures of a large number of plant pathogens and some have been shown to be very long lived (Coley-Smith and Cooke, 1971; Willetts, 1971).

Three broad types of germination are shown by fungal sclerotia. These are distinguished on the basis of the final structure produced. Sclerotia may thus be "myceliogenic" where individual hyphae or hyphal aggregates are formed, "sporogenic" where asexual spores are produced or "carpogenic" where complex carpophores arise from the sclerotium. Examples of these three types of germination are given by Coley-Smith and Cooke (1971). In general, however, it can be said that sclerotia of airborne fungi are either sporogenic or carpogenic whereas those of soil-borne root-infecting fungi (Garrett, 1970) are myceliogenic. It is the latter group to which this review will refer.

III. Host-stimulated Germination of Sclerotia

At the time of Park's (1963) review very few cases had been recorded where fungal propagules germinated in the presence of host plants but not in their absence or in the presence of non-host plants. Much of my own work in this area has involved the fungus *Sclerotium cepivorum* which in its host range is restricted to species of *Allium* in which it causes the disease known as white rot. This work has already been reviewed (Coley-Smith and Cooke, 1971) and I only propose to refer to the main findings here. The sclerotia of *S. cepivorum* are long lived (Coley-Smith, 1959) and do not normally germinate in natural soil because they are inhibited by the microbial population therein (Coley-Smith *et al.*, 1967). This involves the well known phenomenon of fungistasis (Dobbs and Hinson, 1953). The presence of *Allium* species overrides this inhibitory effect in some way and the sclerotia germinate. Germination of sclerotia in natural soil is thus a highly specific response to members of the host genus (Coley-Smith and Holt, 1966). If the fungus is removed from the restraining influence of the soil microflora then germination can occur with or without the presence of host plants or even without an exogenous supply of nutrients. The host-specific effect is a particular feature of natural soil. We have now found that the roots of *Allium* species exude small quantities of alkyl cysteine sulphoxides, methyl and propyl derivatives by onion, and methyl and allyl derivatives by garlic (Coley-Smith and King, 1969). These compounds are metabolized by bacteria in soil (King and Coley-Smith, 1969) to yield a mixture of volatile alkyl thiols and sulphides (Fig. 2). All the volatile propyl and allyl compounds have been shown to be active in stimulating germination (Coley-Smith and King, 1969). These compounds are, of course, particular features of *Allium* species whereas methyl compounds, which are much more widely distributed, are inactive.

With sclerotia produced in pure culture there is variability in the response of different isolates of *S. cepivorum* (Coley-Smith and Holt, 1966), some being much more reactive to the presence of the host plant than others, and some showing a higher level of spontaneous germination in the absence of the host than others. I have recently had an opportunity to compare the response of sclerotia from a field outbreak of disease with the response of ones produced in culture and on artificially infected onion bulbs by an isolate of *S. cepivorum* taken from the same field outbreak (Fig. 3). All three sets of sclerotia had been buried in field soil for six months before use.

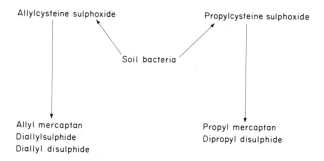

FIG. 2. Production of volatile alkyl thiols and sulphides by microbial decomposition of alkyl cysteine sulphoxides in soil (King and Coley-Smith, 1969).

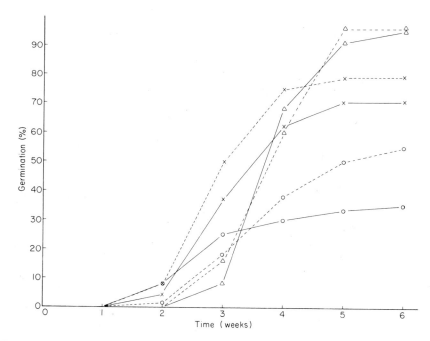

FIG. 3. Reactions of *S. cepivorum* sclerotia from three sources to the presence of onion seedlings (————) and onion bulb extracts (– – –). Sclerotia were from the field, ○; from artificially infected onion bulbs, ×; and from culture, △. Data of Coley-Smith (unpublished).

No germination was recorded in the control series without onion seedlings or onion bulb extracts. On the other hand, high levels of germination were recorded with all three batches of sclerotia in the presence of onion seedlings or bulb extracts, amply confirming the host-stimulatory effect previously found with sclerotia produced in culture. There were differences, however, in both the rate and final amount of germination in the three batches. The field sclerotia, in particular, showed erratic germination and this has been confirmed in several experiments.

This highly host-specific response of *S. cepivorum* has been regarded by many workers as unusual or perhaps even unique amongst plant-pathogenic fungi. Only one other case which appears to involve a degree of host specificity has been reported in sclerotium-forming fungi. This involves *Verticillium dahliae** in which Schreiber and Green (1963) observed that germination of microsclerotia in soil was stimulated more by root exudates from a host plant (tomato) than by ones from a non-host plant (wheat).

In the case of spore-forming fungi the literature certainly suggests (Schroth and Hildebrand, 1964) that responses to exudates from plants are mostly non-specific, germination occurring equally well in the presence of either host or non-host species or in the presence of susceptible or resistant varieties (Schippers and Voetberg, 1969). This seems to be especially true of those fungi in which different physiological races occur. There is no evidence that the reactions of differential varieties to these races are regulated at the spore germination level although such claims have been made in the past.

Outside the fungi there are a few examples of host-specific responses. Hatching of the cysts of the potato cyst nematode (*Heterodera rostochiensis*) is stimulated by root exudates from a few Solanaceous hosts (Shepherd, 1970) and germination of the seeds of parasitic *Striga* and *Orobanche* species is also a fairly specific response to exudates from host plants (Edwards, 1972).

During the last few years a number of sclerotium-forming fungi have been examined in my laboratory for evidence of host-stimulated germination. Only one of these organisms, *Stromatinia gladioli*, is a narrow host-range root-infecting type. This fungus, which occurs commonly in this country and elsewhere, is known principally as the cause of dry rot disease of *Gladiolus* but it also infects *Freesia*, *Crocus* and some other members of the Iridaceae. The few early reports of its occurrence on plants outside the Iridaceae are almost certainly erroneous. Sclerotium-forming fungi occur commonly on bulb and corm producing plants and in the past there has been considerable confusion about the identity of these organisms.

Sclerotia of *Strom. gladioli* can be produced easily in culture (Javed, 1971) and in soil they are long lived. In experiments which are continuing in my laboratory a high proportion of such sclerotia have survived six years burial in field soil. Like those of *S. cepivorum*, the sclerotia of *Strom. gladioli*

---

* This name refers to the microsclerotial form of *Verticillium* which is often called *Verticillium albo-atrum* in the U.S.A.

are subject to fungistasis and do not normally germinate in natural soil in the absence of host plants. Under aseptic conditions however they germinate in the absence of host plants or of any added nutrient.

In experiments reported by Jeves (1974) the effect of host plants on sclerotia was investigated in natural soil with a simple nylon-strip soil-tube technique previously used with *S. cepivorum* (Coley-Smith and King, 1970). The experiments demonstrated a definite host-stimulatory effect. Usually there was little or no germination in controls without host plants but in the presence of *Gladiolus* or *Freesia* seedlings sclerotia gradually germinated over a period of about six weeks (Fig. 4). In the case of *Gladiolus* the host-stimulatory effect was shown in the field as well as in the laboratory. Host-stimulated germination of *Strom. gladioli* sclerotia is much less vigorous than in *S. cepivorum* however (Coley-Smith, 1960). In *Strom. gladioli* only one or two hyphae emerge as a rule but a single hypha is perfectly capable of infecting *Gladiolus* seedlings. A number of plants from a wide range of families were tested by Jeves for their effects on *Strom. gladioli* sclerotia. Of the sixteen members of the Iridaceae tested successfully in this way, fifteen gave higher germination levels than the controls (Table I) whereas none of the thirty-one other plants from different plant families did so. Even within the Iridaceae, however, there were considerable differences in the degree of stimulation observed. Some species gave very little stimulation whereas others induced high levels of germination. In spite of this variation, it is quite clear that the behaviour of sclerotia of *Strom. gladioli* in soil parallels that of *S. cepivorum* in a remarkable way.

The stimulatory factor produced by members of the Iridaceae has not yet been identified. However Jeves found that the effect on germination of

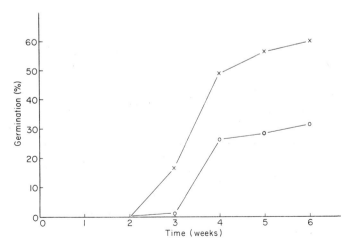

FIG. 4. Response of sclerotia of *Strom. gladioli* to the presence of *Gladiolus* and *Freesia* seedlings (data from Jeves, 1974). *Gladiolus* seedlings = ×, *Freesia* seedlings = ○.

TABLE I

The effect of different plant species on germination in soil of sclerotia of *Strom. gladioli*

Plant	Family	Percentage germination of sclerotia
Control (no seedlings)	—	0·0
Control (no seedlings)	—	3·3
Gladiolus sp.	Iridaceae	73·3
Gladiolus sp. cv. Areta	Iridaceae	53·3
Freesia hybrida	Iridaceae	46·6
Freesia hybrida	Iridaceae	28·3[a]
Dierama pulcherrimum	Iridaceae	51·6
Dierama pulcherrimum	Iridaceae	28·3
Romulea sp.	Iridaceae	16·6
Romulea bulbocodium	Iridaceae	8·3[b]
Tigridia pavonia	Iridaceae	18·3[a]
Tigridia pavonia	Iridaceae	20·0
Lapeyrousia cruenta	Iridaceae	70·0
Sisyrinchium brachypus	Iridaceae	5·0
Sparaxis sp.	Iridaceae	6·6
Curtonus paniculatus	Iridaceae	18·3[a]
Crocosmia masonorum	Iridaceae	40·0
Watsonia hybrida	Iridaceae	0·0
Lactuca sativa	Compositae	3·3
Brassica oleracea	Cruciferae	1·6
Nicotiana tabacum	Solanaceae	1·6
Primula vulgaris	Primulaceae	1·6
Cucumis sativus	Cucurbitaceae	3·3
Allium cepa	Liliaceae	3·3
Viola nigra	Violaceae	1·6
Phaseolus aureus	Papilionaceae	3·3

[a] Germination of seed was less than 33%.
[b] Germination of seed was between 33 and 70%.
All other germination was more than 70%.
Thirty-one other plants from a wide range of families gave no germination of sclerotia.
Data from Jeves (1974).

sclerotia could be produced with water extracts of many Iridaceous species (Table II). Not all extracts were equally effective and *Iris* and *Crocus* extracts in particular gave no evidence of stimulatory capacity. Chromatographic investigations of *Gladiolus* extract gave very poor separation of the stimulatory factor which could possibly indicate that a mixture of compounds is involved.

Jeves' (1974) experiments have demonstrated unequivocally that the host-specific response of *S. cepivorum* is not unique since *Strom. gladioli* has now been shown to respond to the presence of host plants in a similar manner.

TABLE II

The effect of plant extracts on germination in soil of sclerotia of *Strom. gladioli*

Plant	Family	Percentage germination of sclerotia
Control (distilled water)		0·0
Gladiolus cv. Greenwitch	Iridaceae	75·0
Gladiolus cv. Snow Princess	Iridaceae	61·6
Gladiolus cv. Assam	Iridaceae	66·6
Gladiolus cv. Memorial Day	Iridaceae	58·3
Freesia refracta-alba	Iridaceae	23·3
Lapeyrousia cruenta	Iridaceae	68·3
Babiana sp.	Iridaceae	75·0
Acidanthera bicolor var murielae	Iridaceae	80·0
Sparaxis sp.	Iridaceae	68·3
Tigridia pavonia	Iridaceae	8·3
Crocosmia masonorum	Iridaceae	60·0
Tritonia (*Montbretia*) sp.	Iridaceae	46·6
Crocus sp.	Iridaceae	1·6
Iris sp.	Iridaceae	0·0

Data from Jeves (1974).

Although two well documented examples can hardly form the basis for claiming a general rule, it is my opinion that host-stimulated germination of fungal propagules may be more common than it appears at present. It is doubtful whether many more examples will be found amongst sclerotium-forming fungi but the chances of discovering such situations amongst spore formers are much greater. There are many of these which have not yet been examined at all and a number of others which have not been examined using methods which might be expected to reveal such host-specific responses. In my view a systematic search for these effects is required.

IV. THE EFFECT OF FUNGAL SCLEROTIA ON THE SOIL MICROFLORA

The second area of discussion in this review concerns the effect of pathogens upon the microbial population of soils. In Park's review (1953) he said of this system:

"Of the six effects being considered this is almost certainly the least important. The pathogen is commonly less dominant in soil than many members of the saprophytic soil population, and is therefore more often influenced than influencing".

Park, however, did recognize that local "hyphosphere" effects on the microflora could occur in the vicinity of fungal spores and hyphae.

During recent years there has been an increasing number of reports of effects of pathogens on the microbial population in soils. One of the best known examples involves the work of Lingappa and Lockwood (1964) who showed that, when added to soil, spores of several fungi released nutrients. As a result of this leakage microbial activity and numbers increased in the soil. Originally Lockwood thought that antibiotic effects resulting from enhanced microbial activity might account for the phenomenon of fungistasis (Lingappa and Lockwood, 1961) but he has since abandoned this idea in favour of one which suggests that fungistasis results from the depletion, by microorganisms, of nutrients in the sporosphere (Ko and Lockwood, 1967; Hsu and Lockwood, 1973). Fungistasis has recently been reviewed in detail by Watson and Ford (1972). The general opinion at the moment is that leakage of materials from spores is likely to be of short duration and it is difficult to see how it could relate to long-term fungistasis.

In the case of bulkier fungal structures far less information has been published. Sclerotia of many organisms are subject to fungistasis but there appear to be genuine differences in species sensitivity (Hsu and Lockwood, 1973; Javed and Coley-Smith, 1973). It is also known that sclerotia, like spores, can leak substances into soil. In work which we reported some years ago (Dickinson and Coley-Smith, 1970) it was shown that the addition to natural soil of washed and air-dried sclerotia of *S. cepivorum* resulted in considerable increases in bacterial numbers and in levels of "soil respiration". Similar increases occurred if water in which sclerotia had stood was used instead of the sclerotia themselves. The major constituents which leaked into water from sclerotia of *S. cepivorum* were trehalose, glucose and mannitol (Coley-Smith and Dickinson, 1971) but later work with this and with a number of other fungi indicates that a whole range of different substances is involved (Rizki and Gladders, unpublished). Leakage from mature sclerotia has now been demonstrated with many other fungi including *Botrytis cinerea* and *Botrytis tulipae* (Coley-Smith *et al.*, 1974), *Rhizoctonia tuliparum* (Gladders, unpublished), *Sclerotinia minor* and *Sclerotinia sclerotiorum* (Smith, 1972c); *Sclerotium delphinii* (Javed and Coley-Smith, 1973); *Sclerotium oryzae* (Keim and Webster, 1974), *Sclerotium rolfsii* (Smith, 1972a,b) and *Strom. gladioli* (Jeves, 1974).

The small sclerotia of many fungi are difficult to handle when moist and the majority of my early experiments on leakage in *S. cepivorum* involved the use of air-dried sclerotia. Although the effect of air drying on sclerotial viability was always tested the effect on leakage was not. A series of papers by Smith (1972a,b,c), however, demonstrated that short periods of air drying caused nutrients to leak from the sclerotia of a number of fungi during subsequent wet conditions. Leakage into soil was associated with invasion of sclerotia by microorganisms and this resulted in the decay of sclerotia. Smith's results have been substantiated as far as leakage is concerned but of the fungi which we have used (Coley-Smith *et al.*, 1974) viability was only affected in *S. delphinii* (Table III).

TABLE III
Survival of dried and non-dried sclerotia in the field

| Fungus | Percentage viable sclerotia recovered | | | |
| | Dried | | Non-dried | |
	1 month	3 months	1 month	3 months
Botrytis cinerea	100	100	100	100
Botrytis tulipae	98	93	99	98
Sclerotium cepivorum	99	98	98	100
Sclerotium delphinii	4	7	100	99

Data from Coley-Smith *et al.* (1974).

The mechanism of leakage from fungal sclerotia appears to be the same as that recorded for plant seeds. This subject has recently been reviewed by Simon (1974). With sclerotia substances lost during leakage appear to originate from within sclerotial hyphae rather than from hyphal-free space.

Leakage can be simply demonstrated by placing sclerotia on agar seeded with a suitable assay organism such as *Bacillus subtilis*. The agar must be nutritionally weak otherwise bacterial development occurs profusely over the whole plate. On weak agar a zone of stimulated bacterial development occurs around air-dried sclerotia of most fungi (Dickinson and Coley-Smith, 1970). These zones of stimulation vary in size according to the organism used. A better measure of total leakage, however, can be obtained by dry weight determinations after allowing sclerotia to leak into water. The differences between leakage levels of some species are of considerable magnitude (Table IV). Sclerotia can also leak on more than one occasion although the quantity of material leaked declines with each successive cycle. In *S. cepivorum* we have demonstrated leakage on ten successive occasions without appreciable loss of viability. Keim and Webster (1974) have shown a similar capacity in *S. oryzae* although loss in viability was greater than in *S. cepivorum*. It is not yet known whether differences in leakage levels are related to survival capacity nor is there any obvious connection with fungistasis. Sclerotia of many fungi often remain dormant in soil for long periods whether conditions are wet or dry. Release from fungistasis by host-specific stimulators may be the rule in narrow host-range root-infecting sclerotium-forming fungi (Watson and Ford, 1972). In wide host-range types stimulators may also be involved but are unlikely to be host specific. Some evidence for this is available for *S. rolfsii* in which germination in response to alcohols and aldehydes emanating from lucerne residue has been shown (Linderman and Gilbert, 1969, 1973). These and similar compounds are probably produced during the decomposition in soil of many plant materials.

TABLE IV
The effect of successive air dry periods on leakage into water from fungal sclerotia

Fungus	Leakage cycle	Weight loss as a % of original oven dry weight
Botrytis cinerea	1	5·6
	2	1·9
	3	0·8
Botrytis tulipae	1	7·9
	2	5·6
	3	4·1
Sclerotium cepivorum	1	3·1
	2	2·4
	3	0·7
Sclerotium delphinii	1	27·8
	2	2·9
	3	1·0
Rhizoctonia tuliparum	1	42·6
	2	17·1
	3	11·0

Data for *Rhizoctonia tuliparum* are of Gladders (unpublished), the rest are from Coley-Smith *et al.* (1974).

A further complication as far as fungistasis is concerned arises from the fact that sclerotia do not always leak substances which stimulate soil micro-organisms. The *B. subtilis* method gives zones of inhibition around sclerotia of *Claviceps purpurea* and *Rhizoctonia tuliparum* (Gladders, unpublished). Materials having antibiotic properties have been described from continental rye ergots and the same or similar compounds occur in barley and other ergots from this country. Several groups of pigmented compounds are probably involved (Fig. 5) amongst which are ergoflavins, ergochrysins and secalonic acids (Stoll *et al.*, 1952; ApSimon *et al.*, 1965; Aberhart *et al.*, 1965).

In the case of *R. tuliparum* the identity of the inhibitory material is not yet known, nor is it yet known whether this or any of the ergot compounds are

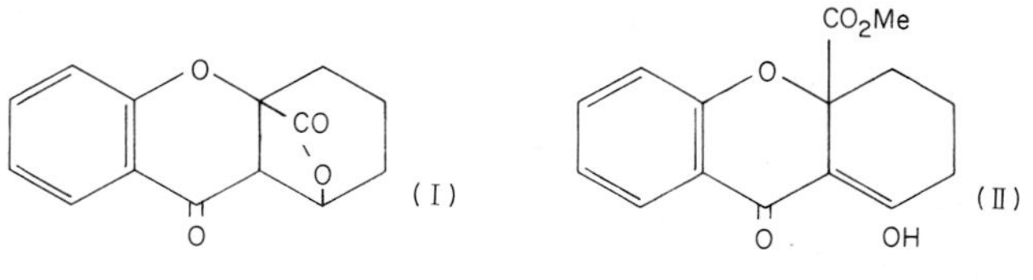

FIG. 5. Structure of ergot pigments. Pigments containing two units of type (I) are termed ergoflavins; those containing one unit of type (I) together with a unit of type (II) are termed ergochrysins and those with two units of type (II) are called secalonic acids (ApSimon *et al.*, 1965).

of importance in the soil. In spite of this one can say that with sclerotium-forming fungi there is a definite effect of the pathogen upon the microbial population of soil. This effect is mediated by the external environment. Under uniformly moist conditions the effect is small or absent but where moist conditions follow dry a considerable increase in the effect occurs. The effect is stimulatory with most fungi but may be inhibitory with some. These effects are transitory in the sense that substances which leak out from sclerotia do so quickly following a change from dry to wet conditions. The effects are long-term, however, in the sense that leakage is possible on more than one occasion and can occur repeatedly with some sclerotium-forming fungi.

ACKNOWLEDGEMENTS

I should like to express my gratitude to the Agricultural Research Council and to the Science Research Council for financial support of many of the investigations reported here.

REFERENCES

Aberhart, D. J., Chen, Y. S., De Mayo, P. and Stothers, J. B. (1965). *Tetrahedron* 21, 1417–1432.

ApSimon, J. W., Corran, J. A., Creasey, N. G., Marlow, W., Whalley, W. B. and Sim, K. Y. (1965). *J. Chem. Soc.* **764**, 4144–4156.

Baker, K. F. and Cook, R. J. (1974). "Biological Control of Plant Pathogens". W. H. Freeman, San Francisco.

Baker, K. F. and Snyder, W. C. (1965). "Ecology of Soil-borne Plant Pathogens. Prelude to Biological Control". University of California Press, Berkeley.

Butler, G. M. (1966). *In* "The Fungi" (G. C. Ainsworth and A. S. Sussmann, eds), pp. 83–112. Academic Press, London and New York.

Coley-Smith, J. R. (1959). *Ann. appl. Biol.* **47**, 511–518.

Coley-Smith, J. R. (1960). *Ann. appl. Biol.* **48**, 8–18.

Coley-Smith, J. R. and Cooke, R. C. (1971). *A. Rev. Phytopath.* **9**, 65–92.

Coley-Smith, J. R. and Dickinson, D. J. (1971). *Soil Biol. Biochem.* **3**, 27–32.

Coley-Smith, J. R. and Holt, R. W. (1966). *Ann. appl. Biol.* **58**, 273–278.

Coley-Smith, J. R. and King, J. E. (1969). *Ann. appl. Biol.* **64**, 289–301.

Coley-Smith, J. R. and King, J. E. (1970). *In* "Root Diseases and Soil-borne Pathogens" (T. A. Toussoun, R. V. Bega and P. E. Nelson, eds), pp. 130–133. University of California Press, Berkeley.

Coley-Smith, J. R., Ghaffar, A. and Javed, Z. U. R. (1974). *Soil Biol. Biochem.* **6**, 307–312.

Coley-Smith, J. R., King, J. E., Dickinson, D. J. and Holt, R. W. (1967). *Ann. appl. Biol.* **60**, 109–115.

Dickinson, D. J. and Coley-Smith, J. R. (1970). *Soil Biol. Biochem.* **2**, 157–162.

Dobbs, C. G. and Hinson, W. H. (1953). *Nature, Lond.* **172**, 197–199.

Edwards, W. G. H. (1972). *In* "Phytochemical Ecology" (J. B. Harborne, ed.), pp. 235–248. Academic Press, London and New York.

Garrett, S. D. (1970). "Pathogenic Root-infecting Fungi". Cambridge University Press, Cambridge and London.

Hsu, S. C. and Lockwood, J. L. (1973). *Phytopathology* **63**, 334–337.

Javed, Z. U. R. (1971). "Studies of Sclerotium-forming Fungi". Ph.D. Thesis, University of Hull.
Javed, Z. U. R. and Coley-Smith, J. R. (1973). *Trans. Br. mycol. Soc.* **60**, 441–451.
Jeves, T. M. (1974). "Studies of the Biology of *Stromatinia gladioli* (Drayt.) Whetz." Ph.D. Thesis, University of Hull.
Keim, R. and Webster, R. K. (1974). *Phytopathology* **64**, 1499–1502.
King, J. E. and Coley-Smith, J. R. (1969). *Ann. appl. Biol.* **64**, 303–314.
Ko, W.-h. and Lockwood, J. L. (1967). *Phytopathology* **57**, 894–901.
Linderman, R. G. and Gilbert, R. G. (1969). *Phytopathology* **59**, 1366–1372.
Linderman, R. G. and Gilbert, R. G. (1973). *Phytopathology* **63**, 359–362.
Lingappa, B. T. and Lockwood, J. L. (1961). *J. gen. Microbiol.* **26**, 473–485.
Lingappa, B. T. and Lockwood, J. L. (1964). *J. gen. Microbiol.* **35**, 215–227.
Park, D. (1963). *A. Rev. Phytopath.* **1**, 241–258.
Schippers, B. and Voetberg, J. S. (1969). *Neth. J. Pl. Path.* **75**, 241–258.
Schreiber, L. R. and Green, R. J. (1963). *Phytopathology* **53**, 260–264.
Schroth, M. N. and Hildebrand, D. C. (1964). *A. Rev. Phytopath.* **2**, 101–132.
Shepherd, A. M. (1970). *In* "Root Diseases and Soil-borne Pathogens" (T. A. Toussoun, R. V. Bega and P. E. Nelson, eds), pp. 134–137. University of California Press, Berkeley.
Simon, E. W. (1974). *New Phytol.* **73**, 377–420.
Smith, A. M. (1972a). *Soil Biol. Biochem.* **4**, 119–123.
Smith, A. M. (1972b). *Soil Biol. Biochem.* **4**, 125–129.
Smith, A. M. (1972c). *Soil Biol. Biochem.* **4**, 131–134.
Stoll, A., Renz, J. and Brack, A. (1952). *Helv. chim. Acta* **35**, 2022–2034.
Toussoun, T. A., Bega, R. V. and Nelson, P. E. (eds) (1970). "Root Diseases and Soil-borne Pathogens". University of California Press, Berkeley.
Watson, A. G. and Ford, E. J. (1972). *A. Rev. Phytopath.* **10**, 327–348.
Willetts, H. J. (1971). *Biol. Rev.* **46**, 387–407.
Willetts, H. J. (1972). *Biol. Rev.* **47**, 515–536.

CHAPTER 3

Development and Use of some Genetically Controlled Lines for Studies of Host–Parasite Interactions

R. JOHNSON

Plant Breeding Institute, Cambridge, England

I. INTRODUCTION

Many aspects of the genetics of host–parasite interactions have been reviewed in recent publications such as those of Hooker and Saxena (1971) on genetics of resistance in hosts, Boone (1971) on genetics of virulence in pathogens and Flor (1971) on the gene-for-gene interaction between host and pathogen. Day (1974) has also considered these topics and in addition has emphasized gene function in the host–parasite interaction and some genetic aspects of disease epidemics.

Rather than attempting to outline the wide range of investigations currently in progress, this chapter will be concerned with recent developments of genetically controlled host and pathogen lines and their use in investigations of host–parasite interactions. The objective of using genetically controlled lines is to help the investigator to distinguish characters which are causes of disease resistance from those which are only correlated with resistance. Two requirements must be fulfilled for the lines to be useful in biochemical and physiological investigations. Firstly the lines should consist of individuals with identical genotypes, i.e. they should be pure lines. This is easily achieved where asexual reproduction occurs as in potato varieties and in pathogen cultures such as uredospore cultures of rusts or clones of pathogens grown in pure axenic cultures. Where the lines can only be reproduced by sexual reproduction as with annual crops, the genotype must be homozygous at all loci in order to be true breeding. The second requirement is that the genetic differences between lines should be known.

II. HOST LINES

A. NEARLY-ISOGENIC LINES

1. *Advantages and Limitations of Nearly-isogenic Lines*

Among the most specialized of genetically controlled lines in the host are nearly-isogenic lines which are available in some inbreeding crops. In any set of such lines differences between lines should be confined to a single locus which, in this case, governs disease reaction.

The advantages of using nearly-isogenic lines, especially in investigations of the mechanisms leading to compatibility or incompatibility between host and pathogen have been emphasized by Daly (1972). Where host lines differ in many characters there is clearly an increased chance that observed differences in biochemical reactions are correlated with but are not the cause of differences in disease reaction. This may be true of events occurring after infection as well as of differences such as levels of phenolic compounds, which exist before infection as demonstrated by comparisons of results obtained with and without the use of nearly-isogenic lines (Daly, 1972; Knott and Kumar, 1972). Whereas tissue concentrations of phenolic compounds were found to be higher in the uninfected wheat variety Khapli which is resistant to an above average number of races of *Puccinia graminis* than in more susceptible varieties, no such differences were found when nearly-isogenic lines with and without gene Sr-6 for resistance were compared.

While the use of nearly-isogenic lines should minimize such problems it cannot entirely eliminate them. This is partly due to some properties of the gene-for-gene relationship which will be considered later, and partly due to limits on the degree to which the lines can be truly isogenic.

2. *Production of Nearly-isogenic Lines*

It is necessary for the most efficient use of nearly-isogenic lines to consider the way in which they are produced. There are two main methods of production which result in important differences in the end-products which, for convenience, will here be referred to as "paired lines" and "grouped lines".

(a) *Paired lines.* Atkins and Mangelsdorf (1942) proposed that pairs of nearly-isogenic lines could be produced by crossing a line possessing the character (resistant) controlled by a single gene, with one lacking the character (susceptible) followed by selfing of plants heterozygous for the resistance locus in each generation. At the end of the series of selfed generations the progeny of a plant heterozygous for the resistance locus is examined and plants homozygous for the allele for resistance and for susceptibility are selected. This produces a pair of lines which, depending on the number of generations of selfing, can be very similar to each other except at the disease resistance locus. There may, however, be large differences between the different pairs of lines which can be selected from such a programme and none may resemble either parent closely, since the occurrence of homozygosity during selfing, at loci other than that selected for resistance, will occur for alleles from either parent with equal frequency. These lines should therefore be treated as pairs, and not compared directly with either original parent.

(b) *Grouped lines.* A resistant parent is crossed with a susceptible parent and, if susceptibility is recessive, plants heterozygous for the gene for resistance are backcrossed in each generation to the susceptible parent. At the end of the series of backcrosses a heterozygous plant is selfed and plants homozygous for the resistance allele are selected. Progeny tests on selfed resistant plants may be necessary if the allele for resistance is completely dominant, to establish that the selected line is homozygous. The selected resistant line is compared with the original susceptible parent. If several lines with different alleles for resistance are developed in this way all may be compared with the common susceptible parent used for backcrossing (Knott and Kumar, 1972).

(c) *The degree of isogeny attained.* The degree to which lines from either type of programme are isogenic will depend on the number of repetitions of the cycles of selfing or backcrossing. Two factors which influence the attainment of isogeny are the behaviour of genes which are independent of, or unlinked to, the gene for resistance and the behaviour of genes linked to the gene for resistance.

(i) Unlinked genes. The graphs of Fig. 1 indicate average rates of return to homozygosity for unlinked genes, calculated from the function $\{(2^m - 1)/2^m\}^n$ where m is the number of generations and n the number of independent gene pairs (Allard, 1960). The curve marked BC shows that, in a backcrossing programme 50% of the heterozygous genes in each generation become homozygous for the genes of the susceptible parent to which the backcrosses are made. The same curve shows that, in a selfing programme, a single heterozygous gene will become homozygous for one or other of the alternative alleles in 50% of the progeny from the individual possessing the

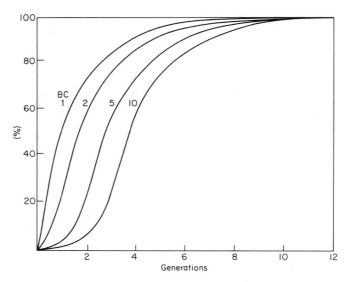

FIG. 1. Proportion of genotype homozygous for unlinked genes from the susceptible parent during backcrossing genes for resistance into a susceptible genotype (BC) and proportion of individuals homozygous at all loci during selfing for 1, 2, 5 and 10 heterozygous independent gene pairs.

heterozygous gene. This rule will apply to all heterozygous, independent genes in each generation so that 50% of all heterozygous genes will become homozygous in each generation. The distribution of genes becoming homozygous will, however, be random so that the proportion of individuals becoming homozygous at all loci will decrease as the number of independent heterozygous genes is increased. Examples are shown for two, five and ten heterozygous genes (Fig. 1). Thus, even after many generations of back-crossing or selfing some heterozygosity can remain. With further selfing during maintenance of the nearly-isogenic lines these heterozygous genes can become homozygous for either of the alternative alleles, and can result in differences between nearly-isogenic lines at loci other than those for disease resistance.

(ii) Linked genes. In the case of the gene for resistance, which is kept heterozygous in each generation, a large block of linked genes can be carried through many generations until they are released by crossing over, as shown by Fig. 2 which is based on calculations by Hanson (1959). Breaking of linkage is quicker with selfing since effective crossing over can occur in both male and female gametes; in backcrossing it can occur only in the male gametes of the heterozygous plants being backcrossed to the susceptible parent. Once linked genes are released from their association with the resistance gene, by crossing over, they can become homozygous for the alleles in the susceptible parent during backcrossing, or for those of either

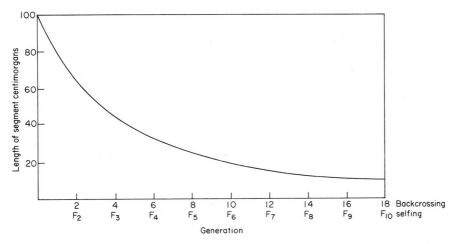

FIG. 2. Probable length of chromosome segment, measured in cross-over units, transferred due to linkage with a gene for resistance which is kept heterozygous during backcrossing or selfing in a chromosome of 100 centimorgans length.

parent during selfing. At the end of the backcrossing or selfing sequence most of the genes from the resistant parent still linked to the resistant gene will become homozygous at the time the homozygous resistant line is selected. When paired lines are selected the alternative alleles at the heterozygous loci will become homozygous in the selected homozygous susceptible line. There are several examples of closely linked genes for resistance such as for rust resistance in flax (Flor, 1971; Shepherd and Mayo, 1972), for powdery mildew resistance in barley (Wolfe, 1972; Moseman and Jorgensen, 1973), for rust resistance in maize (Hooker and Saxena, 1971) and for stem rust resistance in wheat (Green et al., 1960, Loegering and Harmon, 1969). Such genes linked in coupling are likely to remain together in programmes of the type described above and in some cases, depending on the pathogen line used for tests, some alleles may remain undetected until a different strain of the pathogen is used (see below).

3. Some Examples of Nearly-isogenic Lines

(a) Grouped lines. Most of the sets of nearly-isogenic lines which have been developed so far have been produced by the backcrossing technique, resulting in grouped lines. It is evident that there is considerable variation in the number of backcrosses used (Table I) and therefore in the degree of isogeny attained. Backcrossing is a laborious procedure, especially in the case of a crop such as maize which usually consists of hybrid varieties produced by crossing inbred lines, so that it is necessary to produce pairs of inbred lines with each of the alternative alleles for resistance.

TABLE I
Grouped nearly-isogenic lines developed by backcrossing resistance genes into a susceptible parent

Crop	Pathogen	Backcrosses	Reference
Flax	*Melampsora lini*	7	Doubly *et al.* (1960)
Flax	*Melampsora lini*	15	Littlefield (1969)
Flax	*Melampsora lini*	4	El-Gewely *et al.* (1972)
Maize	*Puccinia sorghi*	?	Hooker and Russell (1962)
Maize	*Helminthosporium turcicum*	5	Lim *et al.* (1970)
Maize	*Helminthosporium maydis*	6 to 9	Barrett and Flavell (1975)
Wheat	*Puccinia graminis*	4 to 6	Green *et al.* (1960)
Wheat	*Puccinia graminis*	10	Knott and Kumar (1972)
Wheat	*Erysiphe graminis*	8	Briggle (1969)

(b) *Paired lines.* Two sets of nearly-isogenic lines have been produced partly by backcrossing and partly by selfing, one by Loegering and Harmon (1969) in wheat for resistance to *P. graminis* and the other by Moseman (1972) in barley for resistance to *Erysiphe graminis*.

(i) Lines of Loegering and Harmon (1969). These lines may be said to combine the advantages of both the backcrossing and selfing procedures, since backcrossing for the chromosomes other than those carrying the genes for resistance was carried out for three to eight generations during the production of chromosome substitution lines in the variety Chinese Spring by Sears (1954). The chromosome substitution lines were then crossed to Chinese Spring and plants heterozygous for resistance were backcrossed twice to Chinese Spring. According to Fig. 2 this would leave a fairly large group of genes linked to the resistance locus, but at other unlinked loci the lines would resemble Chinese Spring closely, as indicated by Fig. 1. Heterozygous resistant lines were then selfed for between ten and twelve generations, during which the linkage group round the resistance locus would be greatly reduced, but in which the return to homozygosity could be equally for genes of either Chinese Spring or the donor resistant parent. Thus there could be some divergence between pairs, and Loegering and Harmon (1969) emphasized that comparisons should be made within rather than between pairs. Their estimate of one per cent difference between susceptible lines from different pairs seems to ignore the effect of linkage on the retention of genes from the donor parent and may thus be an underestimate. Lines from this set are being used in pairs for biochemical investigations at laboratories in Nebraska (Daly, 1972) and at Winnipeg (Rohringer *et al.*, 1974; Skipp and Samborski, 1974). The pair of lines with Sr-6 and sr-6 alleles has been especially intensively investigated because the resistance gene Sr-6 is temperature sensitive, being active at 20 °C and inactive at 25 °C (Forsyth, 1956). Thus it is possible to investigate susceptible and resistant reactions using the same pathogen strain which is non-virulent to Sr-6 at low temperatures.

(ii) Lines of Moseman (1972). In the paired lines developed by Moseman (1972) the barley varieties resistant to *E. graminis* were crossed to the susceptible variety Manchuria and plants heterozygous for resistance were backcrossed four times to Manchuria. This was followed by twelve to fifteen generations of selfing of plants heterozygous for resistance. Moseman (1972) calculated that about 12% of the germ plasm would be from the resistant parent after the four backcrosses. Approximately half of this will have remained and be homozygous after the generations of selfing so that quite large differences between pairs may remain, although two members of a pair would be expected to be very similar except at the resistance locus. It would not be surprising, therefore, to find differences between the resistant lines and Manchuria, affecting the host–parasite relationship but not controlled by the resistance locus upon which the nearly-isogenic lines were developed.

An example of this was shown by McCoy and Ellingboe (1966) who studied the influence of resistance genes controlling infection type, on production of secondary hyphal initials of *E. graminis*, using the paired lines developed by Moseman. At the time of their studies the development of the lines was incomplete compared with those described later by Moseman (1972). McCoy and Ellingboe (1966) found that the development of secondary hyphal initials took place over a longer period in lines possessing homozygous resistance genes Ml-a, Ml-g, Ml-p and Ml-k than in Manchuria. They studied the development of secondary hyphal initials in the progeny from selfed plants heterozygous for the resistance genes, presumably derived from the selfing programme of Moseman. They found that segregation for the occurrence of secondary hyphal initials differed from segregation for the resistance genes and concluded that the resistance genes did not control this character. A similar conclusion is suggested by the fact that the homozygous susceptible plants in the progeny and the plants heterozygous for different resistance genes also differ from each other and from Manchuria for this character (Table II).

Such differences between the susceptible lines and Manchuria emphasized the paired nature of the resistant and susceptible lines developed by Moseman and show that a character so intimately concerned with disease development as the development of secondary hyphal initials need not be controlled by genes affecting infection type.

4. *Identification of Genes for Resistance with Observed Effects*

Ellingboe (1972) emphasized the value of establishing an absolute association in segregating populations of characters which may relate to resistance, and their supposed genetic control. This is as important with

TABLE II

Mean percent of conidia of *E. graminis* producing secondary hyphal initials in susceptible plants of the progeny from selfed plants heterozygous for different single genes for resistance to powdery mildew compared with the susceptible variety Manchuria (from McCoy and Ellingboe, 1966)

Allele for susceptibility	ml-a	ml-g	ml-p	ml-k
Percent on line with allele	71	56	29	21
Percent on Manchuria	90	24	24	24
Hours after infection	24	18	18	18

nearly-isogenic lines as in other less genetically controlled lines, as illustrated by McCoy and Ellingboe (1966). Their demonstration of segregation of characters in individual plants of a segregating population would, however, be the envy of those whose investigations require more plant tissue and greater replication as would commonly occur in biochemical studies. For such purposes it would be desirable to establish homozygous lines in, for example, the progeny, from a cross between the partners of a pair of nearly-isogenic lines. This could be achieved by selecting a range of plants from such a cross and selfing without selection for several generations to establish homozygous lines which could be replicated. There is a possibility that such a laborious procedure could, in some cases, be superseded in future by the development of methods permitting the production of haploids from heterozygous individuals, followed by chromosome doubling to produce immediate complete homozygosity (see Kasha, 1974).

B. CHROMOSOME SUBSTITUTION LINES

1. *Some Advantages of Chromosome Substitution Lines*

Hooker and Saxena (1971) drew attention to the importance of modifying genes which alter the expression of resistance. The more highly developed nearly-isogenic lines become, the more likely it is that they will be homozygous for the same modifying genes, whose presence might then remain undetected in comparisons between partners of a pair, or a closely related group. In wheat the use of cytogenetic techniques by which whole chromosomes can be transferred from one variety to another can be useful in studying modifying genes. Law and Johnson (1967) showed that, whereas the wheat variety Chinese Spring was susceptible to an isolate of *Puccinia recondita*, a chromosome substitution line in which chromosome 7B of Chinese Spring was replaced by chromosome 7B of the variety Hope was resistant. Furthermore the Chinese Spring/Hope 7B substitution line was more resistant than Hope itself, suggesting the presence of modifying genes in the Chinese Spring background increasing the expression of resistance carried by chromosome 7B of Hope.

In order to study the location of genetic factors for resistance in chromosome 7B the chromosome 7B substitution line was crossed with Chinese Spring (see Fig. 3). This produced an F_1 in which chromosome 7B was heterozygous for the genes of Chinese Spring and Hope but all other chromosomes were expected to be homozygous for the genes of Chinese Spring. This F_1 was crossed to Chinese Spring monosomic 7B, and monosomic progeny were selected by cytological analysis. Such monosomics would carry the products of meiosis of the heterozygous chromosome 7B as univalent chromosomes. Thus after one generation in which recombination could occur on chromosome 7B further recombination was prevented. The monosomic individuals from this generation were selfed and euploid individuals were selected. These would be homozygous for the particular arrangements of genes on chromosome 7B inherited by the monosomic plants of the previous generation. These homozygous individuals were selfed to produce homozygous lines. Of 70 such lines 41 were susceptible while 29 were resistant to *P. recondita*, which agreed with the 1:1 ratio expected if a single gene controlled resistance. The resistant plants were, however, divided into two classes, one with the high level of resistance shown in the Chinese Spring/Hope 7B substitution line and the other rather less resistant in a 1:1 ratio. It was concluded that chromosome 7B also carried a modifying gene, the allele in Hope increasing resistance and that in Chinese Spring decreasing resistance. An important advantage of the

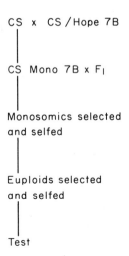

CS x CS / Hope 7B

CS Mono 7B x F₁

Monosomics selected
and selfed

Euploids selected
and selfed

Test

FIG. 3. Crossing programme to determine the control of resistance to *P. recondita* on chromosomes 7B of Hope and Chinese Spring (CS).

technique in the context of the present discussion is that it produced a series of homozygous, true-breeding families differing for the alternative alleles of the resistance gene and of the modifying gene.

2. *Limitations of Chromosome Substitution Lines*

Chromosome substitution lines are produced by backcrossing procedures (Sears, 1953) which permit one chromosome to be transferred without change from a donor into a recipient variety. The original genetic content of the recipient parent for all other chromosomes must be reintroduced by the backcross procedure and this process is subject to the same restrictions as indicated by the line BC in Fig. 1. Thus the question of whether differences between a chromosome substitution line and the recipient parent are controlled by genes on the substituted chromosome or in the background is analagous to that for nearly-isogenic lines. In the study by Law and Johnson (1967) it was noted that there was no segregation for resistance reactions within any of the 70 families produced according to the procedure outlined in Fig. 3, as might have been expected if differences from Chinese Spring affecting resistance existed in chromosomes of the substitution line other than 7B. In addition, both euploid Chinese Spring and the Chinese Spring/Hope 7B substitution line were crossed with Chinese Spring monosomic 7B and monosomic progeny were selected and selfed to produce euploids which were identical to the original Chinese Spring parent and the original Chinese Spring/Hope 7B substitution line respectively, confirming that there were no differences affecting resistance on chromosomes other than 7B.

Since differences can occur for genes not carried by the substituted chromosome, Law and Worland (1973) have pointed out the value of deriving, separately, two substitution lines for each chromosome. If genes affecting resistance derived from the donor parent remain on chromosomes other than the one substituted, it is unlikely that they will be the same in both substitution lines. Thus the magnitude of difference between the two substitution lines can be compared with differences between the substitution lines and the recipient parent to indicate the size of differences due to the substituted chromosome relative to those due to other chromosomes. The same argument could be applied to nearly-isogenic lines and it would therefore be helpful if two separately derived pairs could be developed for each resistance gene. Few such duplicates exist at present but the two pairs of lines carrying gene Sr-5 for resistance to *P. graminis* described by Loegering and Harmon (1969) may be considered as an example.

C. GENETICALLY DIVERSE HOST LINES

In some cases repetition of the same observation on a series of lines less closely related than those described above may serve to increase the probability that an effect is correctly identified with a specific genetic control. Barrett and Flavell (1975) found that the sensitivity of mitochondria from "Texas" cytoplasm in maize to a pathotoxin produced by race T of *Helminthosporium maydis* was reduced by the presence of nuclear restorer genes which negate the male sterility induced by the "Texas" cytoplasm. This effect was observed in nine lines with four different sources of restorer genes and thus increases the probability that it is the restorer genes themselves which modify the sensitivity of the mitochondria. Similar results were obtained by Watrud *et al.* (1975). Kuć (Chapter 12) used several lines of both host and pathogen in studies of phytoalexin production by potato tubers infected with *Phytophthora infestans*.

III. PATHOGEN LINES

1. *The Need for Genetically Controlled Lines*

Relatively little genetic control of the type described above for hosts has been achieved for pathogens. The labour of producing the equivalent of isogenic lines for pathogens such as rusts which are outbreeding and highly heterozygous would be great. Daly (1972) has suggested that strict genetic control is less important in, for example, rust pathogens than in the host because appropriate strains of the pathogen can be produced vegetatively. The production of pure line cultures which, in the host, requires the use of homozygous lines is, however, only one part of the process of development of genetically controlled lines, the other being the development of controlled differences between lines. Person and Mayo (1974) have discussed some implications of the gene-for-gene interaction between host and pathogen which suggest that close genetic control of pathogen lines would be an advantage.

2. *Implications of the Gene-for-gene Concept*

According to the gene-for-gene concept developed by Flor (1956), each gene for resistance in the host interacts with a specific gene for non-virulence or virulence in the pathogen. Person and Mayo (1974), and others, have noted that an interaction preventing growth of the pathogen often occurs between a dominant gene for resistance, "R", in the host and a dominant gene for non-virulence, "A", in the pathogen; they have called this R:A interaction a "stop-signal". With the alternative alleles for susceptibility, "r", in the host and virulence, "a", in the pathogen there are three other interactions at the same related loci which can lead to growth of the pathogen: these are R:a, r:A and r:a. These four interactions have been described as a "quadratic check" by Rowell *et al.* (1963). Slesinski and Ellingboe (1971)

studied such a quadratic group of reactions in wheat to *E. graminis*. They
reported that uptake of ^{35}S by the pathogen was less for the susceptible
reaction, which occurred when the resistance gene Pm-1 was present and
the pathogen carried the corresponding virulence gene p-1, than in the
other two susceptible reactions pm-1/P-1 (host without resistance gene;
pathogen with gene for non-virulence to gene Pm-1) and pm-1/p-1 (host
without resistance gene; pathogen with unnecessary gene for virulence
towards Pm-1). The p-1 gene for virulence seemed unable to overcome
completely the effect of the resistance allele Pm-1. Such differences between
the effects of alleles at the Pm-1 or other loci may be important in the host–
parasite interaction but would not be detected by conventional methods of
observing the interaction.

	Host	
Genotypes	R_1 R_1	R_1 R_1
	r_2 r_2	R_2 R_2
	r_3 r_3	r_3 r_3
	R_4 R_4	r_4 r_4

Parasite

	Host col 1	Host col 2
a_1 a_1 a_2 a_2 A_3 A_3 A_4 A_4	−	+
a_1 a_1 a_2 a_2 A_3 A_3 a_4 a_4	+	+

FIG. 4. Interaction between two host and two pathogen genotypes involving four R and A loci
respectively. Corresponding numerical subscripts indicate corresponding genes in host and
pathogen. In the host upper case letters represent dominant alleles for resistance, lower case
letters represent recessive alleles for susceptibility. In the pathogen upper case letters represent
dominant alleles for non-virulence, lower case letters represent recessive alleles for virulence. In
the interactions, minus (−) represents resistance and plus (+) susceptibility. (Person and Mayo,
1974.)

   Since R genes are only detectable by the presence of the appropriate A
genes in the pathogen and vice-versa they have been called "conditional"
genes (Person and Mayo, 1974). A pathogen strain may therefore carry,
undetectably, any combination of "A" or "a" genes at loci corresponding
with "r" genes in the host, and the host may carry, undetectably, any number
of "R" genes corresponding with "a" genes in the pathogen. This is illustrated
in Fig. 4 taken from Person and Mayo (1974) which shows that only one of
four pairs of alleles in host and pathogen results in a stop-signal. This poses
a problem in selecting pathogen strains which are genetically as similar as
possible.

3. Choice of Natural Variants or Induced Mutants

Alternative sources of pathogen strains seem to be (i) strains selected from wild populations and studied genetically by crosses and by tests of virulence on a wide range of host varieties, preferably choosing those which seem similar at many loci conditioning virulence, but differing at selected loci, and (ii) strains which have been studied genetically and in which artificial mutations have apparently been produced at single loci governing virulence.

The second alternative was adopted by Rowell et al. (1963) who irradiated a culture of P. graminis derived from a cross between races 111 and 36. The culture was known to be heterozygous at eight loci conditioning virulence, and cultures which had become homozygous for virulence at single loci and were true breeding were selected.

Much knowledge of the biochemistry of fungi has been obtained from studies of mutants. Fincham and Day (1971) have, however, pointed out that most of these investigations have been made possible by the occurrence and preservation in the laboratory of mutants which could hardly survive in nature. Although investigations of induced mutations can contribute to understanding of the genetics of pathogenic fungi and of host–parasite interactions (Day, 1974), it should be noted that such mutants may differ from naturally occurring pathogen variants. This may not be evident from preliminary observations of host–parasite interactions.

Person and Mayo (1974) illustrated the apparent segregation for co-dominant alleles at a disease-resistance locus in the host when tested with certain pathogen strains. A similar pattern can be created for segregation of codominant alleles for non-virulence in the pathogen using appropriate host lines (Fig. 5). The non-virulence allele A_1 corresponding with host gene R_1 produces an incompatible reaction, but the non-virulence allele A_2 produces a compatible interaction with the r_2 locus in the host. Thus the A_2 allele could be expressed as a recessive allele for virulence. A similar pattern would arise if the A_2 allele was replaced by a deletion or an inactive allele as might result from artificial mutation. Flor (1960) showed that a culture of race 1 of Melampsora lini was heterozygous for dominant non-virulence on the flax variety Koto. Irradiation with X-rays produced two cultures

	Real Pathogen gene frequency			Apparent Pathogen gene frequency	
	1 :	2 :	1	3 :	1
	$A_1 A_1$:	$A_1 A_2$:	$A_2 A_2$	$A-$:	aa
Host line					
$R_1 - r_2 r_2$	−	−	+	−	+

FIG. 5. Apparent segregation for codominant allelic genes for non-virulence in a pathogen F_2 tested on a host line with one gene for resistance.

virulent on Koto. These were selfed and the 198 cultures obtained were all virulent on Koto. Sixteen of them were also virulent on Leona and Abyssinian to which race 1 was homozygous for non-virulence. Flor suggested that a deletion had occurred removing closely linked genes for non-virulence. The deletions, whether for the locus controlling virulence on Koto only, or on all three varieties, would be non-functional alleles. Day (1974) has described work of G. J. Lawrence demonstrating multiple alleles Ap, Ap1, Ap2 and Ap3 for non-virulence in flax rust and has suggested that this would imply that the Ap locus has some other function in addition to determining virulence since a non-functional allele would apparently cover all eventualities for virulence but not for its other functions. Thus it may be argued that naturally occurring alleles for virulence would not be non-functional and would only appear recessive because of interaction with a corresponding "r" gene for susceptibility in the host as shown in Fig. 5. Comparisons between a mutational deficiency at the Ap locus and a functional allele might enable a specific gene product to be identified with the functional allele but would not allow the differences between alleles at the Ap locus, which control specificity, to be studied.

4. *Changing Pathogen Strains*

It may often be difficult to obtain the same pathogen strains for physiological or biochemical studies as were used in developing genetically controlled host lines. The development of culture collections such as those described by Loegering (1965), including cultures used in genetical studies of hosts and pathogens, will help to alleviate this problem. When it is necessary to use other cultures, however, it should always be remembered that responses may be elicited from previously undetected genes for resistance. The flax variety Bison used by Flor (1954) in developing lines with single genes for resistance to *Melampsora lini* was susceptible to all North American races of this pathogen but Kerr (1960) described Australian races to which Bison was resistant. As mentioned above, this can also occur in nearly-isogenic lines where closely linked genes of resistance remain together during development of the lines, as in the following example. Loegering and Harmon (1969) noted that there was a gene for resistance to *P. graminis* at the Sr-9 locus in the wheat variety Chinese Spring which they used as the susceptible parent in developing lines with genes for resistance. In this case the role of the resistant and susceptible pair of lines could be reversed by appropriate choice of cultures.

IV. CONCLUSIONS

The use of genetically controlled host and pathogen cultures in studies of characters thought to control disease development can provide powerful aids to critical analyses of the physiology and biochemistry of host–pathogen relations and the determination of true causes of resistance. Limits to the power of these aids are imposed by aspects of the gene-for-gene relationship between genes for resistance in the host and for virulence in the pathogen and also by the methods of production of the host and pathogen lines.

Nearly-isogenic lines in the host are especially useful for studying disease resistance mechanisms which are under simple genetic control. Such lines can be produced either in pairs with alternative alleles at a resistance locus, or in groups in which several lines with different genes for resistance can be compared with the susceptible line used in development of the group. These lines can be used most efficiently when the method of production is taken into account and when strains of the pathogen used in their development are available. Chromosome substitution lines, such as those available in wheat, may be useful for investigations of disease resistance which is under more complex genetic control.

The number of repetitions of selfing or backcrossing used in developing nearly-isogenic lines or chromosome substitution lines influences the degree of similarity attained between lines at loci other than the selected resistance gene or chromosome. It also affects the degree of homozygosity achieved and thus the ability of the lines to breed true in subsequent generations. Maintenance of the lines requires care during their reproduction to prevent outcrossing and, if possible, the use of selection against the accumulation of mutations.

In lines which can be reproduced as clones, for example rust uredospores, homozygosity is not necessary for maintenance of a pure line, but the achievement of genetically controlled stocks may still require the use of procedures similar to those used in developing nearly-isogenic lines, or in the use of induced mutations. It may be questioned, in some cases, whether induced mutations represent similar variation to that occurring in wild populations.

In the case of disease resistance, which is under more complex genetic control and which may be of special interest to plant breeders as it is durable under agricultural conditions, the production of genetically controlled lines suitable for the investigations of mechanisms of resistance may be more difficult. The more complex the genetic control of resistance, the less likely that direct observation of a single, genetically diverse pair consisting of a resistant and a susceptible variety will permit a causal role in resistance to be attributed to any character observed in the resistant parent. In such a situation there may be no alternative to observing the character in several varieties of diverse origin, and studying the inheritance of the character and the disease resistance together.

ACKNOWLEDGEMENTS

I am grateful to colleagues in the Cytogenetics and Pathology and Entomology Departments of the Plant Breeding Institute for helpful discussions during preparation of this chapter and to the Director, Professor Ralph Riley for permission to attend the Symposium.

REFERENCES

Allard, R. W. (1960). "Principles of Plant Breeding". John Wiley and Sons, Chichester and New York.

Atkins, I. M. and Mangelsdorf, P. C. (1942). *J. Am. Soc. Agron.* **34**, 667–668.

Barrett, D. H. P. and Flavell, R. B. (1975). *Theoret. appl. Genetics* **45**, 315–321.

Boone, D. M. (1971). *A. Rev. Phytopath.* **9**, 297–318.

Briggle, L. W. (1969). *Crop Sci.* **9**, 70–72.

Daly, J. M. (1972). *Phytopathology* **62**, 392–400.

Day, P. R. (1974). "Genetics of host-parasite interaction". Freeman, San Francisco.

Doubly, J. A., Flor, H. H. and Claggett, C. D. (1960). *Science, Wash.* **131**, 229.

El-Gewely, M. R., Smith, W. E. and Colotelo, N. (1972). *Can. J. Genet. Cytol.* **14**, 743–751.

Ellingboe, A. H. (1972). *Phytopathology* **62**, 401–406.

Fincham, J. R. S. and Day, P. R. (1971). "Fungal Genetics", (3rd edition) Blackwell Scientific Publications, Oxford.

Flor, H. H. (1954). *USDA Tech. Bull.* **1087**, 1–25.

Flor, H. H. (1956). *Adv. Genet.* **8**, 29–54.

Flor, H. H. (1960). *Phytopathology* **50**, 603–605.

Flor, H. H. (1971). *A. Rev. Phytopath.* **9**, 275–296.

Forsyth, F. R. (1956). *Can. J. Bot.* **34**, 745–749.

Green, G. J., Knott, D. R., Watson, I. A. and Pugsley, A. T. (1960). *Can. J. Pl. Sci.* **40**, 524–538.

Hanson, W. D. (1959). *Genetics N. Y.* **44**, 833–837.

Hooker, A. L. and Russell, W. A. (1962). *Phytopathology* **52**, 14 (Abstr.).

Hooker, A. L. and Saxena, J. M. S. (1971). *A. Rev. Genet.* **5**, 407–424.

Kasha, K. J. (1974). "Haploids in Higher Plants, Advances and Potential". Proceedings of the First International Symposium, University of Guelph, Canada, 1974.

Kerr, H. B. (1960). *Proc. Linn. Soc. N. S. W.* **85**, 273–321.

Knott, D. R. and Kumar, J. (1972). *Physiol. Pl. Path.* **2**, 393–399.

Law, C. N. and Johnson, R. (1967). *Can. J. Genet. Cytol.* **9**, 805–822.

Law, C. N. and Worland, A. J. (1973). 1972 Report of the Plant Breeding Institute, Cambridge. pp. 25–65.

Lim, S. M., Hooker, A. L. and Paxton, J. D. (1970). *Phytopathology* **60**, 1071–1075.

Littlefield, L. J. (1969). *Phytopathology* **59**, 1323–1328.

Loegering, W. Q. (1965). *Phytopathology* **55**, 247.

Loegering, W. Q. and Harmon, D. L. (1969). *Phytopathology* **59**, 456–460.

McCoy, M. S. and Ellingboe, A. H. (1966). *Phytopathology* **56**, 683–686.

Moseman, J. G. (1972). *Crop Sci.* **12**, 681–682.

Moseman, J. G. and Helms Jorgensen, J. (1973). *Euphytica* **22**, 189–196.

Person, C. and Mayo, G. M. E. (1974). *Can. J. Bot.* **52**, 1339–1347.

Rohringer, R., Howes, N. K., Kim, W. K. and Samborski, D. J. (1974). *Nature, Lond.* **249**, 585–588.

Rowell, J. B., Loegering, W. A. and Powers, H. R. (1963). *Phytopathology* **53**, 932–937.

Sears, E. R. (1953). *Amer. Nat.* **87**, 245–252.

Shepherd, K. W. and Mayo, G. M. E. (1972). *Science, Wash.* **175**, 375–380.

Skipp, R. A. and Samborski, D. J. (1974). *Can. J. Bot.* **52**, 107–115.

Slesinski, R. S. and Ellingboe, A. H. (1971). *Can. J. Bot.* **49**, 303–310.

Watrud, L. S., Hooker, A. L. and Koeppe, D. E. (1975). *Phytopathology* **65**, 178–182.

Wolfe, M. S. (1972). *Rev. Pl. Path.* **51**, 507–522.

CHAPTER 4

Structural Aspects of Infection by Biotrophic Fungi

D. S. INGRAM

Botany School, Downing Street, Cambridge, England

AND

J. A. SARGENT AND I. C. TOMMERUP

A.R.C. Unit of Developmental Botany, Cambridge, England

I. INTRODUCTION

A biotrophic fungus may be defined as one which, as a parasite, must derive the nutrients it requires for growth and full development from the living tissues of a compatible host. The most important phase in the life history of such an organism occurs during the period immediately following the germination of its spores upon the surface of a potential host plant. During this time a series of characteristic, specialized infection structures are normally elaborated (Bushnell, 1972; Scott, 1972; Figs 1 and 2). The function of these structures is to bring the fungus from the plant surface, where its spores germinate, to a site suitable for haustorium formation.

Most rust fungi (Uredinales) and many of the downy mildews (Peronosporaceae) gain entry to host tissues via the stomata (e.g. Royle and Thomas, 1971). The uredospore germ tube of a rust fungus forms an appressorium over a stomatal pore and an infection thread then grows into the substomatal cavity. This thread gives rise to a substomatal vesicle from which infection hyphae arise and ramify through the intercellular spaces of the leaf (Maheshwari *et al.*, 1967). Haustorial mother cells arise from the infection hyphae and intercellular hyphae, and give rise to haustoria which enter mesophyll cells, invaginating each plasmalemma. These specialized branches then establish an intimate contact with the living protoplasts of the penetrated cells.

In contrast, a few rust fungi (e.g. Zimmer, 1965), all powdery mildews (Erysiphales) (e.g. Bracker, 1968) and some downy mildews (e.g. Sargent *et al.*, 1973) penetrate host organs by direct invasion of epidermal cells. In these cases the germ tube gives rise to an appressorium which adheres firmly to the cuticle of the host. From this structure an infection peg develops and penetrates the cuticle and cell wall. Once inside the penetrated cell *Erysiphe* spp. form a haustorium which invaginates the plasmalemma and immediately enters into an intimate relationship with the protoplast. Vegetative growth of the fungus on the surface of the infected leaf does not commence until this structure has been elaborated (Ellingboe, 1972). Direct invasion of epidermal cells by downy mildew fungi and rusts is normally followed by a brief period of intracellular growth, sometimes involving the elaboration of further infection structures, such as the primary and secondary vesicles of *Bremia lactucae* (Figs 1 and 2). Like haustoria, these structures invaginate the plasmalemma of the host. The fungus then grows out into the intercellular spaces and forms an intercellular mycelium bearing haustoria. Some downy mildew fungi produce an intercellular mycelium directly (Chou, 1970); the infection hypha arising from the appressorium forces its way into the leaf along the middle lamella between adjacent epidermal cells, producing haustoria laterally as it does so.

Direct invasion of host cells by the zoospores of some intracellular, plasmodial fungi such as *Polymyxa betae* (Keskin and Fuchs, 1969) and *Plasmodiophora brassicae* (Aist and Williams, 1971) may involve the injection of the parasite through a puncture in the cell wall. In the case of *P. brassicae* this is achieved when a bullet-shaped projectile, the Stachel, is forced from within the encysted zoospores, through the wall of a root hair cell. Penetration occurs in about one second and, once inside the cytoplasm of the host, the young, uninucleate plasmodium of the invader is surrounded by a complex envelope consisting partly of invaginated plasmalemma.

As infection structures form, a complex series of molecular interactions must occur which determine whether a functional relationship will develop between host and parasite, and what the character of that relationship will be. The more important of these interactions may be: recognition of parasite by host, and vice-versa; the secretion by the parasite of substances causing the modifications of cellular metabolism in the host which lead to its accommodation or rejection; the secretion by the host, in incompatible interactions, of toxins which contribute to the cessation of fungal growth; and the transfer of nutrients and other essential metabolites from host to parasite. Precise investigation of the histology and cytology of infection is an indispensable first step towards biochemical and biophysical definition of these processes.

II. The Infection of Lettuce by *Bremia lactucae*

The sequence of structural and biochemical events which occur during infection may be unique to the particular host and parasite interaction under consideration, and it is dangerous to make generalizations. As a basis for discussion we shall first describe the structural aspects of infection by a single biotroph that we have studied, namely *B. lactucae* race W5, and a susceptible host, *Lactuca sativa* L. (lettuce) cv. Trocadero Improved. *B. lactucae* is responsible for the downy mildew disease of lettuce. The events leading to the establishment of this fungus in the tissues of Trocadero Improved are summarized in Table I and illustrated in Figs 1–20.

A. PRE-PENETRATION

Even before penetration occurs there are a number of interactions between *B. lactucae* and its host. Nutrients may be absorbed from the leaf surface (Andrews, 1975); recognition may occur through the interaction between

TABLE I

The time-course of infection of the lower epidermal cells of detached cotyledons of the lettuce cultivar "Trocadero Improved" by *B. lactucae* race W5 at 15 °C (after Sargent *et al.*, 1973)

Time after release of spores from inhibition[a] (h)	Stage of development of the fungus
0	Spores placed on surface of cotyledon in a droplet of water
1–2	Spore germination
2–3	Elaboration of the germ tube and appressorium
3–4	Penetration and enlargement of the primary vesicle in an epidermal cell
4–13	Enlargement of the secondary vesicle; commencement of nuclear division
13 onwards	Continued enlargement of the secondary vesicle; the first appearance of divided nuclei; invasion of adjacent cells and tissues by haustoria and intercellular hyphae

[a] Spores of *B. lactucae* contain a water soluble inhibitor of germination which is removed by washing (Mason, 1973).

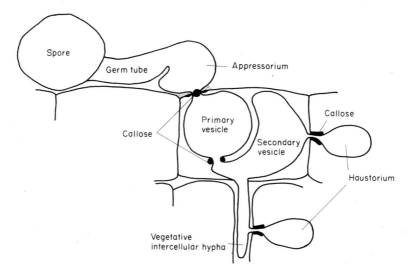

FIG. 1. A diagrammatic representation of the infection of a susceptible lettuce epidermal cell by *B. lactucae* (after Sargent *et al.*, 1973).

fungal secretions, or molecules attached to the wall, and similarly secreted or constitutive host products. This may result in a stimulation of host cell metabolism leading ultimately to accommodation or rejection of the fungus. In addition, fungal enzymes may at this stage begin the processes of preparing the host wall for penetration.

1. *The Germ Tube and Appressorium* (Figs 1 and 2)

Fine structure studies of *B. lactucae* show that before penetration, the outer surfaces of the wall of the germ tube and appressorium are not sharply defined (Fig. 4). This probably results from secretion through the wall of substances which assist in attaching the germ tube and appressorium to the surface of the host (Figs 5 and 6). The structure of the cytoplasm of the germ tube immediately behind the apex is not unusual, mitochondria, endoplasmic reticulum, ribosomes, dictyosomes and lipid droplets being the major components; nuclei are abundant and small vacuoles are present (Fig. 3). However, the cytoplasm at the apex of the germ tube is quite different (Fig. 4), being composed almost entirely of cytoplasmic vesicles, many of which fuse with the plasmalemma, interspersed with a few ribosomes and short profiles of endoplasmic reticulum. The plasmalemma is invaginated and frequently merges into membrane complexes which we call lomasomes. The tip of the appressorium is also occupied by densely packed cytoplasmic vesicles, and here large lomasomes are evident (Fig. 6). Similar aggregations of cytoplasmic vesicles and lomasomes have been noted in the tips of penetrating cells of other fungi, including *Melampsora lini* (flax rust) (Littlefield and Bracker, 1972) and *Olpidium brassicae* (Lesemann and Fuchs, 1970), and in the growing tips of the hyphae of saprotrophic fungi (Grove and Bracker, 1970). The vesicles of both the germ tube tip and the appressorium of *B. lactucae* appear to have their origin in the cisternae of the dictyosomes which are abundant in the cytoplasm of the bodies of the germ tube and appressorium (Fig. 5 in Sargent and Payne, 1974).

In both the germ tubes and the appressoria it is noticeable that the cytoplasmic vesicles show differential staining properties, probably indicating that they contain different substances. It is reasonable to suppose that a proportion of them contain the enzymes and structural molecules which are involved in the rapid extension of the cell wall. Others are likely to contain substances which, as will be seen, prepare the host cells for penetration by causing changes in wall structure, by stimulating metabolism or by interacting with host molecules in the recognition process. The lomasome structures may simply represent the accumulated membranes of vesicles which have discharged their contents into or through the fungal wall. Alternatively, they could be formed by a proliferation of the plasmalemma and be the sites where the contents of different cytoplasmic vesicles mix and become assembled as wall building molecules or as molecules to be involved in the processes of recognition and infection. Quite clearly a precise cytochemical study of the origins and contents of the cytoplasmic vesicles and lomasomes in germ tubes and appressoria is required to ascertain their role in the processes of growth and infection.

2. *Pre-penetration Responses of the Host Cell*

A section through the area of contact between a lettuce epidermal cell and a newly formed appressorium of *B. lactucae* is shown in Fig. 5. At this very early stage the cells below the appressorium differ little in appearance from those remote from germinating spores: the sparse cytoplasm is spread thinly around a large central vacuole and the outer cell wall, which is covered by a thin cuticle, can be divided into an inner zone which stains lightly and a darker staining outer zone. A few minutes later, as the appressorium matures, but before penetration commences, significant structural changes become obvious in the cell which is to be penetrated (Fig. 6). In the region of the periclinal wall of this cell corresponding to the site at which the infection peg will form the cuticle becomes swollen, with

Abbreviations used in Figs 2–20: *B. lactucae*: S = Spore; A = Appressorium; H = Intercellular hypha; HA = Haustorium; PV = Primary vesicle; SV = Secondary vesicle; CP = Callose plug; cv = Cytoplasmic vesicles; d = Dictyosome; e = Endoplasmic reticulum; l = Lipid droplet; lo = Lomasome; m = Mitochondrion; n = Nucleus; v = Vacuole; w = Cell wall; Lettuce: c = Cuticle; ca = Callose deposit; cc = Callose collar; cl = Chloroplast; cm = Crystal-containing microbody; hd = Dictyosome; he = Endoplasmic reticulum; hn = Nucleus; hp = Plasmelemma; hw = Cell wall; mt = Microtubule. Scale bar = 1 μm.

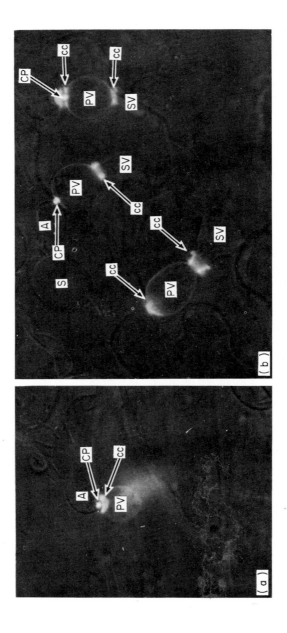

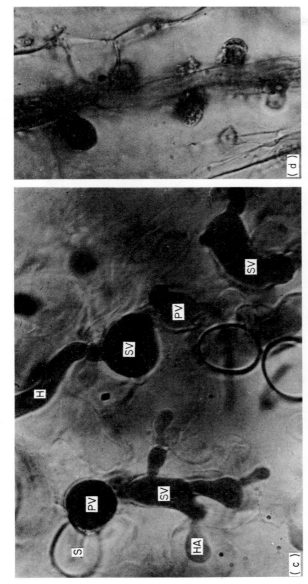

Fig. 2. Light micrographs of the infection structures of *B. lactucae*. (a) Appressorium and primary vesicle (aniline blue/u.v.); note the fluorescent collar around the neck of the primary vesicle and the plug between the primary vesicle and appressorium. (b) Three primary vesicles with developing secondary vesicles (aniline blue/u.v.); note the callose collar around the neck separating each primary vesicle and secondary vesicle. (c) Primary vesicles, secondary vesicles, haustoria and intercellular hyphae (acid fuchsin). (d) An intercellular hypha with haustoria (resorcin blue); photograph supplied by Dr P. A. Mason. Note that the haustoria are in various stages of en-sheathment.

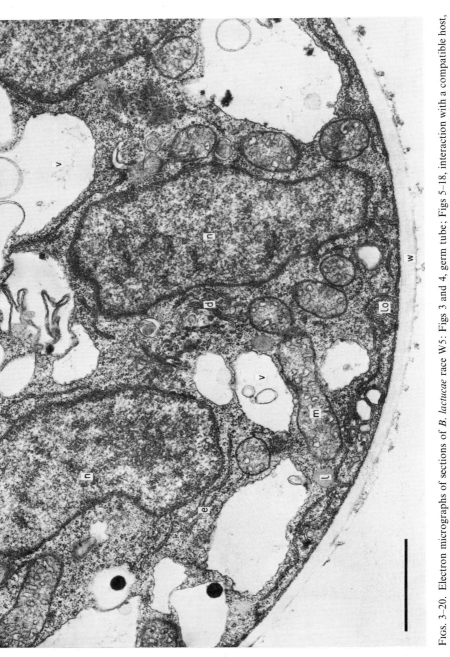

FIGS. 3–20. Electron micrographs of sections of *B. lactucae* race W5: Figs 3 and 4, germ tube; Figs 5–18, interaction with a compatible host, Lettuce cv. "Trocadero Improved"; Figs 19 and 20, interaction with a resistant host, Lettuce cv. "Avondefiance".

FIG. 3. Transverse section through a germ tube a short distance behind the tip. Nuclei, mitochondria, endoplasmic reticulum, dictyosomes, ribosomes and lipid droplets are abundant. Small vacuoles are present and coalescing. Lomasomes are continuous with the plasmalemma (Sargent

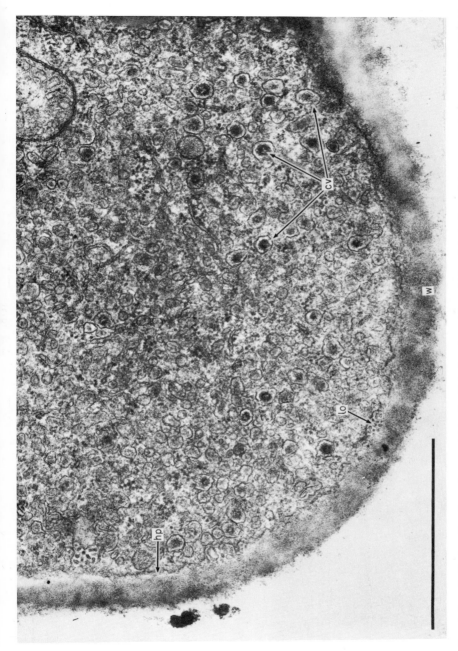

Fig. 4. Longitudinal section through a germ tube tip. Cytoplasmic vesicles, which originate from dictyosomes and whose contents stain at different intensities, accumulate in this region. They fuse with the plasmalemma which, as a result, is highly invaginated and interrupted by lomasomes. The outer surface of the fungal wall is irregular due to the presence of a secretion (Sargent et al., 1973).

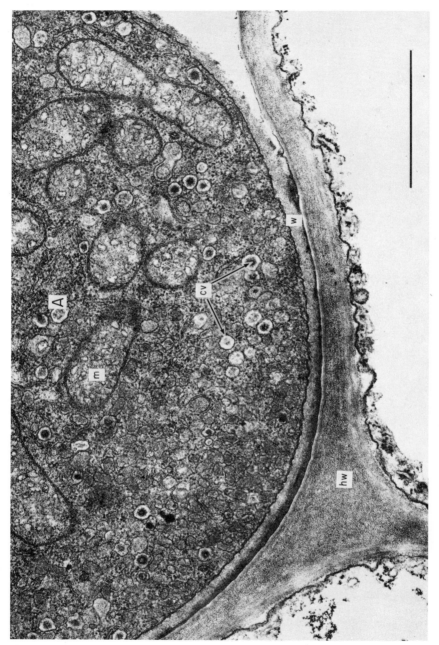

FIG. 5. A section through an area of the appressorium soon after making contact with the epidermal surface of a lettuce cotyledon. Cytoplasmic vesicles are abundant in the cytoplasm of the fungus adjacent to the area in contact with the host. As yet no effect on the host is visible (Sargent *et al.*, 1973).

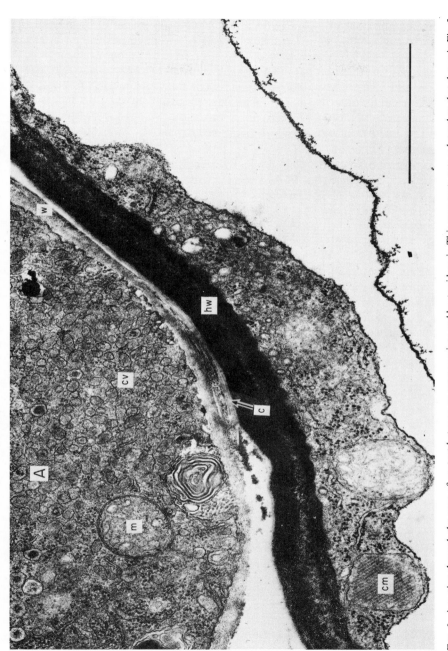

FIG. 6. A section through the zone of contact between an appressorium and host epidermis. This represents a stage later than that shown in Fig. 5 but before the formation of a penetration peg. Fungal cytoplasmic vesicles are discharging their contents through the plasmalemma. The host cuticle is swollen, its wall darkened and its cytoplasm aggregated beneath the area of contact. Host cytoplasm is vesiculated, and a crystal-containing microbody, whose formation is induced by the presence of the fungus, is present (Sargent *et al.*, 1973).

its lamellations separated, while the molecular properties of the wall change
so that it stains abnormally heavily. These changes almost certainly represent
a partial degradation of the cuticle and matrix by enzymes secreted by the
parasite. However, the possiblity of activation of endogenous host enzymes
by the parasite cannot be ruled out. A deposit of amorphous material
between the walls of host and pathogen is probably little more than a fungal
secretion which aids the adhesion of the appressorium. Similar layers of
material have been observed between appressoria or haustoria mother cells
and host walls in a wide range of host–parasite interactions (Bracker and
Littlefield, 1973).

Coincidental with the changes in the cell wall, structural changes indicating a shift in metabolism become evident in the cytoplasm of the host
immediately adjacent to the point of contact, presumably as a result of
fungal influence. One of the most noticeable of these changes is that the
cytoplasm at this point becomes aggregated, partly as a result of migration
to the area and partly because of vesiculation following a stimulation of
dictyosome activity. The vesicles presumably contain molecules which are
important in interactions with the invader. Crystal-containing microbodies
also appear in the cytoplasm of the host at this stage. At a later stage of
appressorium development the changes in the cytoplasm of the host intensify,
and at the same time a pad of material, probably corresponding to the
papilla seen in other host–parasite interactions (Bracker and Littlefield, 1973),
develops between the plasmalemma and cell wall immediately below the point
of contact with the fungus (Fig. 7). This pad probably contains callose since
it fluoresces characteristically with aniline blue and ultra-violet light and
stains blue with resorcin blue (Sargent et al., 1973).

Similar changes in the structure of host cells before penetration have been
observed in other host–biotroph interactions (e.g. Edwards and Allen, 1970;
Stanbridge et al., 1971; Bushnell, 1972; Bracker and Littlefield, 1973). They
point to a significant interference of the metabolism of the host as a prelude
to penetration. This suggests that host organelles are involved in penetration
itself, probably providing additional surface membrane and also secreting
molecules which aid the fungus in its attempt to gain access. The formation
of crystal-containing microbodies is not a response which is peculiar to
lettuce. Similar structures with a lattice spacing of 22 nm have been observed,
for example, in flax and sunflower tissues 11 days after infection by rust
fungi (Coffey et al., 1972). They are particularly interesting since they may
represent a response of the cytoplasm of the host to proteins secreted by the
fungus.

B. PENETRATION

With respect to their function in host cell wall penetration, the appressoria
of powdery mildews, and of those downy mildews which penetrate directly,

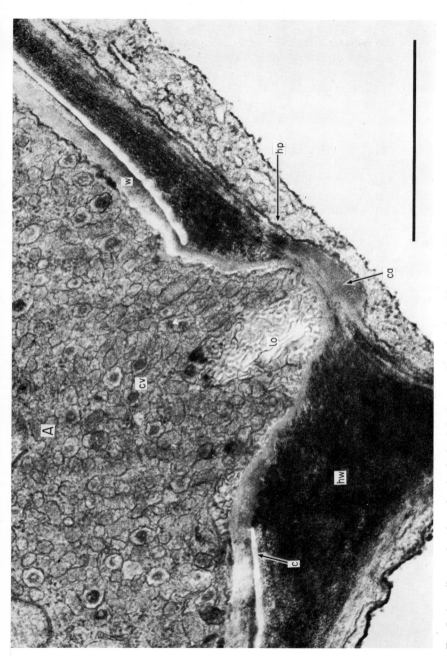

Fɪɢ. 7. A median section through a developing infection peg. Cytoplasmic vesicles are mobilized towards the site of peg formation where they discharge their contents through the plasmalemma so rapidly that a lomasome is formed. The host cuticle and wall microfibrils have been digested but their profiles are undistorted. The outer edge of the infection peg wall is ill-defined and appears to "flow" between the host microfibrils. A callose pad has formed beneath the penetration peg outside the plasmalemma of the host (Sargent *et al.*, 1973).

and the haustorial mother cells of the rust fungi may be regarded as some-what analogous structures. For many years it was assumed that because biotrophic fungi cause little structural damage in their hosts, their capacity for the production of extracellular wall degrading enzymes is limited and that wall penetration must, therefore, be achieved by mechanical pressure. The fact that the appressorium or haustorial mother cell is attached firmly to the host wall by a fungal secretion and would, therefore, provide the necessary anchorage against which the infection peg might generate the force necessary to effect penetration (Dickinson, 1960) lent support to this view, as did the demonstration that fungal infection pegs can penetrate biologically inert barriers such as gold foil (Brown and Harvey, 1927). However, it now seems likely that most biotrophic fungi do possess the appropriate enzymes and that penetration is, therefore, partially or completely by chemical dissolution of walls in very many cases (Peyton and Bowen, 1963; McKeen *et al.*, 1969; Edwards and Allen, 1970; Stanbridge *et al.*, 1971; Heath and Heath, 1971; Littlefield and Bracker, 1972; Coffey *et al.*, 1972; Bracker and Littlefield, 1973; Sargent *et al.*, 1973; Staub *et al.*, 1974). In some instances (e.g. *Erysiphe graminis*) apparent chemical alteration of the wall of the host around the infection site may be extensive; in others (e.g. many rusts) the zone of alteration may be very narrow. It has been suggested that the penetration of the walls of Brassica root hairs by the Stachel of *P. brassicae* may be entirely mechanical (Aist and Williams, 1971), although as Stahmann (1973) has pointed out, enzymatic penetration, mediated by enzymes bound to the Stachel itself, could occur even here.

The kinds of ultrastructural evidence on which most conclusions concerning penetration of walls have been based are shown in Figs 7 and 8. A median section through an almost fully developed infection peg of *B. lactucae* is shown in Fig. 7 and a non-median section in Fig. 8. In both figures it is clear that the infection peg has penetrated the cuticle and outer zone of the wall, leaving the inner zone intact. The profiles of the cuticle and wall microfibrils are not distorted at this stage, and there is no other evidence of the application of mechanical force. The developing wall of the infection peg appears to be in a very fluid state; its surface is irregular, with projections extending between the digested ends of the host wall microfibrils. Large numbers of cytoplasmic vesicles with differential staining properties are concentrated in the appressorium around the site of penetration, and a large lomasome has formed within the infection peg (cf. III A).

Thus it would appear that *B. lactucae* and many other biotrophs have the ability to dissolve enzymatically the waxes (see Staub *et al.*, 1974), cuticle, cellulose, pectic substances, hemicelluloses etc. that comprise the cell wall of a higher plant. However, there is little biochemical evidence that this is so, with the exception of the work of Kunoh and Akai (1969) and McKeen *et al.* (1969) who have shown that the halo of host wall alteration around the infection peg of *E. graminis* contains reduced amounts of cutin,

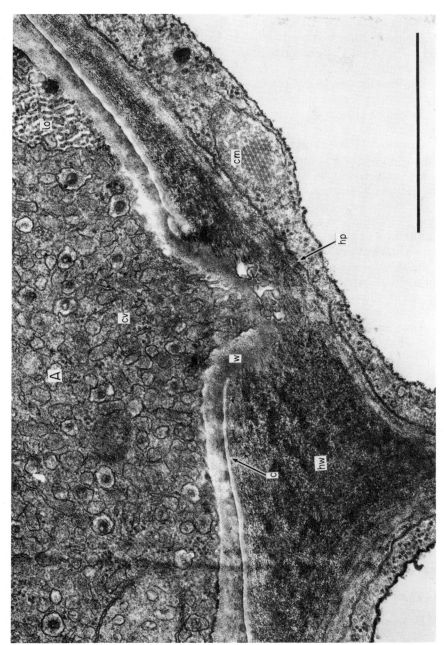

Fig. 8. A non-median section through a penetration peg illustrating the fluid nature of the fungal wall and the presence of a crystal-containing microbody in the host cytoplasm (Sargent *et al.*, 1973).

polysaccharide (including cellulose) and pectin, and increased amounts of reducing sugars, pentose and uronic acid.

A notable feature of penetration by biotrophs is the comparative narrowness of the zone of alteration of host wall around the infection peg, suggesting that diffusible, extracellular enzymes are not produced in large quantities by the invader. It is likely, therefore, that wall-degrading enzymes produced by biotrophs for the purpose of penetration, and presumably subsequent intracellular growth when the middle lamella must be separated, are either of high molecular weight, or more probably bound to the wall of the fungus. Stahmann (1973) has indicated that ways have now been developed for linking enzymes to insoluble supports and that experiments with such insolubilized enzymes have shown that they are more stable than enzymes in solution. The involvement of fungal wall-bound enzymes in penetration would have two major advantages: first it would result in less tissue damage than the production of enzymes into solution, and secondly it would be very economical for the fungus which could penetrate with the least possible expenditure of protein.

Further understanding of cell wall penetration by biotrophs will require biochemical studies of enzyme production by those biotrophs that can be grown in culture, e.g. *Phytophthora infestans* (Friend, 1973), Uredinales and Ustilaginales, and careful cytochemical investigations. It must be determined whether enzymes such as cellulases and pectinases, or their precursors, are located in the cytoplasmic vesicles and lomasomes of the germ tube, appressorium and infection peg; whether certain enzymes are of such high molecular weight that their mobility is limited; whether enzymes are formed in any particular sequence during penetration; and whether enzyme formation is triggered only by contact with particular, appropriate substrates. Some of the methods required are already available, but full investigation will have to wait the development of suitable, sophisticated cytochemical techniques. Such investigations may be important not only in understanding the mechanism of penetration, but in some instances in understanding the mechanism of specificity too, for it has been suggested by Albersheim *et al.* (1969) that it may be through the interaction of wall-degrading enzymes and their substrates that specificity may be determined.

C. INTRACELLULAR INFECTION STRUCTURES

1. *The Primary Vesicle* (Figs 1 and 2)

This structure enlarges very rapidly following penetration by *B. lactucae.* During its formation the host's cuticle and wall become distorted as a result of enlargement of the penetration pore (Figs 19 and 20). At an early stage of development the primary vesicle contains many cytoplasmic vesicles similar in appearance to those observed in the appressorium, and their presence may result from a surge of cytoplasm through the infection peg as the latter penetrates the epidermal cell wall. The irregularly stained cytoplasmic vesicles disperse and finally disappear as the primary vesicle expands and more cytoplasm, rich in mitochondria, nuclei, endoplasmic reticulum, dictyosomes and lipid droplets enters from the appressorium and germ tube. This disappearance of cytoplasmic vesicles, linked with the break-up of the lomasome from the infection peg, suggests that enzyme synthesis may be switched off soon after penetration so that a new and balanced phase of interaction with the host can commence.

Invasion of the epidermal cell by *B. lactucae* does not result in the disruption of the plasmalemma. This membrane is apparently stimulated to increase in area both before and during the rapid expansion of the primary vesicle, and there is no evidence of stretching. Indeed, the interference of host membrane biosynthesis by the fungus is so intense that large folds of surplus plasmalemma may form (Fig. 9). Even in adjacent, non-infected cells metabolism may be affected so that the plasmalemma may proliferate extensively (Fig. 10) and cytoplasm and nuclei migrate to an area adjacent to the infected cell. That the invaginated plasmalemma may be structurally and physiologically different from the normal plasmalemma is suggested by the observation that when infected lettuce cells die rapidly during certain resistance responses, the host membrane which surrounds the primary vesicle remains intact for a considerable time after other host membranes have been destroyed (Fig. 19; see also sections II C 3 and III A).

Noticeably absent from the primary vesicle is any zone of interaction between the fungal wall and the invaginated plasmalemma of the host which may be equated with the sheath (Bracker and Littlefield, 1973) or encapsulation (Peyton and Bowen, 1963) which form following intracellular penetration by the haustoria of biotrophs (see II C 3). If such structures are important in the exchange of molecules between haustoria and host cells (Bushnell, 1972) then one may speculate that the primary vesicle has only a limited function of this kind and may serve primarily as a secondary spore. This idea is supported by the observation that the primary vesicle has only the same number of nuclei as the germinating spore (approximately 16), and that no nuclear divisions occur during its enlargement.

As the primary vesicle ages callose is deposited by the host around its neck and between its wall and the invaginated plasmalemma. In addition a callose plug of fungal origin forms within the infection peg, sealing off the structure from the external appressorium and germ tube (Figs 1, 2 and 20). The microtubules in the cytoplasm of the host close to the primary vesicle in Fig. 11 could be associated with callose deposition and with other aspects of the development of the interface, such as plasmalemma proliferation.

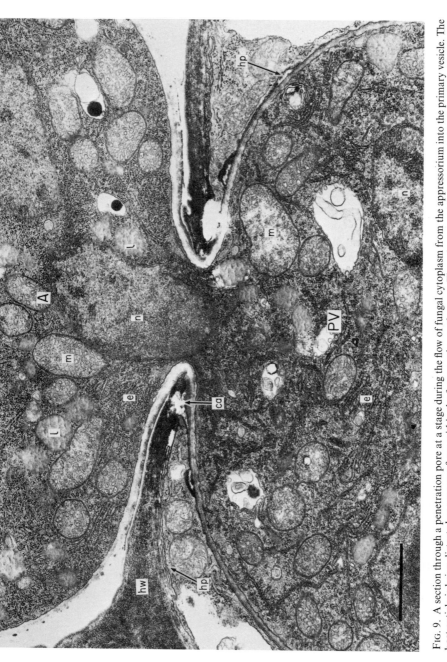

FIG. 9. A section through a penetration pore at a stage during the flow of fungal cytoplasm from the appressorium into the primary vesicle. The host cuticle is being distorted as pressure from within the fungus enlarges the pore. The primary vesicle is surrounded by invaginated host plasmalemma. Rupture of the plasmalemma is apparently avoided by its rapid proliferation in the region of penetration. Areas of proliferated but non-invaginated plasmalemma appear as multilayered membrane profiles surrounding the pore. A callose collar has developed around the neck of the primary vesicle. (Poor embedding of the callose has resulted in some tearing of the section in this region.) (Sargent et al., 1973.)

2. *The Secondary Vesicle* (Figs 1 and 2)

This structure normally becomes visible as a nipple-like projection from the primary vesicle one to two hours after penetration (Fig. 12). It then develops slowly over a period of nine hours or more, invaginating the plasmalemma and almost completely filling the space in the epidermal cell not occupied by the primary vesicle. During this enlargement cytoplasm migrates into it from the primary vesicle which consequently becomes highly vacuolate. The cytoplasm of the secondary vesicle (Figs 12 and 13) is little different from that of the primary vesicle, except that there is an increase in the total volume (i.e. it contains more cytoplasm than that contributed by the primary vesicle alone), and stages of nuclear division (mitosis) occur. These observations indicate that during development of the secondary vesicle the dependence of *B. lactucae* on its host becomes more profound.

A further indication that the relations between the secondary vesicle and the penetrated cell are different from those established by the primary vesicle comes from an examination of the region of interface between the fungal wall and the invaginated plasmalemma of the host (Figs 12 and 13). The most noticeable feature is a dark-staining osmiophilic layer which is continuous with the fungal wall. This layer, which in some ways resembles the encapsulation (cf. sheath, Bracker and Littlefield, 1973) of the haustoria of downy mildew fungi (see below and Peyton and Bowen, 1963; Bushnell, 1972; Bracker and Littlefield, 1973; Beckett *et al.*, 1974), is completely absent from the primary vesicle during all stages of its development. Frequently associated with the osmiophilic layer is an alignment of host endoplasmic reticulum (Fig. 13), although this never fuses with the invaginated plasmalemma. It is not known whether the osmiophilic layer is of host or fungal origin, or both.

Other effects on the cytoplasm of the host include a continuation of membrane proliferation, vesiculation and accumulation of cytoplasm and organelles close to the fungus (Figs 12 and 13), and the stimulation of dictyosome activity (Fig. 12). Vesicles produced by the latter could be involved in the proliferation of membrane, the deposition of dark material at the interface, and in the formation of callose, particularly as a collar around the neck between the primary and secondary vesicles (Figs 1, 2 and 12). A callose plug does not form between the two structures.

3. *The Haustorium* (Figs 1 and 2)

Haustoria of *B. lactucae* normally arise from intercellular hyphae and penetrate mesophyll cells (Fig. 2), although they also arise from the secondary vesicle and occasionally from the primary vesicle (Figs 1, 2 and 14). The portion of an intercellular hypha which initiates a haustorium, like the

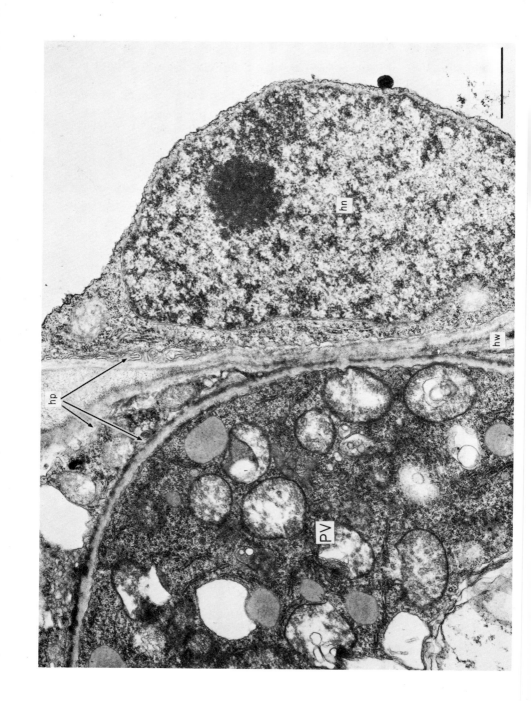

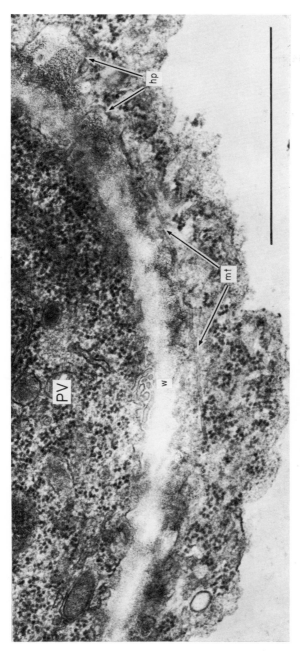

FIG. 10. (Upper) A section through parts of two adjacent cells of the host, one of which contains a mature primary vesicle. The fungus has stimulated proliferation of host plasmalemma in the invaded cell and in the adjacent, as yet uninvaded, cell. Note the close association of fungus and host nucleus albeit they are in separate host cells.

FIG. 11. (Lower) A section through the wall of an enlarging primary vesicle. The invaginated host plasmalemma is highly convoluted and therefore indistinct in profile. Host microtubules are closely associated with the interface between the two organisms.

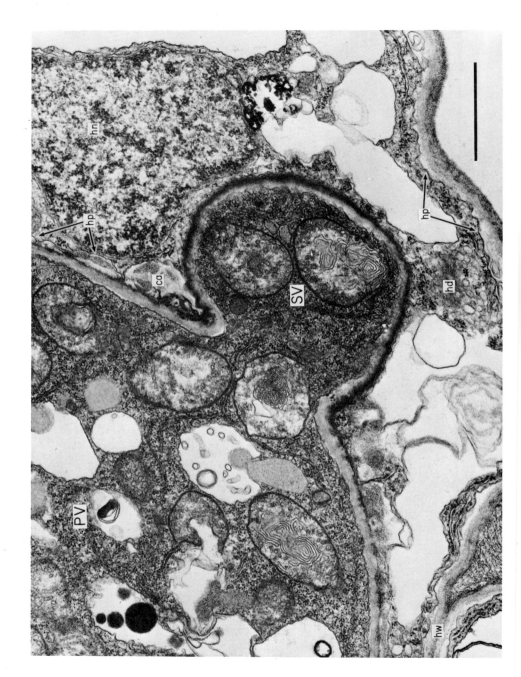

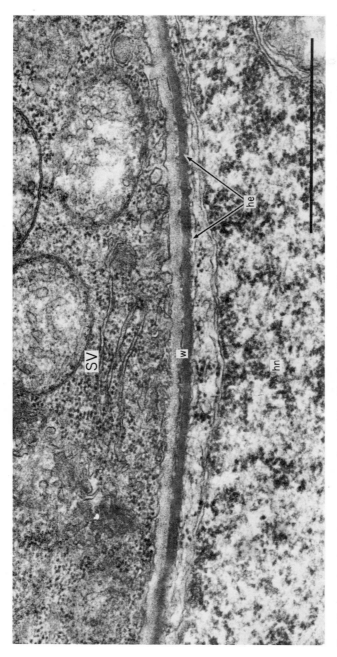

FIG. 12. A section through a young secondary vesicle arising from a primary vesicle. The outer layer of the wall of the secondary vesicle stains darkly like that of the haustorium (Figs 14–18). Host plasmalemma continues to increase in area to accommodate the secondary vesicle, and *in situ* to produce multilayered membrane profiles. Host dictyosomes have become activated and may contribute membrane to the plasmalemma and callose to the enlarging collar around the neck of the secondary vesicle.

FIG. 13. A section through the wall of a mature secondary vesicle. The outer layer of the wall is darkly staining (cf. wall of primary vesicle, Figs 9–12). Host endoplasmic reticulum is clearly associated with the secondary vesicle.

appressorium already described and the haustorial mother cells of rust fungi (Bracker and Littlefield, 1973) is firmly attached by an amorphous secretion to the wall of the cell to be penetrated (Fig. 15). Penetration by haustoria, like the initial penetration, apparently occurs through enzymatic dissolution of the host cell wall (see also Coffey *et al.*, 1972; Littlefield and Bracker, 1972; Bracker and Littlefield, 1973). Penetration of a mesophyll cell by a developing haustorium is preceded by massive proliferation of the host's plasmalemma near the point of attachment, and as the haustorium enlarges to its final club shape within the cell (Fig. 14) this membrane is invaginated without any evidence of stretching. We have already indicated that the plasmalemma invaginated by the primary and secondary vesicles may be different from the normal plasmalemma of the host, although we have no evidence that this is the case here. However, evidence from studies of other host–parasite interactions has suggested that the invaginated host plasmalemma is specialized when it is associated with haustoria. Berlin and Bowen (1964) and Peyton and Bowen (1963) showed an association of this membrane with the secretory apparatus of the host in their studies of *Albugo* and *Peronospora*. Altered thickness of the plasmalemma around haustoria of *Erysiphe* has been noted by Bracker (1968) and Ehrlich and Ehrlich (1963). More conclusively, Littlefield and Bracker (1972) have used the techniques of freeze-etch microscopy and PACP staining to demonstrate that the invaginated plasmalemma around haustoria of *Melampsora lini* is markedly different in structure from that elsewhere in the cell (see also section III C). In view of the probable importance of the membrane surrounding haustoria in the exchange of molecules between the host protoplast and the parasite it is unfortunate that little information regarding its chemical and physical structure, let alone its metabolic potential, is as yet available for any host–parasite interaction.

The cytoplasm of the young developing haustorium of *B. lactucae* is very active and dense, containing many mitochondria, ribosomes, endoplasmic reticulum, lipid droplets, active dictyosomes and 1–3 nuclei (Fig. 4). As the haustorium ages vacuoles become increasingly abundant and fuse, eventually limiting the cytoplasm to a thin peripheral layer (Figs 17 and 18). The nuclei in the ageing haustorium degenerate.

The outer zone of the haustorial wall is, from the time of haustorial initiation, highly osmiophilic (Figs 14 and 15). The host plasmalemma is closely adpressed to this layer which, as the haustorium ages, becomes progressively thicker and deeply lobed on those portions of its outer face which are adjacent to the regions of dense host cytoplasm (Figs 16, 17 and 18).

The cytoplasm of the host shows no evidence of stimulation, and there is little evidence of any adverse influence of the fungus, at least during the earlier stages of haustorium development (Fig. 14). Later, however, degenerative changes do set in (Figs 16, 17 and 18; see also Coffey *et al.*, 1972).

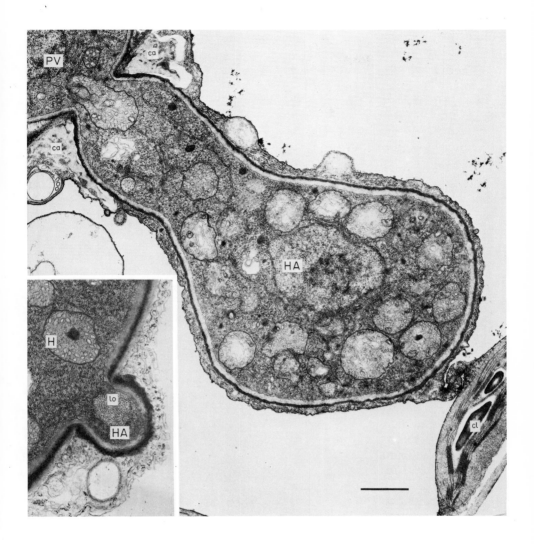

FIG. 14. A longitudinal section through an enlarging haustorium arising directly from a primary vesicle. The wall is characterized by a darkly staining layer which in young haustoria is fairly regular in outline. Host cytoplasm surrounds the fungus which invaginates the host plasmalemma. Around the neck of the haustorium a callose collar has developed between the host wall and plasmalemma.

FIG. 15. A section through a very young haustorium arising from an intercellular hypha. Host plasmalemma is convoluted in the region of penetration but callose has not yet been deposited in appreciable amounts. The darkly staining outer layer of the haustorium wall is already apparent and a large lomasome occupies the penetration pore (cf. Fig. 7).

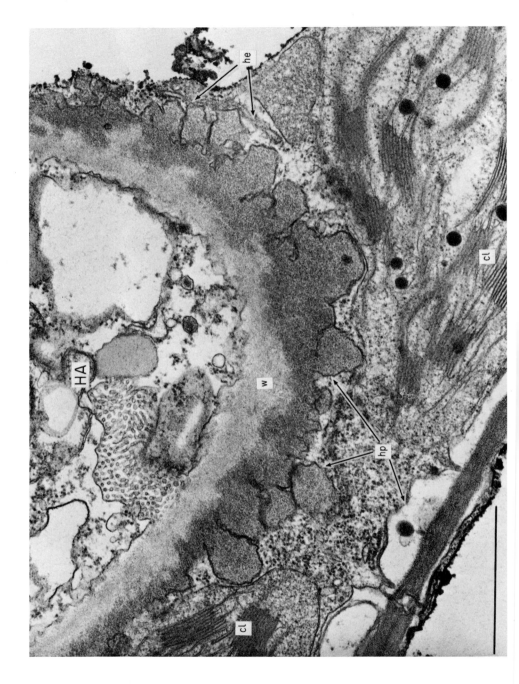

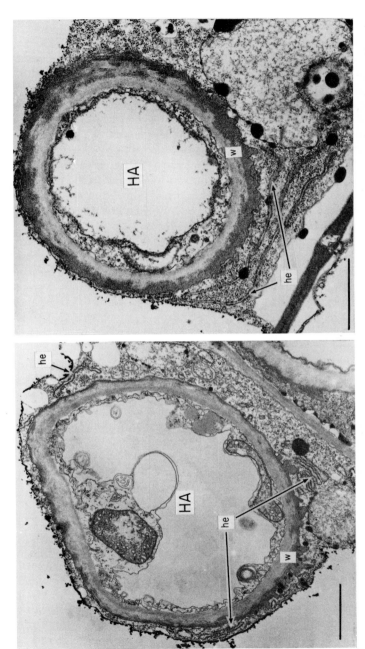

FIG. 16. (Facing page) A section through a mature, vacuolated haustorium from the outer wall of which have developed projections. Since the invaginated host plasmalemma lines these projections their development marks an extension of the fungal/host interface. At this stage of development (qv. Figs 17 and 18) host organelles are beginning to degenerate with the appearance of osmiophilic droplets. Chloroplasts have lost their starch but the grana are largely intact.

FIGS 17 and 18. Sections through haustoria at a stage shown in Fig. 16. Wall projections are not evenly distributed over the haustoria. Host endoplasmic reticulum is closely associated with the haustorium and although it is frequently closely adpressed to the invaginated plasmalemma surrounding the haustorium, the membranes have never been observed in continuity.

An important feature of the cytoplasm of the host is that the endoplasmic reticulum frequently becomes aligned with the haustorium, although it does not fuse with the invaginated plasmalemma (Figs 16, 17 and 18). We have never seen any vesicles or microtubules associated with the dark outer layer of the haustorial wall (see Bracker and Littlefield, 1973 for discussion).

Whether the dark-staining zone of the haustorial wall may be equated with the encapsulation described as surrounding the haustoria of other members of the Peronosporales by Peyton and Bowen (1963) or the sheath around the haustoria of powdery mildews and rusts (Bracker and Little-field, 1973) is a moot point, as is the question as to its origin in the host, fungus or both. The only reliable point of similarity between these various structures is that they occupy a similar position with regard to the haustorium and the protoplast of the host (see for further discussion Berlin and Bowen, 1964; Peyton and Bowen, 1963; Bracker, 1968; Ehrlich and Ehrlich, 1963, 1971; Manocha and Shaw, 1967; McKeen et al., 1966; Littlefield and Bracker, 1972; Bracker and Littlefield, 1973; Bushnell, 1972). The nature and function of the components of the complex interfaces between haustoria and host protoplasts is still almost completely unknown, as is the metabolic function of haustoria themselves (absorption, secretion or both?), despite their obvious importance as the probable sites of molecular exchange, and the intense activity of electron microscopists for more than a decade. This is an area of host–parasite interaction where the application of advanced cyto-logical and cytochemical techniques is urgently required.

Callose is deposited as an extension of the host wall to form a collar around the developing haustorium (Figs 2 and 14). As the haustorium ages this deposit may extend and in most cases eventually completely encloses the haustorial body (Fig. 2). There is no evidence that senescence of haustoria is either a cause or a consequence of callose deposition, although Heath and Heath (1971) did associate the complete encasement of haustoria of Uromyces phaseoli var. vignae with certain types of host resistance. Encase-ment of haustoria with callose in compatible interactions is a comparatively rare phenomenon (Littlefield and Bracker, 1972).

III. DISCUSSION

Description of the structural aspects of infection such as we have outlined in section II and such as are reviewed by Ehrlich and Ehrlich (1971) and Bracker and Littlefield (1973) are numerous, but can only be of value if they are accompanied by correlated biochemical and biophysical studies or are followed by the application of advanced cytochemical and autoradio-graphic techniques to provide a bridge between mere knowledge of the sequence and structure of events and an understanding of the molecular basis of interaction. This has so far happened only rarely. We have already indicated some of the areas where we consider that more research of this kind is urgently required; here we give examples of some such research that is already in progress.

During the establishment of a compatible, intracellular relationship between a biotroph and its host the plasmalemma of the penetrated cell or cells is invaginated to accommodate the parasite. At the same time there occurs a complete reorganization of the metabolism of the infected cell, and sometimes adjacent cells, to meet the needs of the invader (Brian, 1967; Scott, 1972). Such changes may, in some instances, result from the mere physical presence of the parasite (e.g. as a non-specific wound response resulting in the deposition of callose) or, in the majority of cases, as an alteration in the expression of the host's genome in response to regulatory molecules secreted by the parasite, or attached to its surfaces. At the structural level alterations in metabolism may, as we have seen, be reflected in a number of ways, both immediately before and during penetration.

An outstanding example of the application of cytological and cytochemical techniques in the analysis of such alterations is the work of Williams and collaborators who have, for a number of years, been attempting to integrate structural and biochemical studies of the infection of Brassica tissues by the intracellular parasite *P. brassicae*, cause of the club root disease (Williams, 1973). The work on the root hair phase of the life history of the parasite will serve as an illustration. It was shown (Williams *et al.*, 1973), using a combination of interference contrast microscopy, cinematography and electron microscopy, that penetration of cabbage root hairs by *P. brassicae* occurs by the injection of the parasite through a puncture in the host's wall, as already described. Similar techniques were then used to correlate changes in the pattern of nuclear development and division, and in the accumulation of storage lipid in the parasite, thus making it possible to trace the evolution of a functional relationship with the host. The application of the techniques of microdensitometry to studies of the changes in the levels of nucleolar RNA, non-histone protein and histones in the nuclear apparatus of the host led to the suggestion that the parasite may alter the normal transcriptional processes of the host.

Although findings such as these throw some light on the nature of the relations between host and parasite, further progress will only come from investigation of the interface between the two organisms, and cytochemical and cytological studies of interaction during the first few moments of the relationship. Recognizing that the interfacial membranes occupy a key position during the establishment of a compatible relationship between *P. brassicae* and its host, Aist (1974) investigated their origin and structure using the techniques of freeze-etch microscopy. Besides revealing the complex nature of the infacial membrane region, he was able to show that the development of the outer plasmodial envelope, which initially derives from the invaginated plasmalemma of the host, is partially directed by the parasite. It seems reasonable to speculate that the appearance of this altered membrane is a prerequisite for proper exchange of materials between host and parasite.

Studies of interfacial membranes are in great need of further sophisticated biochemical and cytological investigation to uncover the functional significance of the specializations which occur.

B. MOLECULAR EXCHANGE

An understanding of the two-way exchange of informational and nutritional molecules between a biotroph and its host is of fundamental importance. Cytological techniques may be valuable in studying such exchange and in relating it to the continually changing interfaces which occur during the establishment of an infection. Of particular note in this respect is the work of Ellingboe and collaborators (Ellingboe, 1972, 1975) with *Erysiphe* spp. Because of the superficial nature of the mycelium of this group of organisms, the only route for molecular exchange during the establishment of an infection is via the developing haustorium. Mount and Ellingboe (1969), Slesinski and Ellingboe (1971) and Stuckey and Ellingboe (1975) utilized this fact to study the movement of ^{32}P and ^{35}S from host to parasite. By feeding the isotopes to the host and then chemically analysing the superficial mycelium of the fungus during infection they were able to correlate the exchange of molecules with the phases of morphological and cytological development of the haustorium. Perhaps more importantly, they were able to correlate molecular exchange and morphological development of the haustorium and secondary hyphae with the operation of genes for virulence and avirulence in the parasite and for resistance or susceptibility in the host (see III C). Clearly such studies, linked with the techniques of electron microscope-autoradiography and the kinds of microsurgery developed by Bushnell (1971) and Sullivan *et al.* (1974) provide potentially powerful tools for investigating still further the function of the various components of the infection structures of *Erysiphe* spp.

Micro-autoradiography has not been widely utilized in studies of molecular exchange between biotrophs and their hosts, except in the preliminary studies of Sivak and Shaw (1970) and Ehrlich and Ehrlich (1970). Recently, however, Andrews (1975) has used high resolution light and electron microscopy to investigate the movement of the label of ^3H-glucose and ^3H-leucine from lettuce during the sequential development of *B. lactucae* from spore germination to the formation of haustoria. He has shown that the parasite accumulates label from ^3H-glucose both on the surface of the lettuce cotyledon, before and during penetration, and at all subsequent stages during the development of primary and secondary vesicles, intercellular hyphae and haustoria. The label from ^{3}H-leucine, in contrast, does not enter the fungus during infection until after the development of inter-cellular hyphae and haustoria, indicating a significant change in the physiology of the interaction between host and parasite resulting from the development of these structures. This study is an important one, and provides a very valuable basis for further investigations of a similar type involving other hosts and parasites. If the techniques of micro-autoradiography of soluble substances can also be combined with biochemical analysis to indicate the form in which labels move, then we could very rapidly advance our under-standing of molecular exchange between biotrophs and their hosts in the near future.

C. INCOMPATIBILITY

Without question recognition and the determination of specificity are the most important steps in the infection process. Structural studies may be important in helping to define the molecular events which are involved. During the course of time the co-evolution of plants and fungi, particularly biotrophs, has led to the occurrence of specific interrelationships in which species of fungi are able to grow in association only with particular families, genera or species of host plants. Indeed, such specificity may be so refined that it is only a particular, genetically distinct biotype of a fungus which may grow in association with a particular host genotype. Such is the case with *B. lactucae* where genetically distinct but morphologically similar biotypes of the fungus are found on *Lactuca* spp., *Sonchus* spp. and *Senecio* spp. Superimposed on this basic pattern of specificity may be a further level of sophistication whereby a number of physiologic races of the fungus, each carrying different genes for virulence or avirulence may be matched by varieties of the host carrying complementary genes for resistance or susceptibility. That such interactions may have evolved after the basic compatibility between a fungus and its host had been established is suggested by the work of Heath (1974). In a light and electron microscope study of the structural events during the infection of resistant and susceptible cowpea (*Vigna sinensis*) hosts of *U. phaseoli* var. *vignae*, she was able to describe a number of stages during the infection process at which interaction between fungus and host may take place. Each stage represents a "switching point" at which host–parasite specificity may be expressed. Non-host (i.e. species other than *V. sinensis*) responses usually occurred during the earliest stages of infection (i.e. before or during cell penetration), leading to the suggestion that these represent the types of reaction "overcome" by cowpea rust during its evolutionary adaptation to its host. In contrast, varietal resistances of *V. sinensis* did not seem to involve these effective forms of pre-haustorial incompatibility, but were characterized by responses induced after the formation of the first haustorium. The structural studies of Heath have thus laid the foundation for further elucidation at the molecular level of the interaction between cowpea rust and its hosts and non-hosts.

The structural, and therefore biochemical, events leading to accommodation or rejection of a biotrophic fungus by a potential host are likely to be unique for each gene-for-gene interaction, as has been amply demonstrated by the work of Ellingboe (1972) using various races of *E. graminis* and various genotypes of its hosts, wheat and barley, and the studies of Heath (1974) with cowpea rust. However, one structural change which appears to accompany resistance in a wide range of gene-for-gene interactions is the rapid collapse and death of one or more host cells at the penetration site, linked with the death or cessation of growth of the invading fungus. It is generally believed that this reaction of the host leads to a disruption of nutrient supplies to the fungus and the production of fungitoxic metabolites, resulting in the cessation of fungal growth. This traditional view was challenged recently by Brown *et al.* (1966), who suggested that the rapid death of wheat cells in response to infection by rust fungi is not a cause, but a consequence, of the earlier death or inhibition of the invading fungus. Király *et al.* (1972), on the basis of indirect evidence, extended this suggestion to include the resistance of bean to *U. phaseoli* and potato to *Phytophthora infestans*. In contrast Maclean *et al.* (1974), on the basis of light and electron microscopic studies during the very early stages of the infection of lettuce by *B. lactucae*, have clearly shown that in incompatible interactions involving this host and parasite the primary event is the rapid death of the penetrated cells (Figs 19 and 20). The fungus continues to grow for many hours after this, before it too dies. Skipp and Samborski (1974), as a result of structural studies of the effect of the Sr6 gene for host resistance on histological events during the development of the stem rust fungus (*Puccinia graminis* f. sp. *tritici*) in near isogenic lines of wheat have also concluded that rapid death of host cells is closely associated with, and may precede, the death of fungal cells. Finally, Shimony and Friend (1975) have shown that, in the interaction between the potato variety Orion and an incompatible race of *Phyt. infestans*, infected Orion cells die very rapidly, but death of the invading fungus does not occur until 12 hours after inoculation.

These various studies serve to emphasize that each gene-for-gene interaction involving biotrophic fungi is likely to be unique, and that generalizations based on macroscopic observations may be very misleading. In some cases rapid death of host cells may be an integral part of the resistance process, in others it may well be a consequence of resistance only. Microscopic observations cannot reveal the mechanisms of specificity, but can often define very precisely the sequence of events in an interaction, and therefore provide a basis for further molecular investigation. Such is the case with the work of Rohringer *et al.* (1974) who, basing their biochemical investigations on the structural studies of Skipp and Samborski (1974), have suggested that a gene-specific RNA may determine resistance of wheat (where cell

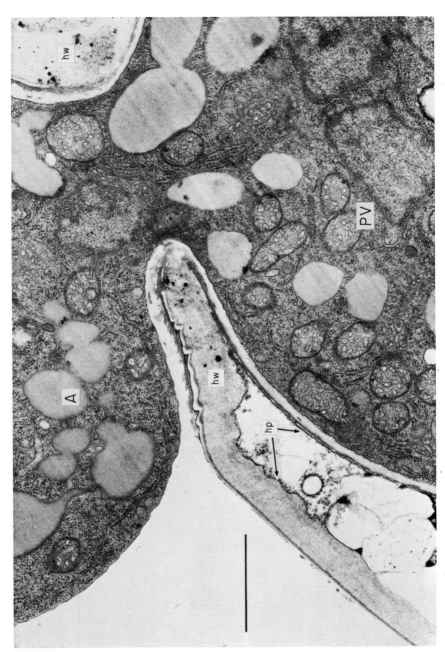

Fig. 19. A section through the penetration pore of the epidermal surface of the resistant lettuce cultivar Avondefiance. Cytoplasm is still entering the primary vesicle from the appressorium. Already the host cytoplasm is severely disrupted but the invaginated host plasmalemma appears intact.

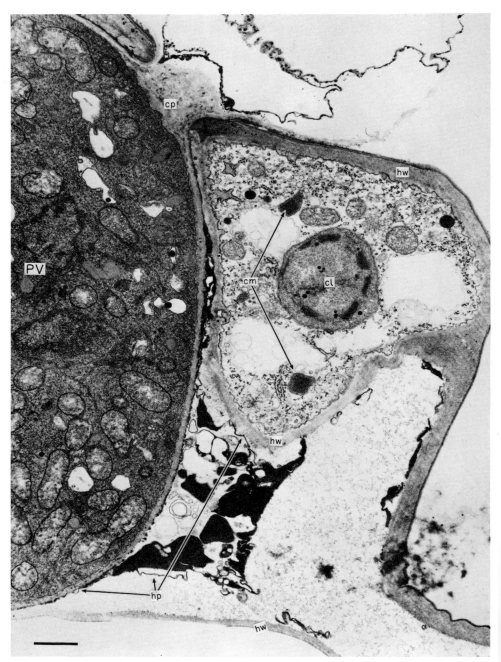

FIG. 20. A mature primary vesicle within an epidermal cell of the resistant lettuce cultivar Avondefiance (4th after inoculation). The neck of the primary vesicle has become plugged with callose and the fungal cytoplasm appears healthy. In contrast the cytoplasm of the invaded cell has become completely disrupted and the invaginated plasmalemma is no longer intact. The cytoplasm of the adjacent, non-penetrated, epidermal cell is little affected but crystal-containing microbodies are present (Maclean *et al.*, 1974).

death is involved) to *Pucc. graminis* f. sp. *tritici*. Cytochemical investigation will probably be important in identifying the sites of specific interaction at the subcellular level.

The correlated genetic, structural and biochemical investigations by Ellingboe (1972) of the interaction between genes for resistance and susceptibility in wheat and barley and for virulence and avirulence in *E. graminis* are outstanding as having made significant contributions to our understanding of specificity at the level of gene expression. The success of Ellingboe is based upon his use of defined genetic material as a basis for his investigations. Such a course must be the basis for any future investigation of the mechanisms of interaction between biotrophs and their hosts, whether it be at the structural or molecular levels.

REFERENCES

Aist, J. R. (1974). *Can. J. Bot.* **52**, 1441–1449.
Aist, J. R. and Williams, P. H. (1971). *Can. J. Bot.* **49**, 2023–2034.
Albersheim, P., Jones, T. M. and English, P. D. (1969). *A. Rev. Phytopath.* **7**, 171–194.
Andrews, J. H. (1975). *Can. J. Bot.* **53**, 1103–1115.
Beckett, A., Heath, I. B. and McLaughlin, D. J. (1974). "An Atlas of Fungal Ultra-structure". Longman, London.
Berlin, J. D. and Bowen, C. C. (1964). *Am. J. Bot.* **51**, 445–452.
Bracker, C. E. (1968). *Phytopathology* **58**, 12–30.
Bracker, C. E. and Littlefield, L. J. (1973). *In* "Fungal Pathogenicity and The Plant's Response" (R. J. W. Byrde and C. V. Cutting, eds), pp. 159–313. Academic Press, London and New York.
Brian, P. W. (1967). *Proc. R. Soc. (Lond.)*, B **168**, 101–118.
Brown, J. F., Shipton, W. A. and White, N. H. (1966). *Ann. appl. Biol.* **58**, 279–290.
Brown, W. and Harvey, C. C. (1927). *Ann. Bot.* **41**, 643–662.
Bushnell, W. R. (1971). *In* "Morphological and Biochemical Events in Plant Parasite Interaction" (S. Akai and S. Ouchi, eds), pp. 229–254. Phytopath. Soc. of Japan, Tokyo.
Bushnell, W. R. (1972). *A. Rev. Phytopath.* **10**, 151–176.
Chou, C. K. (1970). *Ann. Bot.* **34**, 189–204.
Coffey, M. D., Palevitz, B. A. and Allen, P. J. (1972a). *Can. J. Bot.* **50**, 231–240.
Coffey, M. D., Palevitz, B. A. and Allen, P. J. (1972b). *Can. J. Bot.* **50**, 1485–1492.
Dickinson, S. (1960). *In* "Plant Pathology, an advanced treatise" (J. G. Horsfall and A. E. Dimond, eds), Vol. II, pp. 203–232. Academic Press, New York and London.
Edwards, H. H. and Allen, P. J. (1970). *Phytopathology* **60**, 1504–1509.
Ehrlich, H. G. and Ehrlich, M. A. (1963). *Phytopathology* **53**, 1378–1380.
Ehrlich, M. A. and Ehrlich, H. G. (1970). *Phytopathology* **60**, 1850–1851.
Ehrlich, M. A. and Ehrlich, H. G. (1971). *A. Rev. Phytopath.* **9**, 155–184.
Ellingboe, A. H. (1972). *Phytopathology* **62**, 401–406.
Ellingboe, A. H. (1975). *In* "Physiological Plant Pathology" (R. Heitefuss and P. H. Williams, eds). Springer-Verlag, Berlin, Heidelburg and New York (in press).
Flor, H. H. (1971). *A. Rev. Phytopath.* **9**, 279–296.
Friend, J. (1973). *In* "Fungal Pathogenicity and the Plant's Response" (R. J. W. Byrde and C. V. Cutting, eds), pp. 383–394. Academic Press, London and New York.

Grove, S. N. and Bracker, C. E. (1970). *J. Bact.* **104**, 989–1009.

Heath, M. C. (1974). *Physiol. Pl. Path.* **4**, 403–414.

Heath, M. C. and Heath, I. B. (1971). *Physiol. Pl. Path.* **1**, 277–287.

Keskin, B. and Fuchs, W. H. (1969). *Arch. Mikrobiol.* **68**, 218–226.

Király, Z., Barna, B. and Érsek, T. (1972). *Nature* **239**, 456–458.

Kunoh, H. and Akai, S. (1969). *Mycopathol. Mycol. appl.* **37**, 113–118.

Lesemann, D. E. and Fuchs, W. H. (1970). *Arch. Mikrobiol.* **71**, 20–30.

Littlefield, L. J. and Bracker, C. E. (1972). *Protoplasma* **74**, 271–305.

Maclean, D. J., Sargent, J. A. S., Tommerup, I. C. and Ingram, D. S. (1974). *Nature* **249**, 186–187.

Maheshwari, R., Allen, P. and Hildebrandt, A. C. (1967). *Phytopathology* **57**, 885–862.

Manocha, M. S. and Shaw, M. (1967). *Can. J. Bot.* **45**, 1575–1582.

Mason, P. A. (1973). Ph.D. Thesis, University of Cambridge.

McKeen, W. E., Smith, R. and Mitchell, N. (1966). *Can. J. Bot.* **44**, 1299–1306.

McKeen, W. E., Smith, R. and Battacharya, P. K. (1969). *Can. J. Bot.* **47**, 701–706.

Mount, M. S. and Ellingboe, A. H. (1969). *Phytopathology* **59**, 235:

Peyton, G. A. and Bowen, C. C. (1963). *Am. J. Bot.* **50**, 787–797.

Rohringer, R., Howes, N. K., Kim, W. K. and Samborski, D. J. (1974). *Nature* **249**, 585–588.

Royle, D. J. and Thomas, G. G. (1971). *Physiol. Pl. Path.* **1**, 345–349.

Sargent, J. A. and Payne, H. L. (1974). *Trans. Br. mycol. Soc.* **63**, 509–518.

Sargent, J. A., Tommerup, I. C. and Ingram, D. S. (1973). *Physiol. Pl. Path.* **3**, 231–239.

Scott, K. J. (1972). *Biol. Rev.* **47**, 537–572.

Shimony, C. and Friend, J. (1975). *New Phytol.* **74**, 59–65.

Sivak, B. and Shaw, M. (1970). *Can. J. Bot.* **48**, 613–614.

Skipp, R. A. and Samborski, D. J. (1974). *Can. J. Bot.* **52**, 1107–1115.

Slesinski, R. S. and Ellingboe, A. H. (1971). *Can. J. Bot.* **49**, 303–310.

Stahmann, M. A. (1973). *In* "Fungal Pathogenicity and the Plant's Response" (R. J. W. Byrde and C. V. Cutting, eds), pp. 156–157. Academic Press, London and New York.

Stanbridge, B., Gay, J. L. and Wood, R. K. S. (1971). *In* "Ecology of Leaf Surface Micro-organisms" (T. F. Preece and C. H. Dickinson, eds), pp. 367–379. Academic Press, London and New York.

Staub, T., Dahmen, H. and Schwinn, F. J. (1974). *Phytopathology* **64**, 364–372.

Stuckey, R. E. and Ellingboe, A. H. (1975). *Physiol. Pl. Path.* **5**, 19–26.

Sullivan, T. P., Bushnell, W. B. and Rowell, J. B. (1974). *Can. J. Bot.* **52**, 987–998.

Williams, P. H. (1973). *Shokubutsu Byogai Kenkyu* (*Forsch. Gebiet Pflanzenkrankh.*), Kyoto **8**, 133–146.

Williams, P. H., Aist, J. R. and Battacharya, P. K. (1973). *In* "Fungal Pathogenicity and the Plant's Response" (R. J. W. Byrde and C. V. Cutting, eds), pp. 141–155. Academic Press, London and New York.

Zimmer, D. E. (1965). *Phytopathology* **55**, 296–301.

CHAPTER 5

Plant Cell Wall Hydrolysis by Pathogens

D. F. BATEMAN

Department of Plant Pathology, Cornell University, Ithaca, N.Y., U.S.A.

I. INTRODUCTION

Most phytopathogenic microorganisms are capable of producing poly-saccharide degrading enzymes which can alter or degrade a number of the polymeric carbohydrates found in higher plant cell walls. During tissue invasion and ramification by pathogens host cell walls are usually breached repeatedly, particularly by the fungi (Aist, 1976). The enzymes considered to aid or facilitate plant cell wall penetration, tissue maceration and dis-membering of cell wall structure represent one group of "Attacking Mechanisms" found in the arsenal of most phytopathogenic fungi and bacteria (Brown, 1955).

The ability to produce a number of polysaccharide degrading enzymes is a feature plant pathogens share with members of the saprophytic micro-flora and fauna in soil responsible for decay of plant materials, micro-organisms which inhabit the rumen of ruminant animals, and certain herbaceous insects (Bateman and Millar, 1966). It is thus apparent that the ability of microorganisms to secrete an array of polysaccharide degrading enzymes does not in itself enable such organisms to be plant pathogenic.

On the other hand, the ability to produce such enzymes may be an essential feature of paramount importance with respect to the pathogenic capabilities of some pathogens (Zucker and Hankin, 1971; Bateman and Bashman, 1976).

The molecular basis of pathogenicity or susceptibility and/or resistance to disease remains to be defined for any given suscept–pathogen (host–parasite) combination; it is apparent, however, that the ability to cause or resist diseases in most instances involves a number of molecular factors of both pathogen and suscept origin; thus, the diversity of the subject matter of this symposium. Disease and disease resistance each arise and are expressed as a result of the interaction of multiple molecular and physical factors originating in the suscept–pathogen complex; also, the outcome of a given suscept–pathogen encounter is usually dependent upon the nature of the physical and chemical environments in which the encounter takes place. The ability of pathogens to produce plant cell wall degrading enzymes represents one element that should be considered in examining pathogenesis. Those factors that impinge upon the pathogen's ability to secrete cell wall degrading enzymes as well as the environment within the host or suscept where these enzymes function become of equal concern in understanding their significance in pathogenesis. Disease can be viewed as resulting from the interaction of multiple systems, each of which may involve multiple components. The various facets of a suscept–pathogen encounter must ultimately be considered in relation to each other if we are to more fully appreciate the significance of the different elements involved in plant diseases. It is now evident that most plant pathogens produce plant cell wall degrading enzymes (Bateman and Millar, 1966; Albersheim et al., 1969; Bateman and Basham, 1976). What roles do these enzymes play in pathogenesis? In some cases the role may be nil (Reddy et al., 1969; Seemuller et al., 1973) and in others, it would appear to be of great significance (Albersheim et al., 1969; Bateman and Basham, 1976; Beraha and Garber, 1971).

It is perhaps premature to focus attention directly on the above question. The objective of this brief review is to examine certain facets of plant cell wall hydrolysis by plant pathogens that should provide a better appreciation of the problems involved. These include consideration of: plant cell wall structure, the various enzymes that split cell wall carbohydrates, regulation of enzyme synthesis by pathogens, and enzymatic decomposition of intact plant cell walls. Emphasis will also be given to the enzymatic basis of tissue maceration and its consequences in pathogenesis.

II. THE PLANT CELL WALL: A CURRENT CONCEPT

The plant cell wall is a complex yet ordered structure composed primarily of polysaccharides, structural glycoprotein and in certain species lignin

(Keegstra *et al.*, 1973; Lamport, 1970; Northcote, 1972). The plant cell wall may be viewed as a two phase system: a dispersed phase consisting of cellulose fibrils and a continuous matrix made up of other polysaccharides and a hydroxyproline-rich glycoprotein. Functionally, the cell wall may be divided into three regions: middle lamella, primary wall, and secondary wall. The middle lamella serves as an "intracellular cement" that binds cells together in tissue systems. Primary wall is the first formed wall; it is the most dynamic of the wall regions and functions to support the proto-plast in young growing cells. Secondary wall is deposited after cell wall elongation is complete. This wall region adds structural strength and functions as a major element in supporting the plant body. In certain species and/or tissues cell walls are lignified after deposition of the secondary wall region. Lignification is usually associated with cessation of metabolic activity and death of the cells whose walls are lignified. Lignin is generated from the oxidation products of sinapyl, coniferyl, and *p*-hydroxycinnamyl alcohols which condense by a free radical mechanism to form an amorphous 3-dimensional polymer (Brown, 1969; Freudenberg, 1968). Lignin is deposited throughout the cell wall and may be covalently linked to the other wall polymers (Cowling, 1975; Fergus *et al.*, 1969). The polysaccharides in lignified cell walls are believed to be masked and rendered resistant to enzymatic hydrolysis by most plant pathogens.

Plant cell wall polysaccharides have historically been grouped into pectic substances, hemicelluloses and cellulose (Northcote, 1963). This classification is based upon solubilities of the wall polysaccharides in various extractants rather than their chemical compositions. The pectic fraction is extractable in cold and hot water and solutions of chelating agents. The hemicellulosic fraction is extractable in alkali. Cellulose or α-cellulose is the residual polysaccharide which remains after extracting the other components. This means of cell wall fractionation gives rise to mixtures of polysaccharide components.

A major component of the pectic fraction is a high molecular weight polymer of α-1,4 linked α-galacturonopyranose interspersed with 1,2 linked rhamnose (Aspinall, 1970a; Keegstra *et al.*, 1973; Talmadge *et al.*, 1973). The uronic acid carboxyls may be methylated and in some plants the uronide residues may be acetylated at carbons 2 and/or 3 (Deuel and Stutz, 1958; Kertesz and Lavin, 1954). The pectic fraction also contains polymers of neutral sugars, consisting of α-1,3 and α-1,5 linked L-arabino-furanose and β-1,4 linked D-galactopyranose. These neutral polymers are believed to be covalently linked to the rhamnogalacturonide component (Northcote, 1969, 1972).

The hemicelluloses include several polymers. An important hemicellulose in the primary wall region is xyloglucan. This polymer is made up of a β-1,4 linked glucopyranose chain with terminal branches of 1,6 linked D-xylopyranose (Bauer *et al.*, 1973). The xylans represent a widely distributed important group of hemicelluloses in higher plants. These polymers are composed of β-1,4 linked D-xylopyranose chains. Xylans commonly have side branches of 1,3 linked arabinofuranose and α-1,2 linked D-glucuronopyranose (or its 4-methyl ester) (Aspinall, 1970b). Other hemicelluloses include glucomannan, a heteropolymer of D-glucopyranose and D-mannopyranose linked β-1,4, and galactoglucomannan, a glucomannan with 1,6 linked D-galactopyranose side branches (Northcote, 1972). The mannans and galactomannans are β-1,4 linked D-mannopyranose chains with the latter having 1,6 linked D-galactopyranose side branches. These last two polymers occur in higher plants but their significance as structural elements does not appear to be resolved. Xyloglucans, xylans and possibly other hemicelluloses have the capacity to hydrogen bond to cellulose (Bauer et al., 1973).

Cellulose is the major constituent of higher plant cell walls. It is reported to occur as microfibrils which constitute primary structural elements. Chemically, cellulose is a linear polymer of β-1,4 linked D-glucopyranose residues; it exists as flat, band-like molecules. Cellulose fibrils contain both crystalline and amorphous cellulose. Crystalline cellulose is postulated to form as a result of antiparallel deposition of cellulose chains which can result in inter- and intramolecular hydrogen bonding (Frey-Wyssling, 1969; Muhlethaler, 1967; Northcote, 1972; Preston, 1971), but the exact structure of cellulose fibrils is a matter for dispute (Sarko and Muggli, 1974).

The existence of a structural protein in plant cell walls has been a subject of considerable controversy (Colvin and Leppard, 1971; Sadava and Chrispeels, 1969; Steward et al., 1967). It appears that most researchers now agree that a glycoprotein, rich in hydroxyproline, does reside in the plant cell wall as a structural element in the primary cell wall region (Keegstra et al., 1973; Lamport, 1970; Northcote, 1972; Sadava et al., 1973). The carbohydrate moities are arabinose and galactose; the former is glycosidically linked to the hydroxyl of the hydroxyproline residues and galactose is believed to be covalently linked to serine residues (Lamport, 1970; Lamport et al., 1973).

Recent studies of cell walls from suspension cell cultures of sycamore and bean in which specific polysaccharide degrading enzymes were employed to fragment these walls, coupled with methylation analysis of intact walls and wall fragments have given rise to the view that the primary wall can be regarded as a giant molecule (Albersheim, 1975; Keegstra et al., 1973; Wilder and Albersheim, 1973). The composition of the primary wall of sycamore is given in Table I. A proposed model showing the relationship of the various constituents in this wall to each other is presented in Fig. 1. The evidence on which the Albersheim group based this model indicates that the pectic polymers are covalently linked and that the hemicellulosic fraction is covalently linked to the neutral pectic fraction. The hemicellulosic xyloglucan, covalently linked to the pectic fraction, is hydrogen bonded to

TABLE I

Polymer composition of sycamore (*Acer pseudoplatanus*)
primary cell walls (from Talmadge *et al.*, 1973)

Wall component	% of Wall
Arabinan	10
3,6-Linked arabinogalactan	2
4-Linked galactan	8
Cellulose	23
Protein	10
Rhamnogalacturonan	16
Tetra-arabinosides	
(attached to hydroxyproline)	9
Xyloglucan	21
Total	99

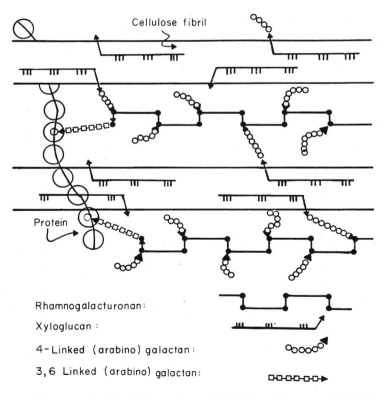

Fig. 1. Proposed model of primary plant cell walls (from Keegstra *et al.*, 1973).

the cellulosic fibrils and serves to unite the amorphous and fibrilar phases of the wall. The Albersheim model proposes that the wall glycoprotein is covalently linked through a minor arabinogalactan component to the pectic fraction. Although aspects of this wall model are based on incomplete evidence, the concept proposed represents a significant advance in our understanding of plant cell wall structure and it should greatly stimulate research on the details of the molecular architecture of plant cell walls. Also, the Albersheim wall model provides a useful context from which to examine our knowledge of the effects of polysaccharide degrading enzymes on plant tissues and cell walls (Keegstra *et al.*, 1973).

III. Enzymes that Cleave Cell Wall Polysaccharides

A. PECTIC ENZYMES

Enzymes that degrade the rhamnogalacturonide portion of cell wall structure have been studied extensively. Pectin esterases, which hydrolyze methyl esters of the uronic acid residues in the pectic chain, have been purified and characterized from a number of sources (Lee and Macmillan, 1968; McColloch and Kertesz, 1947; Miller and Macmillan, 1971). Furthermore, the enzymes that cleave the α-1,4 bonds between the uronide residues in pectins and pectic acids have been characterized. These include both hydrolases and lyases (Bateman and Millar, 1966; Rombouts and Pilnik, 1972). Endo- and exo-enzymes with each mechanism of bond cleavage are known. The exo-hydrolases release monomeric galacturonic acid, whereas the exo-lyases release unsaturated dimers (Hasegawa and Nagel, 1962; Okamoto *et al.*, 1964). Several classifications of the enzymes that split the α-1,4 bonds in the pectic chain have been proposed (Bateman and Millar, 1966; Demain and Phaff, 1957; Deuel and Stutz, 1958); the most recent has been summarized by Rombouts and Pilnik (1972). Little appears to be known about the hydrolytic removal of rhamnose from pectic substances although rhamnose occurs as a reaction product from pectin or pectate incubated with crude enzyme mixtures. *Corticium rolfsii* (*Sclerotium rolfsii*), a pectolytic and plant pathogenic fungus, is known to secrete an α-L-rhamnosidase (Kaji and Ichimi, 1973).

A number of pathogens have been shown to possess the enzymes that hydrolyze β-1,4 galactan as well as araban (Bateman *et al.*, 1969; Bauer and Bateman, 1975; Cole and Bateman, 1969; Fuchs *et al.*, 1965; Kaji *et al.*, 1965; Knee and Friend, 1968; Mullen and Bateman, 1975b; VanEtten and Bateman, 1969). The galactan component is hydrolyzed both by endo-β-1,4 galactanase and by exo-β-1,4 galactanase or β-1,4 galactosidase. Relatively little appears to be known about the enzymes that hydrolyze β-1,3 linked galactans. The araban substrate most frequently used for studies with arabanases contains α-1,3 and α-1,5 glycosidic linkages (Hirst and Jones,

1947). No arabanases of the endo type have been reported; L-arabinose appears to be the only reaction product released and the enzymes are usually designated α-L-arabinofuranosidases. This enzyme has been highly purified from *Sclerotinia fructigena* (Hislop *et al.*, 1974). Also, an α-L-arabino-furanosidase from *Aspergillus niger* which hydrolyzes α-1,5 linked arabino-furanose has been crystallized and characterized (Kaji and Tagawa, 1970).

B. HEMICELLULASES AND CELLULASES

The hemicellulosic xyloglucan chain can be fragmented by endo-β-1,4 glucanase (cellulase, Cx-type), but enzymes that remove the 1,6 linked xylopyranose residues apparently have not been examined (Bauer *et al.*, 1973). Endo-β-1,4 xylanases and xylobiases have been described that convert β-1,4 xylan to xylose (Hashimoto *et al.*, 1971; Mullen and Bateman, 1975b; Strobel, 1963; VanEtten and Bateman, 1969; Walker, 1967). Similarly, endo and exo enzymes which hydrolyze β-1,4 linked mannan have been prepared from a number of fungi (Lyr, 1963; Reese and Shibata, 1965; VanEtten and Bateman, 1969). Terminal branch sugar residues occur in many of the linear cell wall polymers. Specific glycosidases which hydrolyze such constituents are known in many instances. For example, 1,3 linked arabinose residues of xylans are released by α-L-arabinofuransidase (Kaji and Yoshihara, 1970), α-1,6 linked galactose in galactomannan is released by α-galactosidase (VanEtten and Bateman, 1969), etc. Pathogenic organisms such as *Colletotrichum lindemuthianum* and *S. rolfsii* are good sources of a variety of specific glycosidases (English *et al.*, 1971; Jones and Bateman, 1972).

Cellulose in its native state is insoluble and partly crystalline. The conversion of cellulose to glucose involves a complex of enzymes originally designated C_1, C_x and cellobiase (Reese, 1956). The C_x or endo-β-1,4 glucanase splits soluble cellulose chains and cellobiase converts cellobiose to glucose. The C_1 enzyme was proposed long before its discovery; its action renders native cellulose susceptible to enzymes of the C_x-type. The cellulase complex of *Trichoderma koningii* for example has been purified and shown to be composed of four components: (a) cellobiase, (b) endo-β-1,4 glucanase, (c) a "short fiber" forming component, designated C_2, and a C_1 component (Halliwell and Mohammed, 1971). A number of plant pathogens produce the enzymes necessary to convert native cellulose to low molecular weight soluble products or glucose (Bateman, 1964, 1969; Kelman and Cowling, 1965).

IV. REGULATION OF PRODUCTION OF POLYSACCHARIDASES BY PATHOGENS

Cell wall polysaccharide degrading enzymes produced by most pathogenic microorganisms appear to be subject to catabolite repression (Biehn and

Dimond, 1971a,b; Horton and Keen, 1966; Keen and Horton, 1966; Moran and Starr, 1969; Mullen and Bateman, 1971; Mussell and Green, 1970; Patil and Dimond, 1968; Spalding *et al.*, 1973; Weinhold and Bowman, 1974; Zucker and Hankin, 1971); but there are relatively few detailed studies of the factors controlling the synthesis of these enzymes. The increased production of a given polymer splitting enzyme by a pathogen grown in the presence of a substrate of the enzyme in question has usually been termed induction and such enzymes are often referred to as inducible. More critical studies such as those reported on the control of polygalacturonic acid *trans*-eliminase by *Aeromonas liquefaciens* (Hsu and Vaughn, 1969) and certain pathogenic *Erwinia* species (Moran and Starr, 1969) indicate that polygalacturonate *trans*-eliminase may not be inducible in these organisms, but rather it is a constitutive enzyme that is subject to severe catabolite repression. For example, slow feeding of *A. liquefaciens* with glucose, glycerol, or polygalacturonic acid when incubated in a carbon deficient medium permitted up to 500 times more enzyme synthesis than when this bacterium was incubated in a carbon sufficient medium. It seems quite likely that in many of the cases where enzyme induction has been claimed, the actual mechanism involved may be one of derepression or relief of catabolite repression. Catabolite repression results from the accumulation of a given sugar in cells of the organisms carrying on enzyme synthesis; repression of enzyme synthesis may not be specific and sugars released from different polymers by different enzymes may function to repress the synthesis of other polysaccharide degrading enzymes. The extent of repression of enzyme synthesis may depend upon both the specific enzyme and the particular catabolite (Cooper and Wood, 1975).

The induction of specific polysaccharide degrading enzymes in certain pathogens is a real phenomenon, however. This is amply demonstrated by recent studies with *Verticillium albo-atrum* and *Fusarium oxysporum* f. sp. *lycopersici* (Cooper and Wood, 1975). When these fungi were cultured in a carbon deficient medium and various monomeric sugars (or dimers) of cell wall polymers fed to cultures at a rate which prevented accumulation of sugar in the fungal cells, specific sugars served to induce synthesis of specific cell wall degrading enzymes, i.e. D-xylose induced synthesis of xylanase, D-galacturonic acid induced synthesis of endopolygalacturonase and endopectin *trans*-eliminase, cellobiose induced synthesis of cellulase, etc. in *V. albo-atrum* (Table II). These same sugars that serve as specific inducers also function as non-specific catabolic repressors when critical sugar levels in the fungus were exceeded.

There are a number of examples where pathogens are known to produce cell wall degrading enzymes constitutively which are not subject to catabolite repression. These examples include polygalacturonases produced by an isolate of *V. albo-atrum* and *Aphanomyces euleiches* (Ayers *et al.*, 1969; Mussell and Strouse, 1972), the polygalacturonase and xylanase of *Helmintho-*

TABLE II

Specificity of induction of polysaccharidases in *V. albo-atrum* by different sugars
(from Cooper and Wood, 1975)

Carbon Source[b]	Enzyme activities (percentage of maximum[a])					
	Endo-PG	Endo-PTE	Arabinanase	β-Galactosidase	Xylanase	Cellulase (C$_x$)
D-Glucose	0–1·25	0	0–5·8	0·84	5·8	1·46
D-Galactose	0	0	14·4	12·90	5·2	0·37
D-Mannose	0	0	0	0·34	0	0·20
L-Arabinose	0	0	100	100	3·5	0·25
D-Xylose	0	0	3·7	0·26	100	0·08
L-Rhamnose	2·10	9·11	6·6	0	0	0·40
L-Fucose	1·38	0·16	0	0	0	0·26
D-Galacturonic Acid	100	100	15·8	0	16·5	0·30
Cellobiose	3·7	1·34	—	0	—	100

[a] For each enzyme, activities detected in cultures supplied with different sugars for identical periods are expressed as a percentage of the activity in cultures supplied with the specific inducer.
[b] Sugars fed from diffusion capsules at linear rates of *ca* 2·5 to 4 mg 100 ml^{-1} h^{-1} to shake cultures.

sporium maydis (Bateman *et al.*, 1973) and the cellulase produced by *Pseudomonas solanacearum* (Kelman and Cowling, 1965).

There is considerable evidence that regulation of the control systems governing the synthesis and/or excretion of polysaccharide degrading enzymes is of significance in host–parasite relationships (Biehn and Dimond, 1971a; Bugbee, 1973; Horton and Keen, 1966; Keen and Horton, 1966; Patil and Dimond, 1968; Weinhold and Bowman, 1974; Zucker and Hankin, 1971). Regulation of sugar levels in onion tissue infected with *Pyrenochaeta terrestris* has been reported to influence the synthesis of polygalacturonase and cellulase by this pathogen as well as the rate of onion pink root development. Treatments that enhance sugar levels in the host were associated with decreased enzyme synthesis and less tissue breakdown (Horton and Keen, 1966; Keen and Horton, 1966). Studies with *Rhizoctonia solani* on bean (Weinhold and Bowman, 1974) and *Erwinia carotovora* on potato (Biehn *et al.*, 1972) indicate that pathogenicity and subsequent disease development can be regulated through the exogenous supply of sugars which in turn regulates production of cell wall degrading enzymes by these pathogens.

Factors from plant cell walls or tissues are known to enhance production of pectic enzymes in some pathogens. A low molecular weight carbohydrate rich in anhydrogalacturonic acid in tomato cell wall greatly enhances polygalacturonase production by *Fusarium oxysporum* f. sp. *lycopersici* (Jones *et al.*, 1972). An ethanol insoluble fraction from green elm shoots has been shown to stimulate polygalacturonase production by *Ceratocystis ulmi*, some 50 times more than citrus pectin (Biehn and Dimond, 1971b). Polygalacturonic acid *trans*-eliminase production by *E. carotovora* has been reported to be greatly enhanced by a heat labile factor from potato when cultured on a pectin-medium (Zucker and Hankin, 1970).

Isolated plant cell walls can serve as carbon sources and stimulate production of a variety of polysaccharide degrading enzymes by pathogens. The limited studies in which cell walls have been used as carbon sources to investigate polysaccharidase production by pathogens indicate that the cell wall degrading enzymes appear to be produced in a definite temporal sequence. *F. oxysporum* f. sp. *lycopersici* grown on isolated tomato walls (Jones *et al.*, 1972), *C. lindemuthianum* grown on bean walls (English *et al.*, 1971), *F. roseum* "*Avenaceum*" grown on potato walls (Mullen and Bateman, 1975b) and *V. albo-atrum* grown on tomato walls (Cooper and Wood, 1975) all produce enzymes that degrade the pectic polymers first followed by those enzymes which hydrolyze the hemicelluloses and finally cellulase which appears to be last in the sequence. This pattern of enzyme production by pathogens, when cultured on cell walls of their hosts, may reflect the relative accessibility of the various substrates in the plant cell walls. There is mounting evidence that the pectic fraction, particularly the rhamnogalacturonide portion, is a key element in plant cell wall structure (Fig. 1); it is now known that the enzymatic degradation of the rhamnogalacturonide fraction enhances the ability of other polysaccharidases to hydrolyze their substrates in cell walls (Talmadge *et al.*, 1973).

Our knowledge of the factors regulating the control of polysaccharidase production by pathogens is fragmentary and limited to systems involving facultative pathogens. Histological evidence obtained at the electron microscope level indicates that cell wall penetration by biotrophic parasites may be aided or accomplished by enzymatic means (Aist, 1976; Bracker and Littlefield, 1973; McKeen and Rimmer, 1973; Sargent *et al.*, 1973). It is apparent from recent studies of the influence of pectic enzymes on plant tissues, that if cell wall degrading enzymes are involved in establishment of obligate parasites, which have or require intimate contact with living host cells, the production of cell wall degrading enzymes by such parasites must be subject to precise regulation and control, lest the host cells be killed (Basham and Bateman, 1975a,b; Stephens and Wood, 1975). Further elucidation of the systems regulating production and excretion of cell wall degrading enzymes by plant pathogens should add considerably to our understanding of host–parasite relationships.

V. Enzymatic Degradation of Cell Walls

A. DECOMPOSITION OF ISOLATED WALLS

Isolated plant cell walls can serve as substrates for polysaccharide degrading enzymes produced by phytopathogenic organisms (English *et al.*, 1971; Jones *et al.*, 1972; Mullen and Bateman, 1975b). Cell walls from young or parenchymatous tissues are more easily hydrolyzed than walls from older tissues where secondary wall thickening has occurred (Bateman *et al.*, 1969). A study of the hydrolysis of bean cell walls isolated from 10-day-old seedlings by the polysaccharide degrading enzymes in a commercial preparation (Pectinol R-10) revealed that action of a "wall modifying enzyme" may be necessary before many of the enzymes which attack cell wall polysaccharides are able to effectively degrade their substrates within the wall (Karr and Albersheim, 1970). A "wall modifying enzyme" was purified from the Pectinol R-10 preparation and its action shown to be requisite for the release of galactose from isolated walls by an α-galactosidase produced by *C. lindemuthianum*. Purified "wall modifying enzyme" did not attack a wide variety of polysaccharide, glycoside and peptide substrates, but it did degrade polygalacturonic acid. Action of this enzyme on cell walls resulted in both the release of 70% ethanol insoluble polymers as well as alteration of the residual wall.

Subsequent studies with a purified endopolygalacturonase from *C. lindemuthianum* (English *et al.*, 1972) and a homogeneous endopolygalacturonate lyase from *Erwinia chrysanthemi* (Basham and Bateman, 1975b) have demonstrated that the endopectic enzymes that split the α-1,4 galacturonosyl bonds of the rhamnogalacturonide fraction of the plant cell wall are able to cause massive disruption of isolated walls and result in a substantial solubilization of the pectic fraction, including the neutral polymers containing galactose and arabinose. The importance of pectic enzymes in cell wall degradation is further emphasized by the fact that a purified fungal endoglucanase (cellulase) which releases the xyloglucan from isolated sycamore cell walls released only 1% of the wall polysaccharides if it was permitted to act on cell walls prior to treatment with an endopolygalacturonase, whereas this endogluconase released 10–15% of the wall polymers when the wall had been pre-treated with the pectic enzyme. Furthermore, when sycamore cell walls were treated with protease, less than 0·3% of the wall material was solubilized, but walls pre-treated with endopolygalacturonase or endopolygalacturonase-endoglucanase combinations enabled the protease to solubilize 2 and 4%, respectively, of the wall material (Bauer *et al.*, 1973; Keegstra *et al.*, 1973; Talmadge *et al.*, 1973). These results coupled with the demonstrated effects of endopolygalacturonases on isolated bean and tomato cell walls and the effects of endopectate lyase on isolated potato cell walls (Basham and Bateman, 1975b; English *et al.*, 1972; Jones *et al.*, 1972) emphasized the importance of the rhamnogalacturonide fraction in plant cell wall structure (Fig. 1), and the significance of the endopectic enzymes in cell wall disruption. The endopectic enzymes appear to function as "cell wall modifying enzymes".

Crude cell free enzyme preparations from pathogens such as *Rhizoctonia solani* (Bateman *et al.*, 1969), *F. oxysporum* f. sp, *lycopersici* (Jones *et al.*, 1972), *S. rolfsii* (Jones and Bateman, 1972), and *F. roseum* "Avenaceum" (Mullen and Bateman, 1975b) which contain enzymes that split the bonds in the major plant cell wall polysaccharides release or solubilize the various cell wall sugars from isolated cell walls of host species. The degree of solubilization of isolated potato cell walls by the enzymes produced by *F. roseum* "Avenaceum" is shown in Fig. 2. The polysaccharidase mixture produced by *S. rolfsii* has proven to be useful for analyzing the sugar composition of plant cell walls (Jones and Albersheim, 1972).

B. DECOMPOSITION OF HOST CELL WALLS IN INFECTED TISSUES

The occurrence of cell wall polysaccharide degrading enzymes in tissues infected by facultative parasites or colonized by perthophytes is relatively

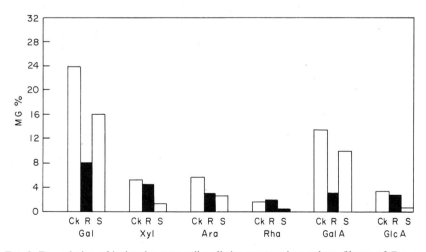

FIG. 2. Degradation of isolated potato cell walls by enzymes in a culture filtrate of *F. roseum* "Avenaceum". The fungus was cultured for 8 days at 22 °C on a sodium polypectate-salts medium. Cell wall degrading enzymes produced included: endopolygalacturonate *trans*-eliminase, two arabanases, endo-β-1,4 galactanase, and xylanase. After treatment of cell walls with culture filtrate buffered at pH 7·0 for 5 h at 30 °C, the wall residue (R) and supernatant (S) were analyzed for constituent sugars and compared with buffer treated walls (Ck). Wall constituents are expressed as mg of a component per 100 mg of walls (MG%). Abbreviations are: galactose (Gal), xylose (Xyl), arabinose (Ara), rhamnose (Rha), galacturonic acid (Gal A) and glucuronic acid (Glc A) (from Mullen and Bateman, 1975a).

easy to demonstrate in extracts of the colonized tissue (Balasubramani *et al.*, 1971; Bateman, 1963a; Bateman and Beer, 1965; Bateman *et al.*, 1969; Garibaldi and Bateman, 1970; Heath and Wood, 1971; Lumsden, 1969; Melouk and Horner, 1972; Muse *et al.*, 1972; Swinburne and Corden, 1969; VanEtten and Bateman, 1969). In some cases in which the diseased tissue contains high oxidase levels precautions are needed during the extraction process to prevent inactivation of the polysaccharidases. For example, *F. roseum* "Avenaceum", a dry-rot pathogen of potato, produces exopolygalacturonase, endopolygalacturonate lyase, endo-β-1,4 galactanase, endo-β-1,4 xylanase, two arabanases, and cellulase (C_x) when cultured on isolated potato cell walls and during pathogenesis. In order to demonstrate these enzymes in extracts of rotted potato tuber tissue, it is necessary to store the intact tissue in a frozen state for an extended period prior to extraction or to extract the tissue in the presence of a reductant such as 2-mercaptoethanol (Mullen and Bateman, 1975a).

Invasion of plant tissues by pathogens such as *Sclerotinia sclerotiorum* and *Sclerotium rolfsii* leads to a very rapid solubilization of host cell wall polymers. This may be due in part to the secretion of oxalate by these organisms (Maxwell, 1973; Maxwell and Bateman, 1968). Oxalate is known to chelate calcium from the insoluble calcium pectates in plant cell walls and this renders the pectic fraction more susceptible to the endopolygalacturonases produced by these two fungi (Bateman and Beer, 1965; Maxwell and Lumsden, 1970). Since endopectic enzymes can serve as "wall modifying enzymes", which by disruption of the pectic fraction render the nonuronide polymers more susceptible to hydrolysis (English *et al.*, 1972; Talmadge *et al.*, 1973), and these fungi produce the full range of enzymes needed to attack the various cell wall polysaccharides (Bateman and Beer, 1965; Hancock, 1966, 1967; Lumsden, 1969; VanEtten and Bateman, 1969), the mechanism of rapid cell wall disruption is apparent.

Analysis of sunflower tissues infected with *S. sclerotiorum* has shown that the methoxyl and pectic acid contents decrease by 74–93% and 64–83%, respectively, within 2 to 4 days of symptom development (Hancock, 1966). The decrease in polymeric arabinose and galactose ranged between 65 and 97%, but the xylan content did not decrease even though xylanase could be extracted from the diseased tissue (Hancock, 1967). Analysis of bean hypocotyl tissues invaded by this same pathogen revealed a decrease in the α-cellulose content of the cell walls within two days after inoculation (Lumsden, 1969). Bean hypocotyl cell walls in tissue colonized with *S. rolfsii* undergo a rapid solubilization of the various cell wall sugars. The depletion of the uronic acid content is particularly rapid; by the time the fungus has fully occupied the tissue, up to 90% of the cell wall galacturonide has been lost (Bateman, 1970). The relative recoveries of the noncellulosic cell wall sugars from bean hypocotyl sections per cm of hypocotyl length three days after inoculation with *S. rolfsii* are shown in Fig. 3. All of the constituents assayed decreased following infection. *S. rolfsii*-infected tissue contains the full range of polysaccharide degrading enzymes needed to attack the various cell wall polymers, including cellulase (Bateman, 1969; Bateman and Beer, 1965; VanEtten and Bateman, 1969).

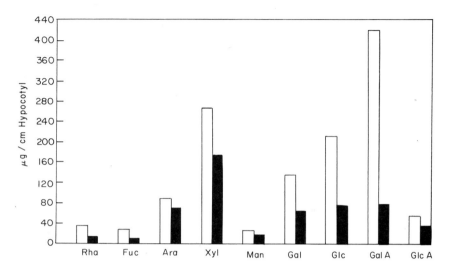

FIG. 3. Noncellulosic-sugar composition of bean hypocotyl cell walls 3 days after inoculating 7-day-old seedlings with *S. rolfsii*. Results are expressed as µg of each cell wall sugar constituent recovered per cm length of hypocotyl for healthy (□) and diseased (■) hypocotyls (adapted from Bateman and Jones, 1976).

Apple fruit infected with *Sclerotinia fructigena*, a pectolytic pathogen that causes a firm rot has been reported to show only about 15% decrease in the pectic fraction while soft-rot fungi like *Botrytis cinerea* and *P. expansum*, both of which are also pectolytic, caused a loss of 50–70% of this cell wall fraction (Cole and Wood, 1961). *R. solani* has the ability to produce a variety of polysaccharide degrading enzymes and this fungus causes an extensive decay of host tissues; it also disrupts host cell walls in invaded tissues and solubilizes the various cell wall polymers during pathogenesis (Bateman *et al.*, 1969). Cellulose represents the most resistant cell wall polysaccharide to enzymatic hydrolysis, a feature attributed to the insoluble, crystalline nature of this wall fraction. A number of pathogens such as *R. solani* (Bateman, 1964), *Pseudomonas solanacearum* (Kelman and Cowling, 1965), *S. rolfsii* (Bateman, 1969), and *Sclerotinia sclerotiorum* (Lumsden, 1969) are known to degrade native cellulose during pathogenesis.

It is now well established that many plant pathogens that cause obvious decays of plant tissues effect solubilization of cell wall polysaccharides in infected host tissues. Our knowledge of cell wall breakdown by obligate parasites is meager, however. The ability of obligate parasites to produce cell wall degrading enzymes apparently has not received attention except for a reported study with *Puccinia graminis* var. *tritici*. Endopolygalactu-ronase, hemicellulases and cellulase were detected when spores of this pathogen were germinated in the presence of appropriate substrates (Van Sumere *et al.*, 1957), but it is not known if these enzymes play a role in infected tissues. Analysis of pine tissues infected with *Cronartium ribicola* showed a 44% decrease in the pectic fraction (Welch and Martin, 1974) but it has not been shown that the pathogen is pectolytic. Certain mycorrhizal fungi are known to be pectolytic (Perombelon and Hadley, 1965), yet they are able to establish compatible, even synergistic relationships with their hosts. It is apparent from work with soft-rot pathogens that pectic enzymes if produced in invaded tissues in an unrestricted, uncontrolled manner macerate host tissues and kill host cells. Yet, recent cytological studies with obligate parasites, indicate that host cell wall penetration in many instances would appear to be aided by enzymes which degrade host cell wall constituents (Aist, 1976). If cell wall degrading enzymes are of importance in the invasion of plant tissues by mycorrhizal and pathogenic biotrophic (obligate) organisms, the sequence and duration of production of cell wall degrading enzymes must be carefully controlled processes.

VI. Enzymatic Basis of Tissue Maceration and its Consequences

A characteristic symptom of many diseases caused by facultative parasites is maceration, separation of cells from each other in invaded host tissue. This symptom is induced by the enzymatic digestion of the middle lamella region of the plant cell wall. The enzymatic basis for this phenomenon as well as the ability of certain pathogens to produce enzymes that cause tissue maceration was well established more than 50 years ago (Brown, 1915; Davison and Willaman, 1927; deBary, 1886; Jones, 1909). The pectic enzymes have long been considered the primary agents responsible for tissue maceration, but proof of their role in the process had to await establishment of the pectic nature of the structural element in the middle lamella of plant cell walls (McClendon, 1964) and development of the technology needed for enzyme purification. In the late 1950s and the early 1960s a number of investigators demonstrated that purified pectic enzymes which split the α-1,4 bonds in the uronide fraction of the pectic substances effect maceration of plant tissues (Bateman, 1963b; Demain and Phaff, 1957; McClendon, 1964; Zaitlin and Coltrin, 1964). Enzymes that attack the other cell wall polymers are not needed to induce cell separation (Bateman, 1968; Sato, 1968). The release of cells from tissue slices treated with pectic enzymes is characterized by a lag followed by a linear rate of cell release (Fig. 4). The length of the lag and rate of cell release are a function of the pectic enzyme concentration and the plant tissue source (Bateman, 1968;

Turner and Bateman, 1968; Tagawa and Kaji, 1966). The endopolygalacturonases and endopectate lyases are by far the most efficient macerating
enzymes. Recent studies of tissue maceration in which homogeneous enzymes
have been employed firmly established the nature of the process and the
enzymes responsible (Basham and Bateman, 1975a; Bush and Codner, 1970).

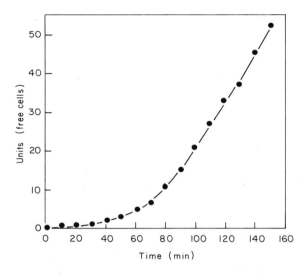

Fig. 4. Release of cells from discs of potato tuber tissue by a purified endopolygalacturonase
from *S. rolfsii* (from Bateman, 1968).

There are a few reports that certain pectic enzymes have failed to macerate
plant tissues (Byrde and Fielding, 1962; Garibaldi and Bateman, 1971;
Kaji, 1958). An endopolygalacturonase produced by *Sclerotinia fructigena*
was separated from the "macerating enzyme" of this fungus (Byrde and
Fielding, 1962); the "macerating enzyme" was subsequently shown to be
an endopectin methyl *trans*-eliminase (Byrde and Fielding, 1968). *Erwinia
chrysanthemi* produced four isozymes of endopolygalacturonate lyase one
of which failed to macerate potato tissue even though it readily attacked
isolated pectic substances from a number of sources (Garibaldi and Bateman,
1971); this isozyme also degraded the pectic fraction in isolated bean cell
walls (Bateman, unpublished). The pectic enzymes mentioned which have
failed to macerate plant tissue represent exceptions to the general rule, and
the reason they do not cause tissue maceration is not understood. There is
no doubt, however, that endopectic enzymes are agents responsible for
tissue maceration (Basham and Bateman, 1975a; Bateman, 1963b, 1968;
Bush and Codner, 1970; Garibaldi and Bateman, 1971; McClendon, 1964;
Mount *et al.*, 1970; Mussell and Morré, 1969; Sato, 1968; Spalding, 1969,
Spalding and Abdul-Baki, 1973).

Arabanases or α-L-arabinofuranosidases were considered at one time to be of possible significance with respect to tissue maceration (Byrde and Fielding, 1962). Subsequent work has failed to indicate a role for these enzymes in the maceration process (Bush and Codner, 1970; Byrde and Fielding, 1968; Cole and Bateman, 1969; Cole and Wood, 1970). A recent report indicates that β-1,4 galactanases from several pathogens of potato were able to macerate and kill potato tuber tissue (Cole and Sturdy, 1973). A purified endo-β-1,4 galactanase from a phytopathogenic *Sclerotinia* sp. has failed to macerate or kill cells in potato tuber tissue (Bauer and Bateman, 1975). Similar negative results have also been obtained with an endo-β-1,4 galactanase from another source (Codner, personal communication). At the present time, the endopectic enzymes that split the α-1,4 galacturonsyl bond in the pectic fraction of plant cell walls remain the only group of polysaccharide degrading enzymes which have been confirmed as agents of tissue maceration.

Plant cells are killed during the process of tissue maceration (Albergbina *et al.*, 1973; Hall and Wood, 1973; Mount *et al.*, 1970; Spalding, 1969; Tribe, 1955). It has been difficult to understand how enzyme preparations responsible for tissue maceration induce cell death. A number of hypotheses have evolved to account for killing of plant cells by enzyme preparations containing pectic enzymes. The major ones have included: (1) pectic enzyme preparations are contaminated with a toxin (Brown, 1915; Tribe, 1955), (2) cell death may be due to the presence of phosphatidases and proteinases present in enzyme preparations causing tissue maceration (Tribe, 1955), (3) some property of pectic enzymes, apart from their catalytic activities, is responsible for the toxic effects on cells (Mount *et al.*, 1970), (4) reaction products of cell wall breakdown by pectic enzymes are toxic to cells (Basham and Bateman, 1975a; Fushtey, 1957), (5) enzymes are released from cell walls as a result of pectic enzyme action that generates products toxic to protoplasts (Mussell, 1973), (6) pectic enzymes may attack a uronide substrate in the cell membrane or within the protoplast (Mount *et al.*, 1970), and (7) pectic enzyme action on the cell wall loosens the wall so that it is no longer able to support the protoplast which is under osmotic stress (Hall and Wood, 1973). It is now possible, based on experimental evidence, to eliminate from serious consideration all of the above hypotheses except for hypothesis seven (Basham and Bateman, 1975a, b; Stephens and Wood, 1975).

Experiments carried out with highly purified or homogeneous pectic enzymes (Basham and Bateman, 1975a) eliminate the first two hypotheses. It has also been demonstrated that phosphatidases and proteinases from pathogens are not toxic to intact plant tissues even though isolated protoplasts may be killed by these enzymes (Stephens and Wood, 1975; Tseng and Mount, 1974). Reports that a variety of proteins with high net-positive ionic charges disrupt the plasma membrane of isolated *Avena* protoplasts (Ruesink, 1971; Ruesink and Thimann, 1965) lead to the development of the third hypothesis

(Mount *et al.*, 1970), but recent studies with a group of purified pectic enzymes, members of which exhibited different ionic properties as well as different mechanisms of α-1,4 bond cleavage (hydrolytic and lyitic), have revealed that the ability of pectic enzymes to macerate and kill plant tissues is not related to the ionic properties of the enzymes but rather to the random (endo) cleavage of the uronide fraction of the cell wall (Basham and Bateman, 1975a). Also, the major tenet of hypothesis four can be eliminated since it has been demonstrated that the hydrolytic and lyitic degradation products of the pectic fraction of cell wall digests are not toxic to cells in plant tissues (Basham and Bateman, 1975a; Fushtey, 1957).

The hypothesis that enzymes are released from cell walls by pectic enzyme action which in turn induce a toxic reaction in tissue seems reasonable in view of studies which have shown that glucose oxidase which generates H_2O_2 as a reaction product is released from cauliflower cell walls by pectic enzyme action (Lund, 1973; Lund and Mapson, 1970), and that peroxidase is a cell wall constituent (Ridge and Oxborne, 1971). The latter enzyme in fact can also be released from isolated tobacco and potato cell walls by pectic enzymes (Basham and Bateman, 1975a). Also, cotton cuttings allowed to take up endopolygalacturonase are induced to yield symptoms similar to those caused by *V. albo-atrum* indicating that the pectic enzyme induces a toxic reaction in this plant (Mussell, 1973). The toxicity of endopoly-galacturonase to cotton may result from some indirect mechanism, but recent studies in which the cell death in tissue discs of potato and tobacco pith were examined in relation to pectic enzyme action, the only toxic entity identified in tissue digests was residual pectic enzyme. Furthermore H_2O_2 concentrations 1 000-fold greater than those reported to accumulate in cotton tissue treated with endopolygalacturonase failed to injure potato tissue; beet tissue, resistant to maceration by pectic enzymes, was not injured when incubated with potato tissue undergoing maceration and cell death caused by a purified pectic lyase (Basham and Bateman, 1975a). It is now apparent from this recent work that one does not have to invoke the view expressed in hypothesis five to account for cell death caused by pectic enzymes.

In tissue exposed to pectic enzymes the rate of cell wall breakdown parallels the rate of cell injury (Basham and Bateman, 1975b; Mount *et al.*, 1970). The only means of separating the two processes involves plasmolyzing the tissue upon exposure to pectic enzymes (Basham and Bateman, 1975a; Tribe, 1955). In the presence of plasmodica, maceration and cell wall break-down proceed, but the protoplasts remain alive. It has also been demonstrated that isolated protoplasts in plasmodica are not injured by purified pectic enzymes (Stephens and Wood, 1975; Tseng and Mount, 1974). In addition, protoplasts subjected to osmotic stretching in the presence or absence of pectic enzymes exhibit similar death rates (Basham and Bateman, 1975a). These observations tend to eliminate the validity of hypothesis six.

Based on the current concept of cell wall structure (Fig. 1) and the available experimental evidence, hypothesis seven is the most rational explanation for the injurious effect of pectic enzymes on plant cells. The random cleavage of the pectic fraction by pectic enzymes results in a dissolution of the middle lamella and a loosening of the primary cell wall structure such that the cell wall is unable to retain the protoplast which is under osmotic stress. The action of pectic enzymes on the middle lamella and primary wall regions occur simultaneously, and in the absence of a plasmodicum tissue maceration and cell death occur at similar rates. Both processes can be attributed to the direct action of pectic enzymes on the rhamnogalacturonide fraction of the plant cell wall.

VII. Conclusions

The plant cell wall is a complex yet ordered structure that represents a barrier that is breached repeatedly during successful infections of plants by most pathogens. Many pathogens possess the ability to produce a number of extracellular enzymes capable of degrading the various polysaccharide constituents of higher plant cell walls. In recent years great progress has been made in our knowledge of the polymeric carbohydrates of cell walls and their relationships to each other as well as in the purification of enzymes of microbial origin that cleave the various glycosidic bonds found therein. The availability of purified polysaccharidases has contributed greatly to our current understanding of primary cell wall structure. The model of primary wall structure which has emerged provides a compatible basis for interpreting and predicting the effects a given polysaccharide degrading enzyme may have on cell wall breakdown. There is a need for further critical studies of cell wall breakdown by the different polysaccharidases individually and in various combinations. Such studies should enable elucidation of further details of cell wall structure and permit clarification of the role of the various enzymes in cell wall decomposition.

The endopectic enzymes which cleave the α-1,4 galacturonide bonds in the rhamnogalacturonide fraction of the cell wall represent the key to enzymatic decomposition of cell walls in herbaceous tissues (Basham and Bateman, 1975a,b; Stephens and Wood, 1975). These same enzymes are responsible for tissue maceration and can cause cell death in tissues; these effects are brought about by dissolution of the middle lamella and weakening of primary wall structure by solubilization of the pectic fraction. The pectic enzymes thus possess the potential of playing a dual role in pathogenesis in addition to rendering the non-pectic polymers in cell walls more accessible to enzymatic hydrolysis.

It is apparent that the problems of cell wall degradation during pathogenesis are complex. A number of factors may determine whether or not a particular enzyme is produced by a pathogen in a given environment. If the enzyme in question is produced, the environment into which it is excreted can be expected to greatly influence its effectiveness and/or its existence in an active form. The molecular environment at the host–parasite interface and in invaded tissues is likely to contain elements that influence enzyme induction or repression in the pathogen as well as activity and stability of the different wall degrading enzymes (Albersheim and Anderson, 1971; Horsfall and Dimond, 1957). Furthermore, the problems of substrate accessibility become relevant since certain wall polymers may be masked by others and deposition of lignin-like materials in cell walls during pathogenesis represents a common phenomenon (Fox et al., 1972; Mullen and Bateman, 1975a) that may also protect cell wall polysaccharides from enzymatic hydrolysis.

Some pathogens produce both pectic hydrolases and lyases (Berndt, 1973; Chan and Sackston, 1970; Garibaldi and Bateman, 1973; Goodenough and Maw, 1974; Hanssler, 1973; Muse et al., 1972; Talboys and Busch, 1970; Wang and Pinckard, 1971), while others produce one or the other of these enzymes (Bateman, 1972; Cole, 1970; Hall and Wood, 1970; Mount et al., 1970; Schulz, 1972). In some cases, the type of enzyme produced appears to be more characteristic of the pathogen strain or isolate than of the species (Garibaldi and Bateman, 1973). The ability of pathogens to produce multiple forms of the different polysaccharide degrading enzymes also appears to be common (Dean and Wood, 1967; Garibaldi and Bateman, 1971; Mullen and Bateman, 1975b), but it is not known whether or not these different isozymes represent different gene products. Isozymes of a given polysaccharidase may exhibit differences with respect to pH optima, molecular size, ionic charge, etc. (Bateman, 1969; Garibaldi and Bateman, 1971; Mullen and Bateman, 1975b). It is tempting to speculate that the existence of isozymes of a given polysaccharidase in the complex environment of a pathogen–suscept interface would enhance the probability that a given substrate would be degraded.

Attempts to correlate production of a cell wall degrading enzyme in vitro by a pathogen with its pathogenicity or virulence appears to be of little value because of the great differences in in vitro and in vivo environments, and the drastic influence a given environmental component can have on production and stability of a given enzyme. For example, where pathogens are capable of producing both pectic hydrolases and lyases, the type of enzyme observed to predominate often is associated with the pH of the environment (Bateman, 1966, 1967; Goodenough and Maw, 1974; Hancock et al., 1970); also the pectic enzymes produced by a pathogen in culture may differ from those found in infected host tissues (Bateman, 1963a; Swinburne and Corden, 1969).

Our knowledge of cell wall decomposition by pathogens has been derived almost exclusively from studies with facultative pathogens that cause extensive decays of invaded suscept tissues (Albersheim et al., 1969; Bateman and Basham, 1976). There is no doubt about the ability of certain pathogens to excrete cell wall degrading enzymes in infected tissues, or the ability of these enzymes to disrupt cell walls during pathogenesis (Bateman, 1970; Bateman and Jones, 1976; Mullen and Bateman, 1975a). The role played by the pectic enzymes in cell wall disruption and the potential of these enzymes to injure plant tissues are reasonably clear (Basham and Bateman, 1975a,b; Stephens and Wood, 1975); our knowledge of the significance of the other wall degrading enzymes in disease processes is at present quite fragmentary.

There is a marked need to focus attention on the involvement of cell wall degrading enzymes in the initial phases of tissue penetration by pathogens and their potential involvement in diseases caused by biotrophic (obligate) parasites. The resolution of these problems, germane to expanding our understanding of the significance of cell wall degrading enzymes in plant diseases, will require merging biochemical and cytological technologies in order to approach the subject in the most meaningful way. Bringing together these areas of research to help resolve the problems of enzymatic degradation of cell walls in relation to pathogenesis represents a significant challenge.

REFERENCES

Aist, J. R. (1976). *In* "Encyclopedia of Plant Physiology" (P. Williams and R. Heitefuss, eds), N.S., Vol. 4. Springer-Verlag, Berlin, Heidelburg and New York (in press).

Albergbina, A., Mazzenchi, U. and Pupillo, P. (1973). *Phytopathol. Z.* **78**, 204–213.

Albersheim, P. (1975). *Sci. Am.* **232**, 80–95.

Albersheim, P. and Anderson, A. (1971). *Proc. Nat. Acad. Sci.* (U.S.A.) **68**, 1815–1819.

Albersheim, P., Jones, T. M. and English, P. D. (1969). *A. Rev. Phytopath.* **7**, 171–194.

Aspinall, G. O. (1970a). *In* "The Carbohydrates—Chemistry and Biochemistry", Vol. 11B (2nd edition), (W. Pigman and D. Horton, eds), pp. 515–536. Academic Press, New York and London.

Aspinall, G. O. (1970b). "Polysaccharides", p. 228. Pergamon Press, Oxford.

Ayers, W. A., Papavizas, G. C. and Lumsden, R. D. (1969). *Phytopathology* **59**, 786–791.

Balasubramani, K. A., Deverall, B. J. and Murphy, J. V. (1971). *Physiol. Pl. Path.* **1**, 105–113.

Basham, H. G. and Bateman, D. F. (1975a). *Phytopathology* **65**, 141–153.

Basham, H. G. and Bateman, D. F. (1975b). *Physiol. Pl. Path.* **5**, 249–261.

Bateman, D. F. (1963a). *Phytopathology* **53**, 197–204.

Bateman, D. F. (1963b). *Phytopathology* **53**, 1178–1186.

Bateman, D. F. (1964). *Phytopathology* **54**, 1372–1377.

Bateman, D. F. (1966). *Phytopathology* **56**, 238–244.

Bateman, D. F. (1967). *In* "The Dynamic Role of Molecular Constituents in Plant-Parasite Interaction" (C. J. Mirocha and I. Uritani, eds), pp. 58–79. Am. Phytopathol. Soc., St. Paul, Minn.

Bateman, D. F. (1968). *Neth. J. Pl. Path.* **74** (Suppl. 1), 67–80.

Bateman, D. F. (1969). *Phytopathology* **59**, 37–42.

Bateman, D. F. (1970). *Phytopathology* **60**, 1846–1847.

Bateman, D. F. (1972). *Physiol. Pl. Path.* **2**, 175–184.

Bateman, D. F. and Basham, H. G. (1976). *In* "Encyclopaedia of Plant Physiology", N.S. (P. Williams and R. Heitefuss, eds), Vol. 4. Springer-Verlag, Berlin, Heidelburg and New York (in press).
Bateman, D. F. and Beer, S. V. (1965). *Phytopathology* **55**, 204–211.
Bateman, D. F. and Jones, T. M. (1976). *Proc. Am. Phytopathol. Soc.* **2**, 131.
Bateman, D. F. and Millar, R. L. (1966). *A. Rev. Phytopath.* **4**, 119–146.
Bateman, D. F., VanEtten, H. D., English, P. D., Nevins, D. J. and Albersheim, P. (1969). *Pl. Physiol.* **44**, 641–648.
Bateman, D. F., Jones, T. M. and Yoder, O. C. (1973). *Phytopathology* **63**, 1523–1529.
Bauer, W. D. and Bateman, D. F. (1975). *Proc. Am. Phytopathol. Soc.* **1**, 78.
Bauer, W. D., Talmadge, K. W., Keegstra, K. and Albersheim, P. (1973). *Pl. Physiol.* **51**, 174–187.
Beraha, L. and Garber, E. D. (1971). *Phytopath. Z.* **70**, 335–344.
Berndt, H. (1973). *Arch. Mikrobiol.* **91**, 137–148.
Biehn, W. L. and Dimond, A. E. (1971a). *Phytopathology* **61**, 242–243.
Biehn, W. L. and Dimond, A. E. (1971b). *Phytopathology* **61**, 745–746.
Biehn, W. L., Sands, D. C. and Hankin, L. (1972). *Phytopathology* **62**, 747. (Abstr.).
Bracker, C. E. and Littlefield, L. J. (1973). *In* "Fungal Pathogenicity and the Plant's Response" (R. J. W. Byrde and C. V. Cutting, eds), pp. 159–318, Academic Press, London and New York.
Brown, S. A. (1969). *Bioscience* **19**, 115–121.
Brown, W. (1915). *Ann. Bot.* **29**, 313–348.
Brown, W. (1955). *Ann. appl. Biol.* **43**, 325–341.
Bugbee, W. M. (1973). *Phytopathology* **63**, 480–484.
Bush, D. A. and Codner, R. C. (1970). *Phytochemistry* **9**, 87–97.
Byrde, R. J. W. and Fielding, A. H. (1962). *Nature Lond.* **196**, 1227–1228.
Byrde, R. J. W. and Fielding, A. H. (1968). *J. gen. Microbiol.* **52**, 287–297.
Chan, Y. and Sackston, W. E. (1970). *Can. J. Bot.* **48**, 1073–1077.
Cole, A. L. J. (1970). *Phytochemistry* **9**, 337–340.
Cole, A. L. J. and Bateman, D. F. (1969). *Phytopathology* **59**, 1750–1753.
Cole, A. L. J. and Sturdy, M. L. (1973). "Second International Congress of Plant Pathology", Abst. 964. Am. Phytopathol. Soc., Minneapolis, Minn.
Cole, A. L. J. and Wood, R. K. S. (1970). *Phytochemistry* **9**, 695–699.
Cole, M. and Wood, R. K. S. (1961). *Ann. Bot.* **25**, 435–452.
Colvin, J. R. and Leppard, G. G. (1971). *J. Microsc.* **11**, 285–298.
Cooper, R. M. and Wood, R. K. S. (1975). *Physiol. Pl. Path.* **5**, 135–156.
Cowling, E. B. (1975). *In* "Cellulose as a Chemical and Energy Resource", (M. Mandels, J. A. Bassham and C. R. Wilkie, eds). Biotechnol. and Bioeng. Symposium No. 5, pp. 163–181. Wiley and Sons, New York and Chichester.
Davison, F. R. and Willaman, J. J. (1927). *Bot. Gaz.* **83**, 329–361.
Dean, M. and Wood, R. K. S. (1967). *Nature Lond.* **214**, 408–410.
deBary, A. (1886). *Bot. Ztg.* **44**, 377–480.
Demain, A. L. and Phaff, H. J. (1957). *Wallerstein Labs Commun.* **20**, 119–139.
Deuel, H. and Stutz, E. (1958). *Adv. Enzymol.* **20**, 341–382.
English, P. D., Jurale, J. B. and Albersheim, P. (1971). *Pl. Physiol.* **47**, 1–6.
English, P. D., Maglothin, A., Keegstra, K. and Albershiem, P. (1972). *Pl. Physiol.* **49**, 293–298.
Fergus, B. J., Procter, A. R., Scott, J. A. N. and Goring, D. A. I. (1969). *Wood Sci. Technol.* **3**, 117–138.
Fox, R. T. V., Manners, J. G. and Myers, A. (1972). *Potato Res.* **15**, 130–145.
Freudenberg, K. (1968). *In* "The Constitution and Biosynthesis of Lignin". (K. Freudenberg, ed.), pp. 47–122. Springer-Verlag, Berlin and Heidelburg and New York.

Frey-Wyssling, A. (1969). *Fortschr. Chem. org. Natstoffe* **27**, 1–30.
Fuchs, A., Jobsen, J. A. and Wouts, W. M. (1965). *Nature Lond.* **206**, 714–715.
Fushtey, S. G. (1957). *Ann. Bot.* (N.S.) **21**, 273–286.
Garibaldi, A. and Bateman, D. F. (1970). *Phytopathol. Med.* **9**, 136–144.
Garibaldi, A. and Bateman, D. F. (1971). *Physiol. Pl. Path.* **1**, 25–40.
Garibaldi, A. and Bateman, D. F. (1973). *Phytopathol. Med.* **12**, 30–35.
Goodenough, P. W. and Maw, G. A. (1974). *Physiol. Pl. Path.* **4**, 51–62.
Hall, J. A. and Wood, R. K. S. (1970). *Nature Lond.* **227**, 1266–1267.
Hall, J. A. and Wood, R. K. S. (1973). *In* "Fungal Pathogenicity and the Plant's Response" (R. J. W. Byrde and C. V. Cutting, eds), pp. 19–38. Academic Press, London and New York.
Halliwell, G. and Mohammed, R. (1971). *Arch. Mikrobiol.* **78**, 295–309.
Hancock, J. G. (1966). *Phytopathology* **56**, 975–979.
Hancock, J. G. (1967). *Phytopathology* **57**, 203–206.
Hancock, J. G. Eldridge, C. and Alexander, M. (1970). *Can. J. Microbiol.* **16**, 69–74.
Hanssler, G. (1973). *Phytopathol. Z.* **77**, 138–156.
Hasegawa, S. and Nagel, C. W. (1962). *J. biol. Chem.* **237**, 619–621.
Hashimoto, S., Muramatsu, T. and Funatsu, M. (1971). *Agric. Biol. Chem.* **35**, 501–508.
Heath, M. C. and Wood, R. K. S. (1971). *Ann. Bot.* **35**, 451–474.
Hirst, E. L. and Jones, J. K. N. (1947). *J. Chem. Soc.* 1221–1225.
Hislop, E. C., Barnaby, V. M., Shellis, C. and Laborda, F. (1974). *J. gen. Microbiol.* **81**, 79–99.
Horsfall, J. G. and Dimond, A. E. (1957). *Z. Pflkrankh. Pfl. Path. Pfl. Schutz.* **64**, 415–421.
Horton, J. C. and Keen, N. T. (1966). *Phytopathology* **56**, 908–916.
Hsu, E. J. and Vaughn, R. H. (1969). *J. Bact.* **98**, 172–181.
Jones, L. R. (1909). *Tech. Bull. Va. agric. Expt. Sta.* **147**, 283–360.
Jones, T. M. and Albersheim, P. (1972). *Pl. Physiol.* **49**, 926–936.
Jones, T. M. and Bateman, D. F. (1972). *Phytopathology* **62**, 767–768. (Abstr.).
Jones, T. M., Anderson, A. J. and Albersheim, P. (1972). *Physiol. Pl. Path.* **2**, 153–166.
Kaji, A. (1958). *Tech. Bull. Fac. Agric. Kagawa Univ.* **9**, 141–145.
Kaji, A. and Ichimi, T. (1973). *Agric. Biol. Chem.* **37**, 431–432.
Kaji, A. and Tagawa, K. (1970). *Biochim. Biophys. Acta* **207**, 456–464.
Kaji, A. and Yoshihara, O. (1970). *Agric. Biol. Chem.* **34**, 1249–1253.
Kaji, A., Tagawa, K. and Motoyama, K. (1965). *J. agric. Chem. Soc. Japan* **39**, 352–357.
Karr, A. L. and Albersheim, P. (1970). *Pl. Physiol.* **46**, 69–80.
Keegstra, K., Talmadge, K. W., Bauer, W. D. and Albersheim, P. (1973). *Pl. Physiol.* **51**, 188–197.
Keen, N. T. and Horton, J. C. (1966). *Can. J. Microbiol.* **12**, 443–453.
Kelman, A. and Cowling, E. B. (1965). *Phytopathology* **55**, 148–155.
Kertesz, Z. I. and Lavin, M. I. (1954). *Food Res.* **19**, 627–632.
Knee, M. and Friend, J. (1968). *Phytochemistry* **7**, 1289–1291.
Lamport, D. T. A. (1970). *Ann. Rev. Pl. Physiol.* **21**, 235–270.
Lamport, D. T. A., Katona, L. and Roerig, S. (1973). *Biochem. J.* **133**, 125–132.
Lee, M. and Macmillan, J. D. (1968). *Biochemistry* **7**, 4005–4010.
Lumsden, R. D. (1969). *Phytopathology* **59**, 653–657.
Lund, B. M. (1973). *In* "Fungal Pathogenicity and the Plant's Response" (R. J. W. Byrde and C. V. Cutting, eds), pp. 69–86. Academic Press, London and New York.
Lund, B. M. and Mapson, L. W. (1970). *Biochem. J.* **119**, 251–263.
Lyr, H. (1963). *Z. Allg. Mikrobiol.* **3**, 25–36.

102   D. F. BATEMAN

Maxwell, D. P. (1973). *Physiol. Pl. Path.* **3**, 279–288.
Maxwell, D. P. and Bateman, D. F. (1968). *Phytopathology* **58**, 1635–1642.
Maxwell, D. P. and Lumsden, R. D. (1970). *Phytopathology* **60**, 1395–1398.
McClendon, J. H. (1964). *Am. J. Bot.* **51**, 628–633.
McColloch, R. J. and Kertesz, Z. I. (1947). *Arch. Biochem.* **13**, 217–229.
McKeen, W. E. and Rimmer, S. R. (1973). *Phytopathology* **63**, 1049–1053.
Melouk, H. A. and Horner, C. E. (1972). *Can. J. Microbiol.* **18**, 1065–1072.
Miller, L. and Macmillan, J. D. (1971). *Biochemistry* **10**, 570–576.
Moran, F. and Starr, M. P. (1969). *Eur. J. Biochem.* **11**, 291–295.
Mount, M. S., Bateman, D. F. and Basham, H. G. (1970). *Phytopathology* **60**, 924–931.
Muhlethaler, K. (1967). *A. Rev. Pl. Physiol.* **18**, 1–24.
Mullen, J. M. and Bateman, D. F. (1971). *Physiol. Pl. Pathol.* **1**, 363–373.
Mullen, J. M. and Bateman, D. F. (1975a). *Phytopathology* **65**, 797–802.
Mullen, J. M. and Bateman, D. F. (1975b). *Physiol. Pl. Pathol.* **6**, 233–246.
Muse, R. R., Couch, H. B., Moore, L. D. and Muse, B. D. (1972). *Can. J. Microbiol.* **18**, 1091–1098.
Mussell, H. W. (1973). *Phytopathology* **63**, 62–70.
Mussell, H. W. and Green, R. J. (1970). *Phytopathology* **60**, 192–195.
Mussell, H. W. and Morré, D. J. (1969). *Anal. Biochem.* **28**, 353–360.
Mussell, H. W. and Strouse, B. (1972). *Can J. Biochem.* **50**, 625–632.
Northcote, D. H. (1963). *Int. Rev. Cytol.* **14**, 233–265.
Northcote, D. H. (1969). *In* "Essays in Biochemistry" (P. N. Campbell and G. D. Greville, eds), Vol. 5, pp. 89–137. Academic Press, New York and London.
Northcote, D. H. (1972). *A. Rev. Pl. Physiol.* **23**, 113–132.
Okamoto, K., Hatanaka, C. and Ozawa, J. (1964). *Agric. Biol. Chem.* **28**, 331–336.
Patil, S. S. and Dimond, A. E. (1968). *Phytopathology* **58**, 676–682.
Perombelon, M. and Hadley, G. (1965). *New Phytol.* **64**, 144–151.
Preston, R. D. (1971). *J. Microsc.* **93**, 7–13.
Reddy, M. N., Stuteville, D. L. and Sorensen, E. L. (1969). *Phytopathology* **59**, 887–888.
Reese, E. T. (1956). *Appl. Microbiol.* **4**, 39–45.
Reese, E. T. and Shibata, Y. (1965). *Can. J. Microbiol.* **11**, 167–183.
Ridge, I. and Osborne, D. J. (1971). *Nature New Biol.* **229**, 205–208.
Rombouts, F. M. and Pilnik, W. (1972). *Chem. Rubb. Comp. Crit. Rev. Food Technol.* **3**, 1–26.
Ruesink, A. W. (1971). *Pl. Physiol.* **47**, 192–195.
Ruesink, A. W. and Thimann, K. V. (1965). *Proc. natn. Acad. Sci. U.S.A.* **54**, 56–64.
Sadava, D. and Chrispeels, M. J. (1969). *Science* **165**, 299–300.
Sadava, D., Walker, F. and Chrispeels, M. J. (1973). *Dev. Biol.* **30**, 42–48.
Sargent, J. A., Tommerup, I. C. and Ingram, D. S. (1973). *Physiol. Pl. Pathol.* **3**, 231–239.
Sarko, A. and Muggli, R. (1974). *Macromolecules* **7**, 486–494.
Sato, S. (1968). *Physiol. Pl.* **21**, 1067–1075.
Schulz, F. A. (1972). *Phytopathol. Z.* **74**, 97–108.
Seemuller, E. A., Beer, S. V., Jones, T. M. and Bateman, D. F. (1973). *Phytopathology* **63**, 207. (Abstr.).
Spalding, D. H. (1969). *Phytopathology* **59**, 685–692.
Spalding, D. H. and Abdul-Baki, A. A. (1973). *Phytopathology* **63**, 231–235.
Spalding, D. H., Wells, J. M. and Allison, D. W. (1973). *Phytopathology* **63**, 840–844.
Stephens, G. J. and Wood, R. K. S. (1975). *Physiol. Pl. Pathol.* **5**, 165–181.

Steward, F. C., Israel, H. W. and Salpeter, M. M. (1967). *Proc. natn. Acad. Sci. U.S.A.* **58**, 541–544.
Strobel, G. A. (1963). *Phytopathology* **53**, 592–596.
Swinburne, T. R. and Corden, M. E. (1969). *J. gen. Microbiol.* **55**, 75–87.
Tagawa, K. and Kaji, A. (1966). *Tech. Bull. Fac. Agric. Kagawa Univ.* **17**, 104–109.
Talboys, P. W. and Busch, L. V. (1970). *Trans. Br. mycol. Soc.* **55**, 367–381.
Talmadge, K. W., Keegstra, K., Bauer, W. D. and Albersheim, P. (1973). *Pl. Physiol.* **51**, 158–173.
Tribe, H. T. (1955). *Ann. Bot.* (N.S.) **19**, 351–368.
Tseng, T. C. and Mount, M. S. (1974). *Phytopathology* **64**, 229–236.
Turner, M. T. and Bateman, D. F. (1968). *Phytopathology* **58**, 1509–1515.
VanEtten, H. D. and Bateman, D. F. (1969). *Phytopathology* **59**, 968–972.
Van Sumere, C. F., Van Sumere-De Preter, C. and Ledingham, G. A. (1957). *Can. J. Microbiol.* **3**, 761–770.
Walker, D. J. (1967). *Aust. J. Biol. Sci.* **20**, 799–808.
Wang, S. C. and Pinckard, J. A. (1971). *Phytopathology* **61**, 1118–1124.
Weinhold, A. R. and Bowman, T. (1974). *Phytopathology* **64**, 985–990.
Welch, B. L. and Martin, N. E. (1974). *Phytopathology* **64**, 1287–1289.
Wilder, B. M. and Albersheim, P. (1973). *Pl. Physiol.* **51**, 889–893.
Zaitlin, M. and Coltrin, D. (1964). *Pl. Physiol.* **39**, 91–95.
Zucker, M. and Hankin, L. (1970). *J. Bact.* **104**, 13–18.
Zucker, M. and Hankin, L. (1971). *Can. J. Microbiol.* **17**, 1313–1318.

CHAPTER 6

Killing of Protoplasts

R. K. S. WOOD

*Department of Botany, Imperial College of Science and Technology, London,
England*

I. INTRODUCTION

This paper will deal mainly with death of protoplasts in soft rots of
parenchyma and chlorenchyma caused by bacteria and fungi and, more
briefly, in certain other types of disease.

The main symptoms of soft rots are as follows:

1. lesions usually grow rapidly to occupy most or all of the available
   tissue;
2. rotted tissue has little or no coherence;
3. cells of rotted tissue are separated along the lines of middle lamellae;
4. after separation, cell walls and cells retain their identity for some time;
5. rotted tissue is water soaked;
6. permeability of protoplasts to electrolytes increases rapidly before cell
   separation;
7. rotted tissue is discoloured in some diseases but not in others;
8. protoplasts of rotted tissue are dead.

Chapter 5 in this book explains how cells are separated by enzymes that
act upon middle lamella/cell wall, thus it is necessary only to refer briefly

to this chapter and to the tentative model of primary walls of cultured sycamore cells proposed by Keegstra *et al.* (1973). In this model cellulose microfibrils are hydrogen-bonded to hemicellulosic xyloglucan which may be linked to galactan side chains of rhamnogalacturonan, a pectic polysaccharide which may be further linked to protein through 3,6-linked arabinogalactan chains. In view of the complexity of this structure it is, perhaps, surprising that cells can be separated solely by polygalacturonases or *trans*-eliminases that rupture α-1,4 linkages between galacturonic acid residues of pectic polysaccharides. It is still not clear how this leads to cell separation in terms of the model which does not include the middle lamella. This does not contain cellulose microfibrils and is assumed to contain larger proportions of pectic polysaccharides than does the matrix of the primary wall. Also, the model does not include divalent ions, especially calcium, which again have been assumed to be important in the structure and insolubility of certain cell wall polymers, particularly the pectic polysaccharides.

The causes of protoplast death in soft rots will now be considered.

II. RELATION BETWEEN CELL SEPARATION AND PROTOPLAST DEATH

The work of many years has established a close association between fractions from culture filtrates or rotted tissue which kill protoplasts and the presence in these fractions of enzymes that cause cell separation and have a mode of action *in vitro* which is confined to the rupture of the α-1,4 glycosiduronic bonds of pectic polysaccharides. (See Brown, 1965 for references to many of the earlier papers.) More recent work has shown for some diseases that cell separation, degradation of the α-1,4 glycosiduronic linkages of pectic polysaccharides *in vitro* and death of protoplasts are all consequences of the action of single polygalacturonases or *trans*-eliminases (pectic enzymes for convenience) produced by plant pathogens *in vitro* and readily extracted from soft-rotted tissue. The rest of this chapter examines the relations between these effects.

III. ACTION OF PECTIC ENZYMES ON PARENCHYMA

The usual procedure is to extract and isolate single pectic enzymes from rotted tissues or from filtrates from cultures of plant pathogens grown so that synthesis of the enzymes is induced and not repressed. Solutions of the enzymes are then applied to slices (*ca* 0·5 mm thick) of various tissues, often medulla of potato tubers. The rate of cell separation is assessed from the time taken for slices to lose coherence. Death of protoplasts is estimated according to loss of ability to plasmolyse or inability to retain neutral red vital stain which accumulates in and is retained by vacuoles of living protoplasts. These assays are no more than semi-quantitative but they give results which can be interpreted with some confidence and which show, with the exception given below, that death of protoplasts is closely associated with cell separation. This applies to enzymes which are polygalacturonases or *trans*-eliminases, to those with very different pH optima, to enzymes activated or unaffected by calcium ions, and when enzyme activity is increased or decreased in various other ways.

Death of protoplasts is preceded by large increases in permeability to ions, as assessed by measurement of $^{86}Rb^+$ or K^+ released into ambient solutions, or by increase in their conductivity. Permeability increases after a few minutes exposure to enzymes; these begin to kill substantial numbers of protoplasts after about thirty minutes. It is not easy to relate these changes in permeability to death of protoplasts because ions may accumulate rapidly and continuously in ambient solutions around discs containing many thousands of cells. This occurs in the following ways: protoplast 1 may be killed quickly and release most of its ions, followed by protoplast 2 and so on; or, more likely, ions are released slowly but continuously and at about the same rate from protoplast 1 and 2, culminating in the death of most of the protoplasts at about the same time. It has been suggested that ions liberated from cell walls by pectic enzymes could contribute significantly to increases in conductivity of ambient solutions (Friedman and Jaffe, 1960). More recent work shows that there is little or no release of ions from discs of tissue exposed to 30% ethanol to kill protoplasts and then leached with water before treating with pectic enzymes (Hall and Wood, 1974; Stephens, 1974).

Protein also is released from tissues treated with pectic enzymes. Thus florets of cauliflower treated with a pectate *trans*-eliminase release glucose oxidase, an enzyme associated with cell walls which causes the release of hydrogen peroxide following the oxidation of glucose (Lund and Mapson, 1970). This enzyme rapidly releases from potato tuber tissue phenoloxidase, which also is associated with cell wall fractions, acid phosphatase and acid phosphodiesterase, alkaline phosphatase and alkaline phosphodiesterase, which are enzymes more usually associated with lysosomes. Surprisingly, it may also release malic dehydrogenase which is usually associated with mitochondria, chloroplasts and cytoplasm. RNase but not DNase also accumulates in the ambient solutions (Stephens, 1974; Stephens and Wood, 1974).

IV. Effect of Other Enzymes on Parenchyma

On suitable substrates plant pathogens may produce in culture enzymes that degrade other polymers of cell walls including proteins. A similar range of enzymes also occur in rotted tissue (Stephens and Wood, 1975). Plant pathogens also produce phosphatidase active on lecithin. Purified preparations of these enzymes either do not cause cell separation and death of protoplasts, or are very much less active than pectic enzymes in these respects. Also, certain pectic enzymes behave anomalously. Thus, an isolate of the bacterium *Erwinia chrysanthemi* produces in culture at least four pectate *trans*-eliminases; each had pH optima in the range 8·2 to 9·8, required Ca^{2+} for activity and degraded pectate *in vitro* in a random manner. Three with isoelectric points at pH 9·4, 8·4 and 7·9 caused cell separation, loss of electrolytes and death of protoplasts, whereas the fourth with isoelectric point at pH 4·6 did not (Garibaldi and Bateman, 1971). Similarly, an endo-polygalacturonase from the vascular wilt pathogen *Verticillium albo-atrum*, which degraded pectate *in vitro* at about the same rate as an endo-pectate *trans*-eliminase from the bacterium *Erwinia carotovora*, was much less active in causing cell separation and death of protoplasts. This is difficult to understand because the enzymes would be expected to have much the same effect on pectic polysaccharides *in vivo* (Cooper, 1974).

V. Effects of Plasmolysis

Tribe (1955) found that when tissue was placed in a plasmolyticum containing pectic enzymes, in conditions in which cell separation was almost as rapid as in hypotonic solutions, killing of protoplasts was greatly delayed. This significant discovery was soon confirmed and is, of course, the basis of methods used to obtain large numbers of protoplasts from various tissues for physiological and other studies. Another important point is that only slight plasmolysis keeps protoplasts alive long after cells have separated.

Obviously, there is little or no leakage of electrolytes from plasmolysed protoplasts in cells of tissue treated with pectic enzymes. But if the toxicity of the ambient solution is decreased slowly, just as it falls below plasmolysing values there is a sudden increase in loss of electrolytes and in the proportion of protoplasts that do not retain neutral red stain. There is little or no leakage of electrolytes during deplasmolysis of protoplasts in untreated tissue. It seems, therefore, that although walls of cells treated with pectic enzymes retain their identity in that they remain coherent, they are altered in some critical way that prevents them from retaining protoplasts in cells exposed to hypotonic solutions (Basham and Bateman, 1975; Stephens and Wood, 1975).

It was mentioned earlier that, as well as ions, protein and certain enzymes are released from unplasmolysed tissue by pectic enzymes. Enzymes are also released from plasmolysed tissue. Thus acid phosphatase accumulates in solutions around plasmolysed tissue at about 40% of the rate at which it accumulates around unplasmolysed tissue. Although acid phosphatase is a typical lysosomal enzyme it has also been reported from cell walls. It seems, therefore, that because plasmolysed protoplasts remain intact, the acid phosphatase is released from cell walls by pectic enzymes. Similar results have been obtained for acid phosphodiesterase and for alkaline phosphodiesterase and, surprisingly, for RNase and malic dehydrogenase which are released from plasmolysed cells at rates of 50–80% of those for unplasmolysed tissue. Again, this implies that these enzymes occur in cell walls. But the proportion of the total in cells that is released by pectic enzymes remains to be determined (Stephens and Wood, 1974).

VI. Effects of Enzymes other than Pectic Enzymes

Proteinase and phosphatidase are produced by certain soft-rot pathogens in culture (Tseng and Bateman, 1968) and they can be obtained from rotted tissue. Singly or together they could cause or contribute to death of protoplasts by degrading plasmalemma or other limiting membranes (Brown, 1965; Tseng and Bateman, 1968). Also, infection of potato tubers by certain soft-rot fungal pathogens causes swelling of cytoplasmic organelles that contain a variety of enzymes including acid phosphatase and proteinase (Pitt and Coombes, 1968, 1969). It has been suggested that disruption of these organelles, mainly lysosomes, in host protoplasts, could release hydrolytic enzymes which cause autolysis of the cytoplasm (Wilson, 1973). However, phosphatidase and proteinase do not cause cell separation or death of protoplasts when applied to unplasmolysed discs of potato tubers in the same concentrations as in rotted tissue (Mount et al., 1970; Garibaldi and Bateman, 1971; Hall et al., 1974). Also, they are inactive on plasmolysed discs even after these have been treated with a pectic enzyme which causes cell separation (Stephens and Wood, 1975). Isolated protoplasts with no cell wall in a plasmolyticum behave quite differently. They are unaffected by solutions of a pectic enzyme which rapidly causes cell separation, but they are killed quite rapidly by phosphatidase and proteinase obtained from rotted tissue and used in the same concentrations as in this tissue (Tseng and Mount, 1974; Stephens and Wood, 1975). Together they kill protoplasts at about the same rate as crude extract. There is, therefore, the anomaly that proteinase and phosphatidase which kill isolated protoplasts in a plasmolyticum do not kill protoplasts within cells in a tissue even after this tissue has been treated with a pectic enzyme to cause cell separation (Stephens and Wood, 1975). These results suggest that the two enzymes do not move inward towards protoplasts through cell walls treated with pectic enzymes. But this treatment does allow rapid loss of protein from cells and, in particular, loss of enzymes from cell walls.

Cellulase from rotted tissue or from commercial samples in various combinations with pectic enzymes, proteinase and phosphatidase, has little or no effect on cell separation or death of protoplasts in plasmolysed or unplasmolysed tissue (Stephens and Wood, 1975).

VII. KILLING OF PROTOPLASTS OTHER THAN BY PECTIC ENZYMES

It is clear from the above that a single pectic enzyme which causes cell separation can also cause striking increases in permeability of protoplasts to ions and to protein and then cause death of protoplasts. Recently it has been shown that extracts from tissue rotted by a soft-rot bacterium contains another factor which kills protoplasts. Thus, an extract from rotted tissue and a pectate *trans*-eliminase from it, at about the same concentration, have similar effects on unplasmolysed tissue of potato tubers. However, whereas protoplasts in plasmolysed tissue treated with the enzyme remain alive for at least 24 h, most of the protoplasts in plasmolysed tissue treated with the crude extract are killed within 5 h; killing by crude extracts is delayed but not prevented (Stephens and Wood, 1975).

Similarly, isolated protoplasts in a plasmolyticum containing pectate *trans*-eliminase live about as long as in water or in a solution of autoclaved enzyme, whereas they die rapidly in a crude extract containing about the same amount of the pectic enzyme. The factor is not proteinase, phosphatidase or RNase from rotted tissue because alone or together these enzymes do not kill isolated protoplasts in plasmolysed tissue. The nature of this factor remains to be established; its effect on infected tissue is, however, likely to be subsidiary to that of the pectic enzymes which act much more rapidly.

VIII. OTHER CAUSES OF DEATH OF PROTOPLASTS

Although proteinase and phosphatide can kill isolated protoplasts, from the evidence cited above it seems unlikely that they cause death of protoplasts in infected tissue. We are left with a pectic enzyme as the direct or indirect cause of death. One possibility is that substances produced by the action of pectic substances on cell walls are toxic to protoplasts. This was investigated by Fushtey (1957) and rejected. More recently, it was reinvestigated by examining the toxicity of fractions obtained from solutions in which discs of potato tuber had been exposed to pectic enzymes for long periods. Fractions containing substances of molecular weight below 10 000 were not toxic to protoplasts in tissue discs. Of fractions containing substances of molecular weight above 10 000, only those containing the pectic enzyme caused death of protoplasts (Stephens and Wood, 1975). Similar results have been obtained with pectate *trans*-eliminase from *E. chrysanthemi* (Basham and Bateman, 1975).

It has also been suggested that oxidases released from cell walls by pectic enzymes may produce hydrogen peroxide in amounts sufficient to kill protoplasts. Toxicity of a polygalacturonase from the vascular wilt pathogen *V. albo-atrum* after absorption by tomato cuttings has been explained on this basis partly because it was suppressed by treating cuttings with catalase (Mussell, 1973). It has recently been found that pectate *trans*-eliminase releases peroxidase and catalase from potato discs and from cell walls isolated from tobacco tissue. However, potato discs treated with the pectic enzyme release little or no hydrogen peroxide, relatively high concentrations of hydrogen peroxide have little effect on permeability to electrolytes and do not kill protoplasts, and catalase does not affect killing by pectate *trans*-eliminase (Basham and Bateman, 1975; Stephens, 1974). Furthermore, damage to protoplasts by a pectate *trans*-eliminase was not affected by mercaptoethanol and diethyldithiocarbamate which are stated to nullify damage caused by hydrogen peroxide (Basham and Bateman, 1975; Siegel and Halpern, 1965). Therefore, although polygalacturonase of *V. albo-atrum* may kill protoplasts of tomato leaf cells by causing the production of hydrogen peroxide, it is unlikely that this mechanism is involved in the killing of protoplasts in soft rots caused by *Erwinia* spp.

Another possible explanation of toxicity is based on the fact that positively charged protein molecules cause protoplasts of *Avena* coloptiles to lyse (Ruesink, 1971; Ruesink and Thimann, 1965). However, although some pectic enzymes have isoelectric points higher than their pH optimum and thus would be positively charged at optimum pH, protoplasts are killed by different pectic enzymes irrespective of their positive or negative charges under the conditions of the tests (Basham and Bateman, 1975).

It has also been suggested that pectic enzymes kill protoplasts by acting directly on their highly specific substrates in membranes or within cytoplasm (Garibaldi and Bateman, 1971; Mount *et al.*, 1970). It is supposed that protoplasts in plasmolysed tissues remain alive for long periods because substrates for the enzyme are less available when the plasmalemma is contracted by plasmolysis or, for substrates within the cytoplasm, because penetration of the contracted membrane by the enzyme is greatly retarded (Mount *et al.*, 1970). These suggestions are difficult to disprove but acceptance of them would depend on a demonstration that suitable substrates do occur in plasmalemma and cytoplasm and that their degradation leads to changes in permeability and death of protoplasts. In this connexion it may be noted that certain cytoplasmic particles in onion protoplasts contain substances that stain with alkaline ferric hydroxylamine which is a reagent for esterified pectic polysaccharides (Albersheim and Killias, 1963). Also, freeze-etching techniques have revealed particles in the matrix of membranes which could be interpreted as intrusions of protein or polysaccharide (Branton, 1969).

Another possible explanation of death involves plasmodesmata which, presumably, become exposed to the ambient solution as pectic enzymes cause cell separation. Does this mean that the plasmalemma, which is continuous with that of the main body of cytoplasm within cells, becomes exposed at an increasing number of points as cell separation proceeds? And what happens to plasmodesmata when the ambient solution is hypotonic? It has been shown that as cell wall is removed in the preparation of isolated protoplasts, plasmodesmata may expand even in hypertonic solutions (Withers and Cocking, 1972). Do they rupture in hypotonic solutions? If so, does this explain the loss of ions that starts soon after tissue is exposed to pectic enzymes and continues until protoplasts lose their ability to plasmolyse? On the other hand, plasmalemma which may be ruptured at plasmodesmata could retract into the cell lumen and reseal (Hall and Wood, 1973); this, presumably, happens readily when protoplasts are plasmolysed. Also, if plasmodesmata are major points of weakness in hypotonic solutions, why do protoplasts in cells immediately below the surface of tissue slices remain alive for long periods as do protoplasts in epidermis? These are removed mechanically from scale leaves of onion in a separation that would seem to resemble closely that caused by a pectic enzyme. A further point is that single callus cells from suspension cultures are also killed by pectic enzymes in hypotonic solutions. Presumably the walls of callus cells do not contain plasmodesmata? On balance, it seems likely that plasmodesmata are not critically involved in cell death caused by pectic enzymes.

Some years ago it was suggested that killing of protoplasts is primarily an osmotic effect (Wood, 1967). Pectic enzymes supposedly cause structural changes in cell walls which allow protoplasts in hypotonic solutions to expand under turgor pressure; elastic stretching of plasmalemma could cause rapid conformational changes which could lead to increased permeability, and when the limit of elasticity is reached, protoplasts would burst to give the effect measured by loss of neutral red from stained protoplasts (Alberghina et al., 1973; Hall and Wood, 1973). Much of the data summarized above, particularly those on effects of plasmolysis, does not refute this hypothesis but there is, as yet, little direct evidence to support it. Transmission electron microscopy of rotted tissue has shown that degradation of wall is associated with disorganization of cytoplasm, enlargement of microbodies and expansion and bursting of cell membranes (Fox et al., 1972; Stephens, 1974). However, these are changes which may occur sometime after the critical changes in cell wall structure referred to above and which will undoubtedly be very difficult to detect by transmission electron microscopy (Stephens, 1974). Also there is the point that Sclerotinia fructigena may kill protoplasts without destroying the physical integrity of the plasmalemma (Calonge et al., 1969), and low molecular weight toxins may greatly alter the permeability of protoplasts without causing changes in the ultrastructure of cell membranes (Wheeler, 1971). Another fact which may be significant is that apart from changes in permeability, the effects of pectic enzymes are usually assessed by loss of neutral red from stained protoplasts. However, neutral red accumulates in vacuoles bounded by tonoplast (Stadelmann and Kinzel, 1972); this membrane is believed to have properties quite different from those of the plasmalemma which is, presumably, the membrane more directly involved in release of electrolytes. Tonoplast remains as a coherent, semipermeable membrane in conditions in which plasmalemma becomes non-functional (Robards, 1970). It may be asked, therefore, whether increases in permeability to electrolytes and loss of ability to retain neutral red are the same or different effects of pectic enzymes on cell walls and cytoplasm.

Scanning electron microscopy of carrot callus cells grown in suspension culture and then treated with pectate *trans*-eliminase revealed several features of interest. Untreated cells are fully expanded with smooth surfaces and with little or no debris other than that clearly associated with mechanically damaged cells. In contrast, the wall of enzyme-treated cells is wrinkled and has much debris, some of which looks like ruptures in the surface and could be interpreted as local extrusion of cell contents. Enzyme-treated cells are often collapsed (Stephens, 1974). Further work of this sort could be profitable.

Finally, after speculating that protoplasts rupture through osmotic effects after the structure of cell walls has been affected critically by pectic enzymes, it is relevant to mention recent work on the regeneration of cell walls by isolated soybean protoplasts which became osmotically stable after four days or so (Hanke and Northcote, 1974). These protoplasts secreted pectic polysaccharides into the ambient solution but none accumulated in the new cell walls which, presumably, have a structure critically different from that proposed by Keegstra *et al.* (1973). It would be of interest, therefore, to determine whether soybean protoplasts with regenerated cell walls are killed by pectic enzymes, as are isolated cells of callus cultures (Stephens and Wood, 1975).

IX. Consequences of Protoplast Death

There is now little doubt that pectic enzymes are critically important in causing cell separation and death of protoplasts for the massive, rapidly spreading lesions in parenchyma caused by typical soft-rot pathogens. Indeed, although final details remain to be established, it is probably safe to assert that we have a better understanding of pathogenesis in this than in any other type of disease. There are, however, a number of questions to be posed about the roles of these enzymes in diseases other than those characterized by cell separation, and about their significance in specificity. First, a very large proportion of all pathogens that grow in culture can synthesize pectic enzymes *in vitro* and they could well be produced *in vivo* notwithstanding the exacting control of their synthesis through induction and catabolite repression (Cooper and Wood, 1975). Also, various enzymes degrade cell wall polymers; pectic enzymes are the first to be synthesized and *in vivo* may well be active in the earliest stages of infection during penetration of the cell wall (English *et al.*, 1970; Cooper and Wood, 1975).

At present we know little about synthesis of pectic enzymes by biotrophic pathogens but there is some evidence that they secrete substances that alter the composition of cell walls through which they grow. Thus, wall material around and some distance from points of penetration stains differently and the physical appearance of walls sometimes suggest that penetration is accompanied by degradation (Sargent *et al.*, 1975).

Clearly, if pectic enzymes are synthesized in the early stages of infection, *prima facie*, conditions are established for the development of a progressive lesion. However, this does not happen in most pathogens that can synthesize these enzymes because the lesion rapidly becomes limited as, for example, in leaf spots caused by a wide variety of pathogens. Soft-rot pathogens also cause only limited lesions under certain conditions. We should aim to explain why synthesis and activity of pectic enzymes does not continue in limited lesions. Further, wilt pathogens also readily produce *in vitro* pectic enzymes and again there is no obvious reason why they should not do so in the xylem elements. If sufficient enzymes were produced, their action on pit membranes and later on cell walls and their effects on protoplasts of xylem parenchyma and cells elsewhere could explain most of the symptoms of vascular wilts, at least as convincingly as the action of low molecular weight toxins.

There are several reasons why pectic enzymes do not seem to act continuously in so many diseases. First, rapid death of protoplasts could lead to changes similar to those associated with the hypersensitive responses to avirulent pathogens which limit or prevent the growth of the pathogen. This would greatly decrease and then stop synthesis of pectic enzymes. Secondly, enzymes released from cell walls by pectic enzymes acting upon substrates released from killed protoplasts could give products that inactivate pectic enzymes or alter substrates so that they are less available or susceptible to degradation and thus progressively limit their effects on host tissues. It is likely that enzymes or other proteins released from cell walls act directly upon the pathogen or its pectic enzymes. Thus, it is known that cell walls contain protein that more or less specifically inhibits the pectic enzymes of a range of plant pathogens (Albersheim and Anderson, 1971). Another possibility is that death of protoplasts will lead to the accumulation of one or more substances that limit synthesis of pectic enzymes by catabolite repression. Clearly, therefore, the action of pectic enzymes on cell walls and protoplasts could activate many mechanisms that would prevent the continued synthesis of pectic enzymes or limit their effects on host tissue. It is also clear that these enzymes and probably other mechanisms often operate soon after plants are infected by most facultative plant pathogens; their significance for biotrophic pathogens cannot yet be assessed.

There are related problems in the different reactions of host tissues to avirulent and virulent races of a pathogen such as *Colletotrichum linde-*

muthianum which readily produces pectic enzymes *in vitro*. Infection by an avirulent race leads to rapid death of host cells in which the role of pectic enzymes is not known; the pathogen is confined to dead cells presumably following the accumulation in them or in adjacent cells of phytoalexins. In susceptible reactions, the protoplasts of infected cells are not killed and the pathogen continues to grow for some time. Then, when it is well established, protoplasts are killed rapidly and a progressive, necrotic lesion develops in which there is good evidence for the activity of pectic and other cell wall degrading enzymes (Mercer *et al.*, 1975; Skipp and Deverall, 1972). There is, therefore, the problem as to why pectic enzymes are not synthesized or not active in the early stages of infection but are synthesized and active after the pathogen has grown in the tissues for some days.

Lastly, there are the implications of the release of protein from cell walls by pectic enzymes probably during the earliest stages of infection by a large proportion of plant pathogens. Little is known about the substances and reactions that determine specific interactions between plants and pathogens, but it is worth speculating that these substances occur in cell walls and are released by the action of pectic enzymes, especially in view of the preliminary evidence for a few diseases that suggests a role for glycoproteins in specificity.

REFERENCES

Alberghina, A., Maccucchi, U. and Pupillo, P. (1973). *Phyt. Zeit.* **78**, 204–213.
Albersheim, P. and Anderson, A. (1971). *Proc. natn. Acad. Sci. U.S.A.* **68**, 1815–1819.
Albersheim, P. and Killias, U. (1963). *Am. J. Bot.* **50**, 732–745.
Basham, H. G. and Bateman, D. F. (1975). *Phytopathology* **65**, 141–153.
Branton, D. (1969). *A. Rev. Pl. Physiol.* **20**, 209–238.
Brown, W. (1965). *A. Rev. Phytopath.* **3**, 1–18.
Calonge, F. D., Fielding, A. H., Byrde, R. J. W. and Akinrefon, O. A. (1969). *J. Exp. Bot.* **20**, 350–357.
Cooper, R. M. (1974). *Ph.D. Thesis, University of London.*
Cooper, R. M. and Wood, R. K. S. (1975). *Physiol. Pl. Path.* **5**, 135–156.
English, P. D., Jurale, J. B. and Albersheim, P. (1970). *Pl. Physiol.* **47**, 1–6.
Fox, R. T. V., Manners, J. G. and Myers, A. (1972). *Potato Res.* **15**, 130–145.
Friedman, B. A. and Jaffe, M. J. (1960). *Phytopathology* **50**, 272–274.
Fushtey, S. G. (1957). *Ann. Bot.* **21**, 273–286.
Garibaldi, A. and Bateman, D. F. (1971). *Physiol. Pl. Path.* **1**, 25–40.
Hall, J. A. and Wood, R. K. S. (1973). *In* "Fungal Pathogenicity and the Plant's Response" (R. J. W. Byrde and C. V. Cutting, eds). Academic Press, London and New York.
Hall, J. A. and Wood, R. K. S. (1974). *Ann. Bot.* **38**, 129–140.
Hall, J. G., Wood, R. K. S. and O'Brien, F. (1974). *Ann. Bot.* **38**, 719–727.
Hanke, D. E. and Northcote, D. H. (1974). *J. Cell. Sci.* **14**, 29–50.
Keegstra, K., Talmadge, K., Bauer, W. D. and Albersheim, P. (1973). *Pl. Physiol.* **51**, 188–196.
Lund, B. M. and Mapson, L. W. (1970). *Biochem. J.* **119**, 251–263.

Mercer, P. C., Wood, R. K. S. and Greenwood, A. D. (1975). *Physiol. Pl. Path.* **5**, 203–214.

Mount, M. S., Bateman, D. F. and Basham, H. G. (1970). *Phytopathology* **60**, 924–931.

Mussell, H. W. (1973). *Phytopathology* **63**, 62–70.

Pitt, D. and Coombes, C. (1968). *J. gen. Microbiol.* **53**, 197–204.

Pitt, D. and Coombes, C. (1969). *J. gen. Microbiol.* **56**, 321–329.

Robards, A. W. (1970). "Electron Microscopy and Plant Ultrastructure". McGraw-Hill, London.

Ruesink, A. W. (1971). *Pl. Physiol.* **47**, 192–195.

Ruesink, A. W. and Thimann, K. U. (1965). *Proc. natn. Acad. Sci. U.S.A.* **54**, 56–64.

Sargent, J. A., Tommerup, I. C., and Ingram, D. S. (1973). *Physiol. Pl. Path.* **3**, 231–239.

Siegel, S. M. and Halpern, L. A. (1965). *Pl. Physiol.* **40**, 792–796.

Skipp, R. A. and Deverall, B. J. (1972). *Physiol. Pl. Path.* **2**, 357–374.

Stadelmann, E. J. and Kinzel, H. (1972). *In* "Methods in Cell Physiology" (D. M. Prescott, ed.), Vol. V. Academic Press, New York and London.

Stephens, G. J. (1974). Ph.D. Thesis, University of London.

Stephens, G. J. and Wood, R. K. S. (1974). *Nature* **251**, 358.

Stephens, G. J. and Wood, R. K. S. (1975). *Physiol. Pl. Path.* **5**, 165–181.

Tribe, H. T. (1955). *Ann. Bot.* **19**, 351–371.

Tseng, T. C. and Bateman, D. F. (1968). *Phytopathology* **58**, 1437.

Tseng, T. C. and Mount, M. S. (1974). *Phytopathology* **64**, 229–236.

Wilson, L. (1973). *A. Rev. Phytopath.* **11**, 247–272.

Withers, L. A. and Cocking, E. C. (1972). *J. Cell Sci.* **11**, 59–75.

Wood, R. K. S. (1967). "Physiological Plant Pathology". Blackwell Scientific Publications, Oxford.

Wheeler, H. (1971). *Phytopathology* **61**, 641–644.

CHAPTER 7

Hormonal Involvement in Metabolism of Host–Parasite Interactions

J. M. DALY AND H. W. KNOCHE

Laboratory of Agricultural Biochemistry, University of Nebraska, Lincoln, Nebraska 68503, U.S.A.

I. INTRODUCTION

For students of plant development and the action of hormones, growth irregularities induced by microorganisms hold the promise of model systems from which insight into modulation and control of normal growth can be obtained. In addition to an obvious analogy with animal cancer, this view undoubtedly stimulated pioneering studies of crown gall caused by *Agrobacterium tumefaciens* and does so today. In certain diseases, such as club root of cabbage, economic factors clearly are at stake. The mechanisms by which overgrowths and irregular growth patterns come about, or can be prevented, in diseased plants are of concern. There is a converse economic interest in the process of nodulation of legumes caused by nitrogen-fixing bacteria, but the physiological and biochemical principles governing antagonistic and symbiotic infectious processes may be quite similar.

Until the middle of the 1950s, understanding of growth alterations of diseased plants relied heavily on findings obtained from studies of normal growth. The independent rediscovery by several laboratories of the growth regulant produced by *Gibberella fujikuroi*, and responsible for elongation of rice plants infected by this organism, provided a new interface between physiologists and pathologists. The history of these findings is summarized in Stowe and Yamaki (1957). Subsequent proof that gibberellins were endogenous hormones of normal plants was an indication that pathogen-induced abnormalities might be useful for discoveries of novel growth regulants, perhaps with roles in normal development. The expectations were nearly realized with another class of hormones, the cytokinins. For many years, the laboratory product, kinetin, was the only representative known. The description of a natural product, zeatin (Letham *et al.*, 1964), was followed by the identification of 6-(3-methyl-2-butenyl) aminopurine in cultures of *Cornyebacterium fascians* (Klambt *et al.*, 1966; Helgeson and Leonard, 1966), causal agent of a growth distortion of peas. The identification was based on the earlier work of Thimann and Sachs (1966) who noted the similarity in disease symptoms with distortions produced by kinetin.

Studies of hormones in diseased plants currently seem to be in quiescent state. There has been a recent review by Sequiera (1973), by design largely a discussion of hormonal metabolism. Some reviews have dealt more directly with the induction of symptoms by hormones (Van Andel and Fuchs, 1972; Veldstra, 1968), while others have provided incidental coverage of disease situations (Hall, 1973; Evans, 1974), including work on mycorrhizae (Meyer, 1974). An earlier review by Brian (1967) includes discussion of the role of hormones in diseases caused by obligate parasites.

Despite the current hiatus, we believe that the interface between studies of normal and abnormal growth will assume interesting new shapes. Helminthosporol, for example, is of interest to pathologists as a toxin of *Helminthosporium sativum* causing damage to oats and wheat (de Mayo *et al.*, 1961). The related alcohol, helminthosporal, independently was shown by Tamura *et al.* (1963) to induce elongation of rice, a non-host for this

Helminthosporal

Helminthosporol

(a)

Fusicoccin

(b)

organism, as well as growth stimulation of other species (Hashimoto *et al.*, 1967; Hashimoto and Tamura, 1967). Equally interesting are the growth effects of fusicoccin, found originally to be responsible for transient early wilt symptoms in a canker disease of almond and peach by an action on stomates (Turner and Graniti, 1969). Fusicoccin has growth promoting activities in species other than hosts for *Fusicoccum amygali* (Lado *et al.*, 1973), activities which include a fast induction of coleoptile elongation in a manner analogous to IAA (Cleland, 1974). A short-term stimulation of growth of susceptible plants has been reported for the host-specific toxins, victorin (Evans, 1973) and race-T toxin (Evans, 1974). Finally, the history of ethylene should be recalled. It was for a long time considered to be a curious emanation of diseased or injured plants (Williamson, 1950), but its role in normal plant growth processes has gradually emerged.

These results with agents that cause deleterious effects in certain plants but growth responses in other species have several types of potential applications. It seems unlikely that the great morphological and ontogenetic diversity of higher plants will be explained completely by a universal interaction, admittedly complex, between four or five known hormones. The existence of additional endogenous, chemical modifiers of growth is a possibility which might be investigated through pathological phenomena. Second, toxins responsible for deleterious effects may function at, or close to, metabolic sites of hormonal interaction. Their value as probes for under- standing modes of hormonal action may be a reflection of their role in disease.

II. HORMONES AND PATHOGENESIS

The comments above illustrate some past and potential usefulness of infectious diseases as tools for analysis of hormonal functions. For an understanding of disease, however, there is an additional facet that requires a different sort of experimental approach. The crucial and unique question for phytopathology is whether an abnormality is merely an unhappy consequence of a host–parasite interaction or whether it is a manifestation of a change, presumably hormonal, which is essential in some way for the establishment of a parasite or for the initiation of disease. To pose the question in another way, is hormonal imbalance one of the biochemical devices which leads to disease or is it a biochemical side effect, that is, only a symptom of other, pertinent events in pathogenesis?

At first glance, it would appear that hormonal imbalance is an important factor causing disease symptoms in those instances where pathogens produce growth regulators in culture which mimic some symptom of a disease or symbiotic relationship, as is the case with *C. fascians*. Production of growth regulators would be a part of pathogenic armoury, much as toxins are acknowledged to be a part of an arsenal for virulence. Even in such situations, however, one cannot rule out innate metabolic differences of the pathogen in the host and in artificial culture, or artifacts of extraction. Rathbone and Hall (1972), for example, have questioned whether the high titers of cytokinin in culture fluids of *C. fascians* (Klambt *et al.*, 1966; Helgeson and Leonard, 1966) were not the product of an acid treatment during extraction which released cytokinin from bacterial RNA. The work of Kuo and Kosuge (1969, 1970) has demonstrated with *Pseudomonas savastanoi*, the incitant of olive knot, the possibilities of two pathways for IAA (indoleacetic acid) synthesis as well as the possibility of allosteric control of IAA accumulation in culture. Allosteric control of hormonal synthesis limits ready interpretation of the relative contributions of host and microorganism during disease or in symbiotic relationships. A further complication is the potential regulation of pathogen-produced hormones by the host through catabolic degradation.

Even in instances where the pathogen obviously is responsible for a hormonal imbalance *in vivo*, the significance for pathogenesis is difficult to determine. It is clear that the role of gibberellins in "foolish-seedling" disease of rice is not crucial for pathogenesis. Only a certain proportion of infected rice plants show the exaggerated growth response, and the ability of strains of *Gibberella fujikuori* to cause disease appears to be independent of their ability to produce gibberellins.

The alternative to imbalance caused by hormones or regulators produced by the pathogen is a modulation of host control of hormonal production. Sakai *et al.* (1970) have reported that the accelerated ethylene production by tissues of sweet potato tubers infected with *Ceratocystis fimbriata* is dependent on the TCA cycle, which apparently is not the case for normal tissue. The chief evidence is the kinetics of labeling from acetate and pyruvate. The use of cell free systems showing the same differences would alleviate the problems of differences in uptake of substrate.

In view of these difficulties, it would appear foolhardy to ask similar questions for diseases in which growth abnormalities are not obvious. Yet, hormonal imbalance may be a primary determinant of disease reaction in tissue incapable, perhaps through maturity, of morphological change but metabolically responsive to changes in endogenous levels of growth regulants. The lengthy catalogue of enzymatic and metabolic change induced by each of the known hormones of higher plants has a companion volume in the literature of host–parasite relations. Even if the inducing agents are not chemically identical, the metabolic control points may be. Future emphasis in growth regulation may place less weight on the concept of a circulating hormone acting on a remote tissue target to induce differentiation (Hall, 1973). Instead, localized perturbation through several interacting endogenous regulators may be more fruitful. Perturbation may be initiated at the nucleic acid level, as with the so called "long-term" effects of hormones. The growth distortions seen in some virus diseases, such as wound tumor virus, should be noted. More recently, attention has shifted to membranes as a possible site for the rapid, short-term responses to hormones and the role of membranes in metabolic phenomena associated with disease is now becoming better documented (Gardner *et al.*, 1974).

The intriguing possibilities for the induction of metabolic events that may be primary determinants of disease, as well as the difficulties in providing the necessary experimental proof, is illustrated by the relationship between ethylene and the black rot disease of sweet potato tubers caused by *Cerato-cystis fimbriata*. A number of years ago, it was noticed that uninoculated control discs of sweet potato tubers housed for 48 h in a closed chamber with infected discs did not become diseased when subsequently inoculated (Clare *et al.*, 1966). Ethylene produced by inoculated discs was responsible for the induced resistance. Subsequent work by Stahmann *et al.* (1966), as well as the group at Nagoya (Imaseki *et al.*, 1968; Shannon *et al.*, 1971; Imaseki, 1970) showed that ethylene could cause increased activity of some enzymes of aromatic biosynthesis as well as the accumulation of chlorogenic acid. It was believed by Clare *et al.* (1966) and Stahmann *et al.* (1966) that induction of these activities was responsible for a chemical barrier to invasion, although the Japanese workers do not mention an increase in resistance in their papers. Chalutz *et al.* (1969a,b) reexamined the question of induced resistance and found that even though ethylene treatment, when started at the time of inoculation, caused accumulation of phenolic constituents (notably isocoumarin), the disease reaction of several sweet potato varieties and pathogens was not appreciably altered. In this instance, it appears that the biochemical changes were only incidental to disease reaction or perhaps involved in the formation of a non-specific morphological barrier which became effective *after* 48 h of treatment. The results do underline the additional experimental burden of providing a direct link between observed metabolic changes and pathogenesis, especially studies of resistance.

III. HORMONAL CHANGES IN DISEASES CAUSED BY BIOTROPHIC ORGANISMS

At some stage in nearly all diseases caused by biotrophic organisms an alteration of growth patterns can be found. In some, such as bean rust, they are not dramatic. Infected primary leaves become thicker with some curling, the leaves do not abscise at the usual time, infected petioles are distorted and eventually the entire plant becomes stunted. In others, such as rusts of pine, thistle, and *Euphorbia* and mildews of apple or cucumber, the alterations are so extensive as to be diagnostic. It is not surprising, therefore, that research on the metabolism of diseases caused by obligate parasites and some of the concepts of resistance and susceptibility have been concerned with hormonal imbalance, even with monocotyledonous hosts where growth effects are lacking.

It would be immensely satisfying if the following account dealt with definite studies dealing quite directly with hormone synthesis and degradation and with the ways in which hormones determine pathogenesis. Because the evidence is fragmented among a number of diseases, even *in toto* it is only suggestive of a potential role of hormones.

A. RESPIRATION AND HOST GROWTH

A number of years ago, Allen (1953) proposed that the increased respiratory rate and the abolition of the Pasteur effect in rust and mildew disease was due to the production by the pathogen of a diffusible toxin which acted as an uncoupler of mitochondrial oxidation in host cells. Evidence was advanced from several laboratories in support of the hypothesis (Millerd and Scott, 1956; Samborski and Shaw, 1956; Pozsar and Király, 1958). Non-photosynthetic safflower hypocotyl infected with *Puccinia carthami* was chosen as a system in which the hypothesis might be readily checked, but certain observations in the course of these experiments caused a revision of the original rationale for the research (Daly and Sayre, 1957). The increases in respiration were accompanied by significant elongation of infected hypocotyls during vegetative mycelial growth of the parasite. The correlation between vegetative fungal development and elongation was noteworthy. When fungal sporulation occurred, the additional elongation ceased. This also was the time when the elongation of normal, uninfected hypocotyls stopped. Whether conversion from vegetative to reproductive activities of the fungus halted additional diseased host elongation or, conversely, whether the incapacity of the host to elongate triggered sporulation was not resolved. The general results, however, appeared inconsistent with the action of a toxin uncoupling respiration from cellular work (Daly and Sayre, 1957).

In the same experiments, several lines of evidence suggested that the abolition of the Pasteur effect with the onset of higher respiratory rates was due to the development of a highly active oxidative pentose pathway. As a consequence of these and related studies, it was proposed that, instead of a pronounced change in respiratory metabolism of the host during infection, the major increases were due to the activity of the parasite (summarized by Daly, 1967). A reexamination of the early concepts by Bushnell (1967, 1970) has indicated that the degree of host involvement is much less than originally thought, although this view apparently has not entirely disappeared.

If a typical oxidative pentose pathway were operative, there should be a decline with time of the ratio of C_1 to C_6 of glucose appearing as CO_2, but the decline was very small. Furthermore, intermediates such as sedo-heptulose phosphate and possibly erythrose phosphate should be detected, yet previous studies in other laboratories had demonstrated no unusual alteration in carbohydrate of diseased plants. With the finding of large pools of sugar alcohols (Daly et al., 1962a), particularly arabitol, it became apparent that the abolition of the Pasteur effect was the result of the conversion of hexoses to pentitols which served as reserve carbohydrates in uredospores. Subsequent studies revealed extensive utilization of polyols during germination, rather than the use of lipid as an exclusive endogenous reserve as generally had been believed (Daly et al., 1967).

The elongation of infected hypocotyls was considered at that time to be only a consequence of an initial successful colonization of a susceptible host and not a critical factor in the initial establishment of infection. Thus elongation was a manifestation of metabolic change of the host conducive to, but not absolutely required for, parasite growth. Daly and Inman (1958) showed that elongation most probably occurred through an increase in IAA very early during infection. The system seemed useful for attempts to learn whether hormonal changes in diseased tissue resulted from increased synthesis in hormone or a change in the rate of degradation. Attempts to determine rates of synthesis with postulated precursors of IAA remain unpublished because microbial contamination inevitably obscured the significance of any findings, despite the use of sterile solutions and combinations of various antibiotics (Deverall and Daly, 1964). Further work with this host–parasite system was halted when seed of varieties, both susceptible to rust and free of an *Alternaria* species causing seedling blight, was no longer available.

B. IAA DECARBOXYLATION AND DISEASE RESISTANCE

Unlike safflower hypocotyls, cereal leaves infected by rusts and mildew show no pronounced growth distortions, but Shaw and his co-workers in Canada obtained some evidence for involvement of IAA in stem rust infections of susceptible Little Club wheat. A limited examination indicated, as with safflower rust, an increase in auxin (Shaw and Hawkins, 1958). Additional indirect evidence was the observation that decarboxylation of IAA in tissue sections from susceptible, infected Little Club leaves was greater and occurred at an earlier stage of infection than degradation by the more resistant Khapli wheat (Shaw and Hawkins, 1958). A third line of evidence was the report that IAA caused a shift in respiration to an oxidative pentose pathway in wheat leaves (Shaw *et al.*, 1958). Because of the implications of this result in assigning metabolic roles to host or parasite, an examination of the interactions between IAA, other hormones and modifying factors was undertaken (Daly *et al.*, 1962b). It failed to show any significant effects of IAA on respiratory pathways although slight increases in respiration rates (10–30%) were consistently observed.

Daly and Deverall (1963) and Deverall and Daly (1964) reexamined IAA degradation by susceptible Little Club wheat and did observe extensive degradation in the first few days after inoculation. The exact time at which maximum rates of degradation occurred appeared to depend on environmental factors, and especially on the infection density. A comparison with Khapli wheat was not attempted because Khapli wheat is genetically and morphologically quite distinct from Little Club wheat. Further, it gives an infection of type 1 in which rust fungi develop to the extent that sporulation occurs.

The availability of near-isogenic lines of Chinese Spring wheat, differing only in the temperature-sensitive Sr6 allele for incompatibility or the corresponding sr6 allele for compatibility helps to resolve this type of problem (Daly, 1972). At 20–21 °C, lines with the Sr6 allele shows an infection type with no sporulation while the sr6 line gives infection type 3 to 4. The early patterns of IAA degradation reported as characteristic of compatibility or incompatibility for Little Club and Khapli were not obtained with these infection types by Antonelli and Daly (1966). There was a marked increase, however, in rates of decarboxylation of IAA in resistant, but not susceptible, tissue starting the third or fourth day after inoculation. This period coincided with the time during infection when resistance mechanisms were believed to become operative, based on the effects of temperature transfers at intervals after inoculation (Antonelli and Daly, 1966).

C. PEROXIDASE CHANGES

It is tempting to derive plausible connections between IAA decarboxylation and the induction of purposeful hormonal changes in resistant plants, but IAA decarboxylation by tissue slices is, as yet, an incompletely defined process. As shown by Daly and Deverall (1963), there may be differences between susceptible and resistant infected tissues in rates of uptake of IAA and perhaps in its conversion to metabolites of IAA (Deverall and Daly, 1964). There is no available evidence to indicate that the process is mediated by a specific enzyme as might be expected for an important regulatory molecule. IAA degradation *in vitro* can be accomplished by oxidative action of peroxidase, with monophenols as activators and polyphenols as inhibitors of the reaction. Rather than invoking a regulated destruction of IAA in incompatible tissue, the results were as easily explained by general activation of peroxidase, either through *de novo* synthesis or an increase in regulatory phenolics. The second possibility was examined first because Farkas and Király (1962), Király and Farkas (1962) and Fuchs *et al.* (1967) in Canada had indicated that infected plants had increased rates of aromatic biosynthesis. Under our conditions, however, none of the near-isogenic lines with either the Sr6 or Sr11 alleles for disease showed any significant change in the quantity of types of phenolics during infection (Seevers and Daly, 1970a).

Peroxidase activity in extracts of the infected compatible and incompatible lines of wheat did correlate reasonably well with changes in IAA decarboxylation, both in terms of differences observed between resistant and susceptible wheats and in the time during infection when increases were first observed (Seevers and Daly, 1970b). The selection of the Sr6 and sr6 alleles for these studies was based on the experimental advantage that, after an incompatible reaction develops at 20–21 °C, a compatible reaction can be initiated when plants are transferred to a temperature of 25–26 °C. If a biochemical event is responsible for resistance at 20–21 °C, it must either disappear or become inoperative at 26 °C. If the event persists, there is the strong probability that it is incidental to resistance, not a cause of resistance (Daly, 1972). When Sr6 plants with high peroxidase activity and showing visible signs of incompatibility were transferred to high temperature, peroxidase activity did not decline even after sporulation of the compatible interaction was well advanced (Seevers and Daly, 1970b).

To reinforce these findings by ways other than temperature transfer of the Sr6 line, attempts were made to induce peroxidase in the susceptible sr6 line. If peroxidase, acting as an IAA oxidase, is important any increase should lead to resistance. Treatment with ethylene at 80 μl/l at 20–21 °C did cause an increase in peroxidase in both compatible and incompatible lines (Daly *et al.*, 1970). That the sr6 line remained susceptible was not surprising; the fact that the resistant Sr6 line became completely susceptible with ethylene treatment at 20–21 °C was surprising.

The existence of several electrophoretically distinct forms of peroxidase precludes an easy interpretation of the role of peroxidase in the disease reaction. Activation of new isozymes and repression of specific isozymes by temperature or ethylene treatments could invalidate the conclusion that peroxidase *per se* was not a factor in incompatibility. Fourteen peroxidase isozymes were detected in normal and infected wheat leaves by gel electrophoresis, but only two of these (isozymes 9 and 10) were consistently higher in activity during infection. In most, but not all infections, isozyme 10 increased in susceptible as well as in resistant reactions. Isozyme 9 consistently increased in activity only in incompatible reactions and to the extent and at the time expected from peroxidase assays of homogenates

(Seevers *et al.*, 1971). The main issue to be resolved was the effect of temperature, in that isozyme 9 either was maintained at the high level caused by incompatibility or actually increased slightly in the first few days when a compatible reaction obviously was being initiated (Seevers *et al.*, 1971).

The sum of these experiments indicated that peroxidase could not be the gene product for resistance controlled by the Sr6 allele. This conclusion is reinforced by the observation that disease resistance controlled by the Sr 11 allele also involves activation of isozyme 9 (Daly *et al.*, 1971). It seems unlikely, although not impossible, that alleles on different chromosomes, which respond differently to temperature and to ethylene (Daly *et al.*, 1971), are responsible for an identical gene product. At present, the most reasonable explanation is that peroxidase activity becomes high as a result of non-specific stress or injury associated with, but not a cause of, incompatibility (Daly, 1972).

There are several aspects which remain to be explored. First, although changes in peroxidase and peroxidase isozymes are of common occurrence in variety of diseases and stresses, the substrates for physiological functions of individual isozymes *in vivo* is not known. Second, the purported activity of an isozyme in a gel is assumed to reflect the relative contribution of that isozyme in homogenates and *in vivo*. The conditions of assay on gels are not kinetically equivalent to any other conditions (Seevers *et al.*, 1971). Therefore, the role of specific isozymes in disease requires their isolation in order to resolve these uncertainties.

D. PROPERTIES OF ISOZYME 9

The findings of Catedral and Daly (1976) with purified individual isozymes are encouraging in that isozyme 9 has a specific activity 3–4 fold higher on phenylenediamine, the substrates used in previous work, than does isozyme 10. Thus, the relatively large increase in peroxidase activity observed in homogenates could be accounted for by only a relatively small increase in isozyme 9. We have not observed any significant increase over healthy plants in total protein of homogenates, despite the general belief that protein synthesis is necessary for incompatibility (Catedral and Daly, 1976), presumably to account for increases in enzymatic activity of infected plants. Attempts to show synthesis of isozymes 9 and 10 by ^{14}C or D_2O labeling were inconclusive because of the very small amount of protein required. From the kinetic properties of purified isozyme 9, it can be calculated that the observed increase in enzyme activity would require the total synthesis of less than 1 μg of protein per leaf spread over a minimum of 4 days. The detection, by techniques other than enzymatic activity, of synthesis of that amount amidst existing leaf protein appears remote.

It is of interest to know whether peroxidase isozyme 9 has a specific physiological function or cellular location that is connected to the events in the incompatible disease reaction. Peroxidase has been implicated in a variety of reactions that bear on disease: aromatic biosynthesis, ethylene production, lignin formation, etc. None of these functions including IAA oxidation, has been shown to be associated with specific isozymes. Very preliminary results with small amounts of purified isozymes 6, 8, 9 and 10 have given contradictory results with IAA as substrate. With dichlorophenol and Mn^{2+} as co-factors, decarboxylation of solutions of carboxyl-labeled ^{14}C-IAA was approximately the same for each isozyme, yet only isozyme 10 showed any activity in Endo's method (1968) for detecting IAA oxidase on gels (Catedral and Daly, 1975). The nature of the reaction involved in Endo's method is unknown, but is believed to involve a complexing of dye with an as yet unknown intermediate of the oxidation. It is now clear that the products of the reaction of IAA with peroxidase are dependent on pH and substrate concentration (Hinman and Lang, 1965; Ricard and Job, 1974). Work by Yamazaki and Yamazaki (1973) also suggests that individual isozymes of peroxidase have different reaction pathways for IAA oxidation. The enzyme intermediate appears to be specifically formed only with isozymes B and C of horseradish peroxidase.

E. CYTOKININS AND TRANSLOCATION

In experiments attempting to "break resistance" of wheat leaves to stem rust, application of the growth regulator, maleic hydrazide, was found to cause higher infection types with more sporulation (Shaw et al., 1958). A similar effect could be caused by floating detached leaves on water (Person et al., 1958), but normal resistance was retained with solutions of kinetin or benzimidazole. The results subsequently were of unusual interest because of the demonstration that drops of kinetin (Mothes and Engelbrecht, 1961) cause accumulation of metabolites at the site of application, as do isolated pustules of wheat rust fungi (Shaw, 1961). The retention of chlorophyll by kinetin treatments of detached leaves also is reminiscent of the "green island" effect frequently seen in mildew and rust infections (Bushnell, 1967).

Short- and long-distance movement of metabolites appears to be important for the development of obligate parasites. Individual pustules and entire leaves act as sinks in translocation processes (Livne and Daly, 1966). Pozsar and Király (1966) and Király et al. (1967) prepared extracts of leaves of rusted beans which caused typical cytokinin responses in several bio-assays. A more detailed study by Dekhuizen and Staples (1968) included a bioassay based on cytokinin-directed flow of metabolites in detached leaves. Bushnell and Allen (1962) and Bushnell (1967) also were able to mimic the green-island effect with water-soluble components obtained from the conidia of powdery mildew.

All these observations are consistent with the occurrence of increased concentrations of cytokinin-like factors whose physiological function would be retention of active metabolism in host cells. Two comments seem appropriate. Bioassays based on chlorophyll retention have been criticized for a lack of specificity (Skoog and Armstrong, 1970). Dr S. Mayama, working in our department, recently observed prolonged retention of chlorophyll in detached wheat leaves floated on mannitol and arabitol, known to be storage products in both rust and mildew fungi. It is important, therefore, to show cytokinin-like activity in leaf extracts before sporulation occurs, an appropriate time if cytokinin is important in disease development. Second, translocation may be controlled by factors other than cytokinins. Bowen and Wareing (1971) have reported IAA-directed movements. Livne (1964) demonstrated that ^{14}C-importing rusted leaves had lower rates of photosynthesis than comparable healthy leaves while exporting non-infected leaves of the same plants had significantly stimulated rates of photosynthesis. A source-sink relation could be developed from this effect, as well as by hormones.

F. OTHER HORMONES

As with *Puccinia carthami* infection of safflower hypocotyls, infection of *Cirsium arvense* by *Puccinia punctiformis* causes very large increases in stem length during vegetative growth of the parasite (Bailiss and Wilson, 1967). Unlike the former disease, however, there are only minor increases in IAA. The principal agents in this early growth appear to be gibberellins A_1 and A_2. Later during infection, GA_3 appears to be the dominant growth regulator. The cytological events of infection, as well as the elongation phenomenon, can be mimicked effectively by as little as 1·0 µg of gibberellin applied at weekly intervals to uninfected plants.

IV. PRESENT OUTLOOK

Most of the work at our laboratories has been based on the generally held view that specificity of host–parasite interactions in rust diseases is primarily the result of incompatibility controlled by single complementary, dominant alleles in host and parasite. In this view, biochemical reactions during an early stage (1–2 days) are induced in the host by invasion causing restriction or death of the parasite. The striking effect of ethylene in inducing susceptibility, coupled with the data on phenolic compounds and peroxidase, led to a consideration of other possibilities.

A survey of the literature indicates that despite sporadic suggestions there is no clear evidence for the involvement during incompatible reactions of compounds which are inhibitory to the parasite. A promising instance seemed to be the synthesis of *N*-feruly and *N*-p-coumaryl derivatives of hydroxyputrescine amides in the Sr6 line of wheat at 20–21 °C (Stoessl

R—〜〜CO.NH.CH₂.CH(OH)(CH₂)₂NH₂

(formula with HO— substituent)

R = H = N - (p -Coumaryl)-2- hydroxyputrescine
R = OCH₃ = N -(Ferulyl) - 2- hydroxyputrescine

(c)

et al., 1969). As with peroxidase, however, these compounds also were synthesized in the same line at 26 °C during a compatible disease reaction and they could be induced by chemical injury or stress (Samborski and Rohringer, 1970). Failure to find inhibitory compounds may either be a result of some unusual chemical properties or due to a lack of sustained effort. The first possibility seems unlikely if experience with the variety of compounds obtained in other diseases, and described in subsequent chapters of this volume, is a gauge. The second possibility is difficult to judge, except for the realization that unsuccessful attempts in science do not usually obtain favorable peer reviews.

Biotrophic organisms, by definition, appear to depend on the metabolic activity of their hosts. In rust and mildew diseases, susceptible interactions are distinguished from resistant by high rates of metabolism only during the period when sporulation is initiated, usually 4 to 5 days after inoculation (Antonelli and Daly, 1966). In disease where metabolic events can be assayed at early stages, such as powdery mildew of barley, acceleration of processes requiring energy are characteristic of compatible, not incompatible, interactions (Slesinski and Ellingboe, 1971). Much of this metabolic activity is associated with biosynthesis of components which clearly belong to the parasite. As indicated earlier, carbohydrates which are present in high concentration in rusted and mildew leaves are typical fungal storage material.

7,(Z)-24 (28) stigmastadien-3β- ol

$CH_3(CH_2)_7CH_2- CH (CH_2)_7 COOH$ *cis*-9, 10 - epoxyoctadecanoic

Knoche (1971) demonstrated appreciable biosynthesis in rusted leaves of 9,10 epoxyoctadecanoic acid and Lin *et al.* (1972) demonstrated synthesis of 7,(Z)-24(28) stigmastadien-3β-ol, both of which were shown to be unique components of uredospores (Knoche, 1968; Lin and Knoche, 1974). It has been argued that the synthesis occurs in the host followed by transport into the mycelium, but germinating uredospores are capable of sterol and phospholipid synthesis for as long as 12 h (Lagenbach and Knoche, 1971a,b).

In view of the metabolic demands of the parasite, some activation of host metabolism is required to ensure a minimum flow of substrate. The distinction between compatibility and incompatibility may lie in the ability to induce a reversion of host cells from a relatively quiescent condition to one resembling a juvenile tissue rather than the induction of the active host process of parasite restriction. It seems reasonable that hormones can fulfil such functions, perhaps in concert and in a complex manner.

It is clear from the literature reviewed above that considerable research effort will be required to support either of these alternatives. At least a portion of this effort must be directed at establishing the validity of assumptions concerning the developments of fungi in their hosts. For example, it has long been held that the hypersensitive response of cells of resistant wheats in the early stages of disease (Skipp and Samborski, 1974) is in some way a deterrent to fungal development. In recent studies, Mayama *et al.* (1975a) have demonstrated that the hypersensitive response can occur with equal, or greater, intensity when resistant wheats are induced to susceptibility by certain treatments.

There is also the general assumption that rust fungi in incompatible hosts are markedly inhibited in growth or killed. Histological observations reveal only that development in both compatible and incompatible disease reactions is similar for as long as 60 h (Skipp and Samborski, 1974), after which time colonies in compatible hosts undergo a more rapid development. Significantly, growth apparently accelerates slightly in the incompatible situations.

Histological examinations are tedious and recently we have attempted to resolve these questions by chemical analysis of constituents of fungal origin. A method for glucosamine analysis was of enough sensitivity to detect germ tube growth on the surface of a single wheat leaf prior to penetration (Mayama *et al.*, 1975b). Surprisingly, no further increases were detected until 55 h later when, in keeping with histological studies, increases were detected in both compatible and incompatible infections, with much more rapid development in compatible tissue. Failure to detect growth may have been due to a low content of chitin in young vegetative mycelium and in haustoria (Mayama, *et al.*, 1975b), although germ tubes (Trocha and Daly, 1974) and an axenic culture of rust fungus several months old (Mayama, *et al.*, 1975b) contained approximately 20% glucosamine in wall components.

Currently, we are examining, with the use of tracer techniques, the bio-synthesis of other components, such as epoxydecanoic acid and fungal sterols. The latter may be of considerable utility since they would probably be a measure of synthesis of membranes of the fungus, regardless of location in vegetative or reproductive structures.

In conjunction with data on hormonal metabolism, it appears that a general requirement for progress is the development of objective quantitative measures of parasitic growth in host tissue. Only then will it be possible to determine whether hormonal changes are primary or secondary determinants of disease.

ACKNOWLEDGEMENTS

We wish to acknowledge the help of Mr H.-K. Chin and Mr P. Ueng during preparations of this manuscript.

J. M. Daly and H. W. Knoche were supported in part by grants from the Nebraska Wheat Commission.

Published with the approval of the Director, Nebraska Agricultural Experiment Station, as paper no. **4006**, journal series.

REFERENCES

Allen, P. J. (1953). *Phytopathology* **43**, 221–229.
Antonelli, E. and Daly, J. M. (1966). *Phytopathology* **56**, 610–618.
Bailiss, K. W. and Wilson, I. M. (1967). *Ann. Bot. Lond.* **31**, 195–211.
Bowen, M. R. and Wareing, P. F. (1971). *Planta* **99**, 120–132.
Brian, P. W. (1967). *Proc. Roy. Soc. Ser. B.* **168**, 101–118.
Bushnell, W. R. (1967). *In* "The dynamic Role of Molecular Constituents in Plant Parasite Interaction". (C. J. Mirocha and I. Uritani, eds), pp. 21–39. Bruce Publishing Co., St. Paul.
Bushnell, W. R. (1970). *Phytopathology* **60**, 92–99.
Bushnell, W. R. and Allen, P. J. (1962). *Pl. Physiol.* **37**, 50–59.
Catedral, F. F. and Daly, J. M. (1976). *Phytochemistry* **15**, 627–631.
Chalutz, E. and DeVay, J. E. (1969a). *Phytopathology* **59**, 750–755.
Chalutz, E., DeVay, J. E. and Maxie, E. C. (1969b). *Pl. Physiol.* **44**, 235–241.
Clare, B. G., Weber, D. J. and Stahmann, M. A. (1966). *Science* **153**, 62–63.
Cleland, R. (1974). *Pl. Physiol.* (*Suppl.*) **53**, 242.
Daly, J. M. (1967). *In* "The dynamic Role of Molecular Constituents in Plant Parasite Interaction". (C. J. Mirocha and I. Uritani, eds), pp. 144–164. Bruce Publishing Co., St. Paul.
Daly, J. M. (1972). *Phytopathology* **62**, 392–400.
Daly, J. M. and Deverall, B. J. (1963). *Pl. Physiol.* **38**, 741–750.
Daly, J. M. and Inman, R. E. (1958). *Phytopathology* **48**, 91–97.
Daly, J. M. and Sayre, R. M. (1957). *Phytopathology* **47**, 163–168.
Daly, J. M., Inman, R. E. and Livne, A. (1962a). *Pl. Physiol.* **37**, 531–538.
Daly, J. M., Krupka, L. R. and Bell, A. A. (1962b). *Pl. Physiol.* **37**, 130–134.
Daly, J. M., Knoche, H. W. and Wiese, M. V. (1967). *Pl. Physiol.* **42**, 1633–1642.
Daly, J. M., Seevers, P. and Ludden, P. (1970). *Phytopathology* **60**, 1648–1652.

Daly, J. M., Ludden, P. and Seevers, P. (1971). *Physiol. Pl. Path.* **1**, 397–407.
Dekhuizen, H. M. and Staples, R. C. (1968). *Contrib. Boyce Thompson Inst.* **24**, 39–52.
Deverall, B. J. and Daly, J. M. (1964). *Pl. Physiol.* **39**, 1–9.
de Mayo, P., Spencer, E. Y. and White, R. W. (1961). *Can. J. Chem.* **39**, 1608–1612.
Endo, T. (1968). *Pl. Cell. Physiol.* **9**, 333–341.
Evans, M. L. (1973). *Bioscience* **23**, 711–718.
Evans, M. L. (1974). *Ann. Rev. Pl. Physiol.* **25**, 195–223.
Farkas, G. L. and Király, Z. (1962). *Phytopathl. Z.* **44**, 106–150.
Fuchs, A. R., Rohringer, R. and Samborski, D. J. (1967). *Can. J. Bot.* **45**, 2137–2154.
Gardner, J. M., Scheffer, R. P. and Higinbotham, H. (1974), *Pl. Physiol.* **54**, 246–249.
Hall, R. H. (1973). *Ann. Rev. Pl. Physiol.* **25**, 415–444.
Hashimoto, T. and Tamura, S. (1967). *Pl. Cell. Physiol.* **8**, 35–45.
Hashimoto, T., Sakurai, A. and Tamura, S. (1967). *Pl. Cell. Physiol.* **8**, 23–34.
Helgeson, J. P. and Leonard, N. J. (1966). *Proc. natn. Acad. Sci. U.S.A.* **56**, 6–63.
Hinman, R. and Lang, J. (1965). *Biochemistry* **4**, 144–149.
Imaseki, H. (1970). *Pl. Physiol.* **46**, 172–174.
Imaseki, H., Uchiyama, M. and Uritani, I. (1968). *Agric. Biol. Chem.* **32**, 387–389.
Kiràly, Z. and Farkas, G. L. (1962). *Phytopathology* **52**, 657–664.
Király, Z., El Hammady, M. and Pozsar, B. I. (1967). *Phytopathology* **57**, 93–94.
Klambt, D., Thies, G. and Skoog, F. (1966). *Proc. natn. Acad. Sci. U.S.A.* **56**, 52–59.
Knoche, H. W. (1968). *Lipids* **3**, 163–169.
Knoche, H. W. (1971). *Lipids* **6**, 581–583.
Kuo, T. T. and Kosuge, T. (1969). *J. gen. appl. Microbiol.* **15**, 51–63.
Kuo, T. T. and Kosuge, T. (1970). *J. gen. appl. Microbiol.* **16**, 191–194.
Lado, P., Caldogno, R., Pennacchioni, A. and Marré, E. (1973). *Planta* **110**, 311–320.
Langenbach, R. J, and Knoche, H. W. (1971a). *Pl. Physiol.* **48**, 728–734.
Langenbach, R. J. and Knoche, H. W. (1971b). *Pl. Physiol.* **48**, 735–739.
Letham, D. S., Shannon, J. S. and McDonald, I. R. (1964). *Proc. Chem. Soc.*, 230–231.
Lin, H. K. and Knoche, H. W. (1974). *Phytochemistry* **13**, 1795–1799.
Lin, H. K., Langenbach, R. J. and Knoche, H. W. (1972). *Phytochemistry* **11**, 2319–2322.
Livne, A. (1964). *Pl. Physiol.* **39**, 614–621.
Livne, A. and Daly, J. M. (1966). *Phytopathology* **56**, 170–175.
Mayama, S., Daly, J. M., Rehfeld, D. W. and Daly, C. R. (1975a). *Physiol. Pl. Path.* **7**, 35–47.
Mayama, S., Rehfeld, D. W. and Daly, J. M. (1975b). *Physiol. Pl. Path.* **7**, 243–257.
Millerd, A. and Scott, K. (1956). *Austr. Jour. Biol. Sci.* **9**, 37–44.
Meyer, F. H. (1974). *A. Rev. Pl. Physiol.* **25**, 567–586.
Mothes, K. and Englebrecht, L. (1961). *Phytochemistry* **1**, 58–62.
Person, C., Samborski, D. J. and Forsyth, F. R. (1958). *Nature* **180**, 1294–1295.
Pozsar, B. I. and Király, Z. (1958). *Nature* **182**, 1686–1687.
Pozsar, B. I. and Király, Z. (1966). *Phytopath. Z.* **56**, 297–309.
Rathbone, M. P. and Hall, R. H. (1972). *Planta (Berl.)* **108**, 93–102.
Ricard, J. and Job, D. (1974). *Eur. J. Biochem.* **44**, 359–374.
Sakai, S., Imaseki, H. and Uritani, I. (1970). *Pl. Cell. Physiol.* **11**, 737–745.
Samborski, D. J. and Rohringer, R. (1970). *Phytochemistry* **9**, 1939–1945.
Samborski, D. J. and Shaw, M. (1956). *Can. J. Bot.* **24**, 601–619.
Seevers, P. M. and Daly, J. M. (1970a). *Phytopathology* **60**, 1322–1328.
Seevers, P. M. and Daly, J. M. (1970b). *Phytopathology* **60**, 1642–1647.
Seevers, P. M., Daly, J. M. and Catedral, F. F. (1971). *Pl. Physiol.* **48**, 353–360.
Sequiera, L. (1973). *A. Rev. Pl. Physiol.* **24**, 353–380.

Shannon, L. M., Uritani, I. and Imaseki, H. (1971). *Pl. Physiol.* **47**, 493–498.

Shaw, M. (1961). *Can. J. Bot.* **39**, 1393–1407.

Shaw, M. and Hawkins, A. R. (1958). *Can. J. Bot.* **36**, 1–16.

Shaw, M., Samborski, D. J. and Oaks, A. (1958). *Can. J. Bot.* **36**, 233–237.

Skipp, R. and Samborski, D. J. (1974). *Can. J. Bot.* **52**, 1107–1115.

Skoog, F. and Armstrong, D. J. (1970). *A. Rev. Pl. Physiol.* **21**, 359–384.

Slesinski, R. S. and Ellingboe, A. H. (1971). *Can. J. Bot.* **49**, 303–310.

Stahmann, M. A., Clare, B. G. and Woodbury, W. (1966). *Pl. Physiol.* **41**, 1505–1512.

Stoessl, A. Rohringer, R. and Samborski, D. J. (1969). *Tetrahdron Letters* **33**, 2807–2810.

Stowe, B. B. and Yamaki, T. (1957). *A. Rev. Pl. Physiol.* **8**, 181–216.

Trocha, P. and Daly, J. M. (1974). *Pl. Physiol.* **53**, 527–532.

Tamura, S., Sakurai, A. and Kainuma, K. (1963). *Agric. Biol. Chem.* **27**, 738–739.

Thimann, K. V. and Sachs, T. (1966). *Am. J. Bot.* **53**, 731–739.

Turner, N. C. and Graniti, A. (1969). *Nature, Lond.* **223**, 1070–1071.

Williamson, C. E. (1950). *Phytopathology* **40**, 205–208.

Yamazaki, H. and Yamazaki, I. (1973). *Arch. Biochem. Biophys.* **154**, 147–165.

Van Andel, O. M. and Fuchs, A. (1972). *In* "Phytotoxins in Plant Disease". (R. K. S. Wood, A. Ballio and A. Graniti, eds) Academic Press, London and New York.

Veldstra, H. (1968). *Neth. J. Pl. Path.* **74** (Suppl. 1.), 55–66.

CHAPTER 8

Toxins of Plant Pathogenic Bacteria and Fungi

GARY A. STROBEL

Department of Plant Pathology, Montana State University, Bozeman, Montana, U.S.A.

I. INTRODUCTION

The concept that plant pathogens may produce substances that are toxic to their respective host plants arose in the late nineteenth century. Since that time numerous compounds have been isolated from pathogenic fungi and bacteria and shown to be phytotoxic. The toxic compounds generally belong to or are derivatives of one or a combination of the following classes of organic substances: glycoside (or polysaccharide), terpenoid, phenolic, and amino acid. In some cases, the chemical structure of the toxin is known. In still other cases, we only know that the crude cultural filtrate of the pathogen possesses some toxic activity. The rarest case of all is the integrated knowledge on the structure, biosynthesis and biological function of the toxin. Beside the interesting chemistry that they exhibit, the phytotoxins have served as useful probes into the normal function and structure of plants, including the molecular basis of plant disease resistance.

There have been numerous reviews on the chemical nature and biological activity of phytotoxins (Strobel, 1974; Scheffer and Samaddar, 1970; Patil, 1974; Wheeler and Luke, 1963; Wood, 1972). Rapid progress is being made on the chemistry and mode of action of certain toxins from previously uninvestigated fungi and bacteria. This review emphasizes the most recent progress made on a number of toxins produced by plant pathogenic fungi and bacteria and briefly summarizes some of the past work on toxins.

II. Helminthosporoside

Steiner and Byther (1971) reported that a host-specific toxin was produced by *Helminthosporium sacchari*, the causal agent of eye spot disease of sugar-cane. The fungus causes eye-shaped lesions on leaves followed by the development of long reddish brown streaks or runners extending from the lesion towards the leaf tip. Since the fungus could be isolated only from the lesion and not the "runner" areas, the suggestion was made that a toxin was involved in the symptomatology. These workers partially purified a substance that was capable of producing runners only on susceptible clones of sugarcane. It could, therefore, be called a host-specific toxin. Steiner and Strobel (1971) isolated this toxin from cultures of *H. sacchari* and named it helminthosporoside. The structure was determined on the basis of mass, infrared, and nuclear magnetic resonance spectroscopy plus numerous chemical and biochemical tests. The structure proposed for helmin-thosporoside is 2-hydroxycyclopropyl-α-D-galactopyranoside (Fig. 1). Organic synthesis has not been carried out, nor has the structure been confirmed by X-ray analysis. There may be some reason to doubt the aglycone structural assignment, but hopefully additional studies will provide more data on this point. These workers devised a reliable biological

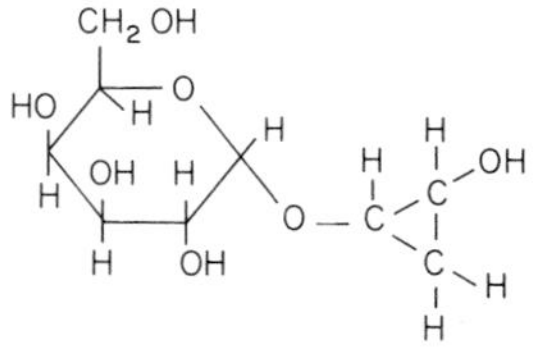

FIG. 1. Proposed structure of helminthosporoside.

assay for the quantitative determination of the toxin based on the degree of symptom expression on susceptible cane leaves (Fig. 2). To show the actual involvement of the toxin under natural field conditions, it was isolated from infected leaves in amounts that could account for symptoms of eye spot (Strobel and Steiner, 1972).

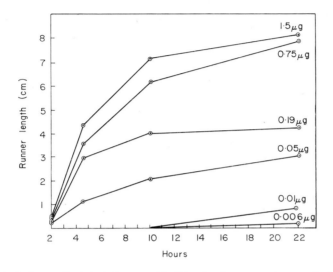

FIG. 2. Biological assay for helminthosporoside. The plot shows symptom production (runner length) as a function of time and the amount of toxin in 1 μl of solution applied to the leaf.

One of the first visible symptoms on susceptible leaves treated with helminthosporoside is water clearing and the formation of water droplets on both leaf surfaces within 30 min after toxin application. Studies by Strobel *et al.* (1972a), however, showed that no visible ultrastructural effects were noticeable until at least 1 h after toxin application. Those changes that were detected after 1 h were slight alterations in chloroplast membranes. Furthermore, after the administration of ^{14}C-toxin to leaves of both resistant and susceptible clones, there was no difference in the amount of ^{14}C recoverable as helminthosporoside from these two clones of cane. This suggested that toxin breakdown was not a factor in resistance to toxin action. In addition, the labeled toxin moved as well in the vascular system of susceptible as in resistant sugarcane leaves.

Strobel (1973a) demonstrated that membrane preparations from susceptible clones of sugarcane bound ^{14}C-helminthosporoside and that there was no binding by similar preparations from resistant clones. Furthermore, *in vivo* binding of the toxin occurred in susceptible clones and not resistant ones. Membranes from clones giving an intermediate reaction to the toxin likewise had only intermediate amounts of binding activity (Table 1). Symptom

TABLE I

Reaction to helminthosporoside of various clones of sugarcane, according to groups, relative to their ability to bind helminthosporoside

Group	Clone	Helminthosporoside bound nmoles/mg protein	Mean for group	Standard error of mean
1 Resistant	H50–7209	0·00		
	H52–4610	0·00	0·028	0·02
	222	0·08		
2 Intermediate	Fiji–62	0·73		
	57NG100	0·73		
	675	0·95		
	CP63–588	1·10		
	117	0·73	0·79	0·055
	CP44–101	0·85		
	51NG127	0·85		
	CP47–193	0·70		
	57NG134	0·53		
3 Susceptible	51NG97	1·58		
	CP50–28	1·80		
	606	1·30		
	Fiji	1·17		
	CP52–68	2·40	1·43	0·13
	CP57–603	1·30		
	736	1·30		
	CP55–30	1·06		
	CP62–258	1·10		
	619	1·38		

production by the toxin, as well as binding of the toxin to membrane preparations, was reduced in the presence of a number of α-galactosides. The binding activity was associated with a membrane bound protein that was purified and shown to have molecular weight of 48 000 and to consist of four identical sub-units. Scatchard plots derived from equilibrium dialysis experiments were used to demonstrate the presence of at least two toxin-binding sites on the protein. The protein has a K_m for helminthosporoside of 6×10^{-5} and an isoelectric point near pH 5·0. Resistant clones have a comparable protein with the same molecular weight, which is immunologically indistinguishable from the binding protein of susceptible clones (Strobel, 1973b). The two proteins, however, differ slightly in their mobility on disc gel electrophoresis. The protein from resistant clones is also a four sub-unit protein, but its amino acid composition differs by four residues. While freshly prepared protein from resistant plants does not bind helminthosporoside, binding can be observed upon treatment of the protein with mild detergent. It has been suggested that the protein from resistant clones possesses helminthosporoside binding sites, but these were most likely masked by virtue of a change in the tertiary or quaternary structure of the protein resulting from alteration in amino acid composition.

The available evidence strongly suggests that the binding protein is primarily associated with the plasma membrane (Strobel and Hess, 1974). A susceptible clone of sugarcane took up an antiserum prepared to the binding protein and subsequently became completely protected from the effects of helminthosporoside. The binding protein was successfully pyridoxylated and reduced with $NaBH_4$, *in vivo*, which also supports its association with the plasma membrane. Further, sugarcane protoplasts were agglutinated by the antiserum to the binding protein, whereas a control serum did not cause agglutination (Fig. 3). After fractionation of the cellular membranes on sucrose density gradients, the toxin-binding activity was associated with the enriched plasma membrane fraction.

The role of the binding protein to the function of the normal susceptible cell is in α-galactoside transport. The affinities of the common α-galactosides, raffinose and the melibiose for the binding protein are slightly greater than that of the toxin (Strobel, 1974). Further, ^{14}C-raffinose was taken up by susceptible sugarcane protoplasts with a K_m approximately that of raffinose binding to the purified protein. No active uptake of the ^{14}C-raffinose was noted in preparations of resistant sugarcane protoplasts. ^{14}C-raffinose was associated with the contents of the susceptible cell, along with sucrose and monosaccharides. Helminthosporoside inhibited the uptake of raffinose by protoplasts. In addition, helminthosporoside was not detected inside the protoplasts.

Since resistant cells function quite normally without the capacity to actively transport α-galactosides, the binding protein must function in some secondary manner to cause cell death. Protoplasts treated with the toxin show immediate swelling, protrusions, and rupture suggesting an effect of the toxin on ion transport (Fig. 4). Using giant cells of susceptible sugarcane, Van Sambeck *et al.* (1975) demonstrated a depolarization of the plasma membrane upon toxin treatment. Toxin treated tissues show an immediate uptake of $^{86}Rb^+$ which also supports the idea of an interference of the toxin, via the binding protein, with membrane function. An enriched plasma membrane fraction from susceptible plants showed a toxin activation of the membrane K^+, Mg^{2+} ATPase (Strobel, 1974).

The plasma membrane K^+, Mg^{2+} ATPase is attributed to have ion regulation activities for the cell. The toxin-binding protein does not have ATPase activity, and the ATPase does not bind the toxin. The membranes of the resistant clone do not show toxin activation of the ATPase. The binding protein does undergo a conformational change upon binding the toxin (Strobel, unpublished). Presumably this change could be amplified via a

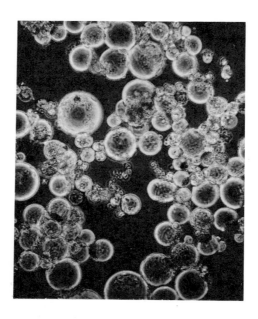

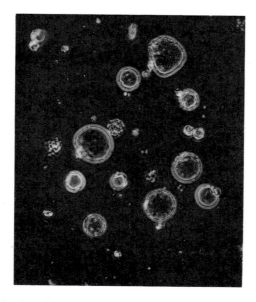

FIG. 3. The agglutination of sugarcane protoplasts by the antiserum to the binding protein (upper). A control serum did not cause agglutination (lower).

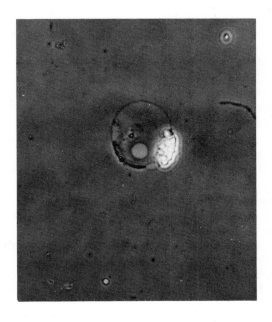

FIG. 4. Susceptible sugarcane protoplasts treated with helminthosporoside. The protoplasts become distorted and eventually burst. Resistant protoplasts remain unaffected (not shown).

phase shift in the membrane bilayer to activate the K^+, Mg^{2+} ATPase. The effect is not limited to the activation of the K^+, Mg^{2+} ATPase since Thom and Maretzki (unpublished) observed a toxin activation of a membrane glycosyl transferase.

Is the binding protein the primary site for the toxin? This question has been answered recently via some experiments in which the purified binding protein was incubated with protoplasts of resistant sugarcane and also tobacco (Strobel and Hapner, 1975). After the cells were treated with helminthosporoside they showed from 50 to 70% mortality after 3 h (Fig. 5). Protoplasts treated with the binding protein also were able to take up ^{14}C-raffinose. The success of the experiment was related to a different procedure to acquire the purified binding protein, that is without detergent extraction. This was effectively accomplished with 1 M trichloroacetate, pH 7·2, as the extraction agent and subsequent protein purification by affinity column chromatography on Bio-gel P-150 linked to melibionate.

The binding protein was labeled with ^{14}C-acetic anhydride, repurified and then incubated with tobacco protoplasts. After 3 h the protoplast membranes were fractionated on a sucrose density gradient column and the labeling peaked in the fraction enriched with plasma membranes. These results strongly support the concept that the binding protein is the key factor in disease susceptibility. This argument is also supported by mutation studies on sugarcane clone H54-775 which is intermediate in its reaction to the toxin. Seed pieces were irradiated with 3 Krγ-radiation, planted and subsequently checked for their reaction to the toxin. All of the totally resistant plants that arose and that were examined lacked toxin-binding protein activity (Strobel et al., 1975). In at least two cases, the modified

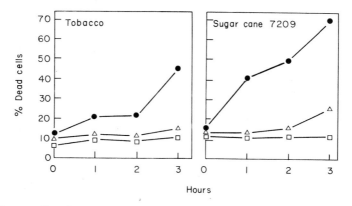

FIG. 5. The mortality of resistant sugarcane protoplasts and tobacco protoplasts having been treated with the binding protein and helminthosporoside in combination and alone. Treatments include: ●————————● protoplasts plus binding protein and helminthosporoside, △————————△ protoplasts plus binding protein alone, □————————□ protoplasts plus helminthosporoside alone (see Strobel and Hapner, 1975 for details).

binding protein in the mutant cane was immunochemically and electrophoretically related to the binding protein from the parent clone, but possessed no binding activity.

H. sacchari can be made non-pathogenic by transferring it for two-week intervals on six separate occasions to a completely synthetic medium (Pinkerton and Strobel, 1976). The organism does not produce any detectable toxin and it does not infect susceptible sugarcane under normal conditions of inoculation. Material from washed leaves from a susceptible cane plant when placed with the attenuated culture will cause toxin production and the pathogenic characteristic of the culture is reinstated. These observations strengthen the notion that the toxin is vital to the pathogenicity of the fungus. Serinol has been isolated and identified from the leaf wash material and shown to be an activator of toxin production at μmolar concentrations. At least one other larger molecular weight activator is also present in crude leaf wash preparations.

III. Some Other Host-specific Toxins

Aside from the first example given in this report of a host-specific toxin, helminthosporoside, there are several others known. These compounds do not belong to a single class of organic substances, but to several including glycoside, terpenoid, and peptide. Historically, the Japanese worker Tanaka (1933) observed what is considered to be the first host-specific toxin from *Alternaria kikuchiana* on pear. In the United States, the outbreak of victoria blight of oats signaled the advent of an enormous amount of work on the toxin of *H. victoriae*. Since the host-specific toxins seem to play such a critical role in their respective diseases, it has been long realized that in depth studies on such compounds and their mode of action could lead to an understanding of the mechanism of disease resistance. The best known examples of host-specific toxins are found in the pathogenic fungi, although some specificity seems to exist in the polysaccharide toxin of the causal agent of fire blight, *Erwinia amylovora* (Goodman *et al.*, 1974).

A. *H. VICTORIAE* AND *P. CIRCINATA* TOXINS

Scheffer's group at Michigan State has worked with the host-specific toxins of *Periconia circinata* and *H. victoriae*. The host-specific toxin from *H. victoriae* causes an immediate loss of electrolytes and other material from susceptible but not resistant oat tissue (Samaddar and Scheffer, 1971) and thus affects the permeability of the plasma membrane (Wheeler and Black, 1963; Keck and Hodges, 1973). The toxin also causes the rapid disruption and lysis of protoplasts from susceptible but not from resistant plants (Samaddar and Scheffer, 1968). Further, there is no demonstrable effect of the toxin on isolated organelles (Scheffer and Samaddar, 1970). Cycloheximide-treated susceptible oat leaves are protected from the effects of the toxin (Gardner and Scheffer, 1973). By deduction, Scheffer's group suggested that the plasma membrane is the site of action of the toxin and that a protein serves as a toxin receptor. The toxin of *P. circinata* also seems to have its effect on the susceptible host, grain sorghum, in a manner that resembles the toxin of *H. victoriae* (Gardner *et al.*, 1972). Additional direct supportive evidence for the plasma membrane as the site of action of both the *H. victoriae* and *P. circinata* toxins has come from electropotential measurements on plant cells. In both cases, in their respective hosts, these toxins caused immediate but gradual decreases in the negative electropotential of the single cells being monitored (Gardner *et al.*, 1974). Nevertheless, incontrovertible evidence for the actual involvement of membrane receptors in susceptible oats on susceptible sorghum will come when the receptor protein is isolated and characterized.

The *H. victoriae* toxin has not been completely characterized and this has precluded many advances toward understanding its precise mode of action and its possible protein receptor. It had been suggested that the toxin has two components, a peptide and a heterocyclic nitrogen containing moiety (Pringle and Braun, 1958). The peptide component contains aspartic acid, glutamic acid, glycine, valine and one of the two leucines. Some dispute has existed for a number of years over the nature of the heterocyclic moiety; however, recently Dorn and Arigoni (1972) have confirmed the fact that victoxinine has the empirical formula of $C_{17}H_{29}NO$ and they propose a structure for it as shown in Fig. 6. It occurs as a free base in fungal cultures and also is found in non-toxin producing isolates of *H. victoriae*.

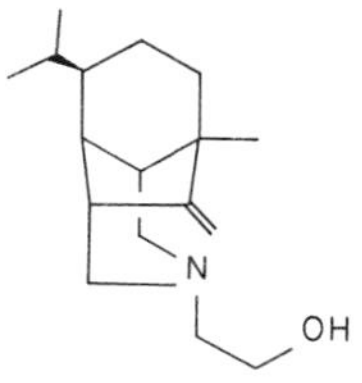

FIG. 6. Victoxinine—the terpenoid moiety of victorin.

Apparently victoxinine has only 1/7500 the toxicity of the intact toxin and is non-host specific. It is well known that the intact *H. victoriae* toxin, also known as victorin (Wheeler and Luke, 1954), affects only oat varieties bearing the Victoria gene. Susceptibility to the toxin is dominant and is inherited in a simple 3:1 mendelian ratio. This suggests that susceptibility to the toxin is controlled by the action of a single gene. An excellent recent review on the subject of *H. victoriae* has been prepared by Luke and Gracen (1972). The historical aspects of the disease are discussed and numerous references are given relative to the early work not only on this interesting toxin but other toxins of *Helminthosporium* as well.

Some information is available on the nature of the host-specific toxin of *P. circinata*, the causal agent of milo disease of sorghum (Pringle and Scheffer, 1963, 1967). There appears to be two host-specific toxins of *P. circinata* and at least one of these was crystallized (Pringle and Scheffer, 1966). Satisfactory criteria for homogenity of the crystalline toxin preparation were not presented, yet approximately 80% of the weight of the crystalline sample after acid hydrolysis could be accounted for as alanine, aspartic acid, glutamic acid, and serine, suggesting a peptide nature of the toxic component.

B. *A. KIKUCHIANA* TOXIN

Alternaria kikuchiana produces a host-specific toxin that is extremely toxic to susceptible Japanese pear, *Pyrus serotina* cv. Nyisseiki and its derivatives, but has no effect on resistant pear cultivars (Tanaka, 1933). Otani *et al.* (1974) have shown that this toxin causes loss of electrolytes from treated tissue. A mild heat treatment (55 °C–2 sec) results in partial reversible protection of leaves to the effects of the toxin. Further, disulfide reagents gave partial protection against toxin induced loss of electrolytes. This protection was completely reversed by an oxidizing agent. The data suggest that disulfide groups in the cell may be involved in the reaction to the toxin and that a toxin receptor site may be involved. The Japanese group is currently working on the structure of this interesting host-specific toxin.

C. *H. CARBONUM* TOXIN

Pringle and Scheffer (1967) isolated the host-specific toxin of *H. carbonum*, and Pringle (1971) described it as being a cyclic peptide containing proline and alanine in a ratio of 1:2 with two unstable amino acids, one of which was identified as α-amino-2,3-dehydro-3-methylpentanoic acid. Since the molecular weight of the toxin was estimated to be about 700, it is conceivable that proline and alanine occur as one and two residues per mole, respectively. The remaining unknown compound was suggested to be a hydroxy amino acid, and its breakdown during hydrolysis may have accounted for such compounds as ammonia, glycine, and unknown products. Although Pringle quantitatively accounted for alanine and proline in the toxin preparations, data on the remaining 70% of the weight of the samples were not presented. Recently, Pringle (1973) has shown that electrolytic reduction of the toxin renders it non-specific, and decreases its relative specific biological activity. Presumably, this loss in activity is due to the reduction of the dehydroamino acid to isoluecine.

Yoder and Scheffer (1973) demonstrated that toxin treated susceptible corn roots took up and retained more NO_3^-, Na^+, Cl^-, 3-o-methylglucose and leucine than did control roots. Their data suggested that the toxin does not cause derangement of the plasmalemma but instead results in specific changes in permeability to certain solutes. Other effects of the toxin on the physiology of the affected plant have been described (Kuo *et al.*, 1970).

D. *H. MAYDIS* TOXINS

In 1970, *H. maydis* caused an epidemic of Southern corn leaf blight in the United States (Tatum, 1971). The pathogen was devastating on corn carrying Texas male sterile cytoplasm, while corn with normal cytoplasm was resistant. A host-specific toxin had been associated with this organism (Smedegård-Petersen and Nelson, 1969) and a number of workers suggested that the toxin was responsible for the rapid graying, chlorosis, and necrosis associated with this disease. Using a leaf bioassay test, Karr *et al.* (1974) isolated four host-specific toxins from *H. maydis* and showed numerous criteria of purity for at least three of these compounds. Two toxins, labeled I and II, reacted postively with all of the reagents commonly used in terpenoid identification and gave UV, IR, NMR, and mass spectral data comparable to those of known tetracyclic triterpenoids. Toxins I and II differed only slightly in spectral properties and molecular weight, whereas toxin III appeared to be a glycoside of either toxin I or II. All four compounds were recovered from leaf and stem tissues infected with *H. maydis*.

Karr *et al.* (1975) developed a colorimetric assay for the *H. maydis* toxins which utilized a modification of the Liebermann-Burchard procedure for the determination of cholesterol. Their data showed that still another toxin (Toxin V) was present in cultures of *H. maydis*. Toxin V appeared in culture before the other toxins, thus it was suggested as a precursor to them. *H. maydis* also causes blighting symptoms on a line of spring wheat, "Fortuna". All of the toxins could be isolated from infected wheat. This observation may ultimately prove useful to those interested in investigating the mechanism of susceptibility of corn carrying Tms cytoplasm to the *H. maydis* toxins since susceptible wheat may carry the same susceptibility factor(s).

Numerous bioassays have been developed for the toxins for *H. maydis* including root inhibition (Lim and Hooker, 1972; Hooker *et al.*, 1970), leaf symptom production (Karr *et al.*, 1974) and pollen germination (Laughnan and Gabay, 1973). Thus far, virtually all of the work on the mode of action of the *H. maydis* toxins has been done with crude fungal culture preparations. The results of such work can only be taken as a possible key for more comprehensive studies with purified compounds. Studies on corn mitochondria by Miller and Koeppe (1971) gave the first indication that the toxins of *H. maydis* affect cell membranes. Crude toxin preparations inhibited phosphorylation, altered electron transport rates and caused swelling of mitochondria of Tms corn and not normal cytoplasm corn. Watrud *et al.* (1975) suggested that the removal of the outer membrane of mitochondria of normal cytoplasm corn results in a sensitive response follow-ing the addition of a toxin preparation. They also suggest that there may be some nuclear control in modifying the host response to the toxins. Nevertheless, since the mitochondrial functions were not affected *in vivo* after toxin treatment, these investigators explored other possibilities as the site for initial toxin effect. A number of physiological functions of susceptible corn were affected including: (1) an inhibition of CO_2 fixation probably related to stomated functioning (Arntzen *et al.*, 1973), (2) an inhibition of K^+ uptake by guard cells (Arntzen *et al.*, 1973), (3) a de-polarization of the electrochemical membrane potential of epidermal corn root cells (Mertz and Arntzen, 1973), (4) a decrease in the content of cellular ATP (Arntzen *et al.*, 1973), (5) an increase in the rate of ion leakage (Arntzen *et al.*, 1973, and Gracen *et al.*, 1972), and (6) an inhibition of microsomal K^+, Mg^{2+} ATPase (Tipton *et al.*, 1973). It is apparent that one of the primary effects of the toxin is on the inhibition of K^+ accumulation and subsequent depolarization of membrane potential since these phenomena are observed immediately upon toxin treatment. The use of purified toxins, and isolated microsomes should be tried to substantiate this notion and determine the nature of the membrane site involved.

E. *P. MAYDIS* TOXIN

The causal agent of yellow leaf blight of corn, *Phyllosticta maydis* produces a substance(s) which selectively inhibits seeding root growth, induces leaf chlorosis, and causes an increased leakage of electrolytes from corn leaves carrying Tms cytoplasm (Comstock *et al.*, 1973). Crude toxin preparations also cause irreversible swelling, inhibition of O_2 uptake and an uncoupling of oxidative phosphorylation on mitochondria from Tms corn. Comparable results on the *P. maydis* toxin were obtained by Yoder (1973).

F. ALTERNARIOLIDE

Alternaria blotch on apple is caused by *Alternaria mali*. Okuno *et al.* (1974) isolated and characterized a depsipeptide which causes characteristic necrotic brown spots on leaves and fruits of the plant. The toxic peptide was shown to be host-specific for certain apple varieties.

IV. STEMPHYLIN

Leaf spot of lettuce caused by *Stemphylium botryosum* has symptoms associated with it that are reminiscent of those caused by toxins. Barash and Strobel (1975) investigated *S. botryosum* and found three phytotoxins produced in defined liquid culture. An anionic compound called stemphylin was isolated and characterized via chemical and spectroscopic analyses. The toxin is an amorphous yellow solid with absorbance maxima at 218, 268 and 427·5 nm and exhibits a bathochromic shift in alkaline pH. It has a molecular weight of 370 and an empirical formula of $C_{17}H_{22}O_9$. Glucose is detected after acid hydrolysis. The proposed structure of the compound is 3-hydroxy-2,2-dimethyl-5-α-D glucopyranoside, 2,3-dihydrochromone (Fig. 7). A

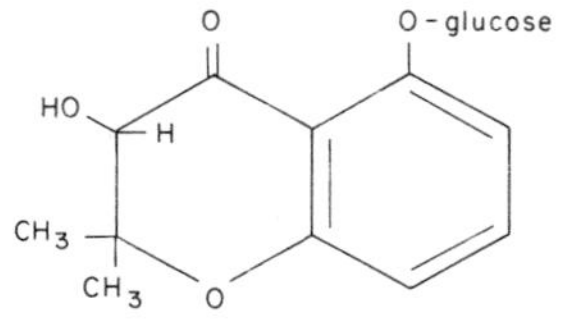

Fig. 7. Stemphylin—a phenolic toxin from *S. botryosum*.

linear relationship exists between the area of a lesion produced on a susceptible lettuce leaf and the amount of toxin applied to it. The role of this toxin in symptom production, and its mode of action, are unknown.

V. Toxic Glycopeptides and Polysaccharides

For years pathologists have worked with cultures of *Erwinia amylovora* trying to extract compounds that are toxic to apple and pear and reproduce some or all of the symptoms of fire-blight. However, Goodman *et al.* (1974) showed that they could recover a toxic compound from apple fruit tissue infected by *E. amylovora* that not only caused symptoms of the disease, but was selective. That is, the toxin caused wilting in plants that were the most susceptible to the bacterium, and only produced wilting in resistant plants after a long duration of exposure to the toxin (Table II). It did not cause symptoms on non-host plants. The toxic compound has a molecular weight of approximately 165 000 and is primarily a polymer of galactose. Traces of protein are also present. Similar results were found by Eden-Green (1973).

Future study will undoubtly show the origin of the toxin, the role of the host plant in its production, its mechanism of action, and its biosynthesis.

One of the best characterized phytotoxic glycopeptides is that of *Coryne-bacterium sepedonicum* (Strobel *et al.*, 1972b). *C. insidiosum*, another wilt causing pathogen, also produces a phytotoxic high molecular weight glycopeptide (Ries and Strobel, 1972a). The organism produces profuse quantities of this toxin in culture media. A molecular weight of 5×10^6 was calculated for this toxin by both light scattering and gel column techniques. The molecule is blue due to the chelation of approximately 75 moles of copper/mole of toxin. It consists primarily of L-fucose, mannose, glucose, galactose, and has an unidentified keto-deoxy acid. A single

TABLE II

Specific activities of the phytotoxic polysaccharide against rosaceous species. The activity is expressed as the length of time in hours required for a plant cutting to show wilting when its base is placed in toxin solution at a concentration of 100 µg/ml.

Plant tested	Specific activity (h)	Susceptibility to *E. amylovora*
Pyrus malus		
Jonathan	1	Susceptible
Malling-26	1	Susceptible
Red Delicious	18	Resistant
Pyrus communis		
Bartlett	2	Susceptible
Starkrimson	3	Susceptible
Magness	12	Resistant
Moonglow	12	Resistant
Kieffer	12	Resistant
Starking Delicious	14	Resistant
Old Home	15	Resistant
Cydonia oblonga		
EM-A	8	Moderately resistant
Province	12	Resistant
Polish 1	12	Resistant
Polish 2	24	Very resistant
Polish 3	18	Resistant
Crataegus crusgalli	1	Susceptible
Spiraea corymbosa	1	Susceptible
Cotoneaster pyracantha	6	Susceptible
Sorbus americana	6	Susceptible

peptide with glycine as the sole NH_2-terminal amino acid is covalently linked through the hydroxyl of threonine to a sugar residue in the molecule. There appear to be at least 77 peptide chains per mole of toxin. L-Fucose comprises about 40% of the toxin with 20% as glucose, 20% as galactose, and 4% as mannose. Sadowski and Strobel (1973) successfully demonstrated the presence of a guanosine diphosphate-L-fucose glycopeptide fucosyltransferase in membrane preparations of *C. insidiosum*. The enzyme transferred L-fucose from DGP-L-fucose to preparations of a partially hydrolyzed toxin which served as a primer. This toxic glycopeptide is present in alfalfa plants infected with *C. insidiosum* (Ries and Strobel, 1972b). It is also a wilt-inducing compound, and it not only affects alfalfa, but a wide range of plant species. Interestingly enough, however, if used in low concentrations on alfalfa cuttings, it selectively causes wilting in susceptible but not in bacterial wilt-resistant clones (Straley *et al.*, 1974). This fact should enable plant breeders to rapidly sort out wilt-resistant clones in alfalfa breeding projects.

Salemink *et al.* (1965) reported on the presence of toxic glycopeptide from cultures of *Ceratocystis ulmi*. These toxins varied in molecular weight from 25 000 upward. The main sugar components were reported to be mannose and galactose. The toxins induced wilting, browning and formation of gums and tyloses in the vascular elements. Recently, Van Alfen, Strobel, Hapner and McNeil (unpublished) obtained a glycopeptide from cultures of *C. ulmi* by traditional ion exchange chromatography to remove contaminants followed by affinity chromatography. Since the glycopeptide contains α-mannose linkages, it easily absorbed to a concanavalin-A-Sepharose column and was eluted with methyl-α-mannoside. The toxin represents about 50% of the high molecular polymers produced by *C. ulmi*. It exists as a polydisperse substance with a multitude of molecular weights in the vicinity of 300 000. Using the procedure of Bjorndal *et al.* (1967) the glycopeptide was methylated, hydrolyzed, reduced and acetylated and subjected to gas chromatography combined with mass spectroscopy. In conjunction with experiments involving mild acid hydrolysis, it was learned that the glycopeptide consists primarily of 1 ⟶ 6 linked mannose, with virtually every mannose residue containing a 1 ⟶ 3 linked rhamnose. A peptide composed of residues of aspartic acid, threonine, serine, glutamate, protein, glycine, alanine, valine isoleucine and leucine is present. β-Elimination, followed by reduction showed that both serine and threonine are the amino acid residues to which sugar residues of the polysaccharide chains are attached. Van Alfen and Turner (1975) have done extensive studies on the mode of action of the *C. ulmi* glycopeptides. Stem conductance and leaf water potential are reduced in elm seedlings treated with crude toxin preparations. Further, stomata close and transpiration decreases in toxin-treated seedlings. Disruption of the water conducting system of the plant may be the factor responsible for wilting symptomatology.

VI. FUSICOCCIN

Without a doubt, some of the most comprehensive structural and physiological work on a fungal toxin is that on fusicoccin and its derivatives. *Fusicoccum amygdali* causes wilting and desiccation of shoots on infected almond and peach trees. Since these foliar symptoms appear on leaves remotely located from fungus infected nodes, toxins elaborated by the pathogen have been designated as the cause of such damage (Graniti, 1962). Cultures of the actively growing fungus produced a phytotoxin which was isolated, partially characterized, and called fusicoccin by Ballio *et al.* (1964). It was later shown by Ballio *et al.* (1968) to be a glucoside of a carbotricyclic terpene with a molecular weight of 680 and a structure shown in Fig. 8. The introduction of fusicoccin into the xylem of peach and almond shoots produces symptoms closely resembling those following infection by *F. amygdali*. The toxin affects a wide range of plants in very dilute solution (2 μg/ml). Graniti (1962) established a bioassay test for the toxin using four-leaved cuttings from young tomato plants. This test has served other investigators in biological studies with the toxin and its derivatives.

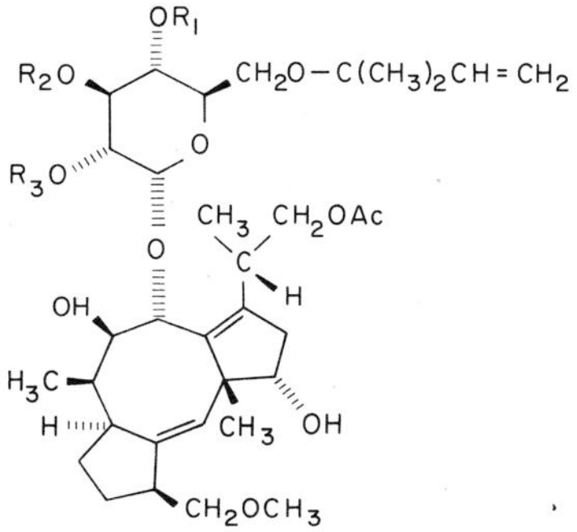

FIG. 8. The structure of fusicoccin and its derivatives

$R_1R_3 = H$ $R_2 = Ac$ fusicoccin
$R_1R_2 = H$ $R_3 = Ac$ allofusicoccin
$R_2R_3 = H$ $R_1 = Ac$ isofusicoccin

Ballio *et al.* (1973) have described a number of minor metabolites of fusicoccin. Of the diacetyl derivatives of *F. amygdali*, fusicoccin is the most toxic, followed by allofusicoccin which is 11 times less active and isofusicoccin which is 25 times less active (Fig. 8). Other modifications have been made on fusicoccin and the derivatives studied. Several deacetyl-fusicoccins were identified by Ballio's group and shown to have less biological activity than fusicoccin (Ballio *et al.*, 1972, 1974).

Ballio *et al.* (1971) showed that fusicoccin produces a pronounced increase in the wet weight of pea internodes, an increase in the rate of tissue elongation, and enhances tissue deformability. These observations led Ballio's group to suggest that fusicoccin possesses auxin-like activity or promotes such activity.

Lado *et al.* (1972) confirmed the observations that toxin-treated pea segments showed increased water uptake over nontreated controls in leaf discs of tomato, clover, and tobacco. Since the osmotic pressure values of treated tissues were lower than those of the control, the authors concluded that fusicoccin acts to irreversibly extend the cell wall. They showed that fusicoccin lacks the ability possessed by auxins to inhibit cell elongation in root tissues. Nevertheless, fusicoccin promotes the extrusion of protons into the medium containing pea tissues which accompanies cell enlargement (Marré *et al.*, 1974a,b). The same observation is made when tissues are treated with either auxin or cytokinins. Such a mechanism would explain the appearance of a negative intercellular potential in fusicoccin treated tissues. A mechanism that accounts for a coupling of proton extrusion to K^+ uptake in toxin-treated tissues has been proposed by Marré *et al.* (1974c). Turner (1972) originally demonstrated the toxin stimulated the uptake of potassium by guard cells in both the light and dark, confirming the idea that potassium is necessary for stomatal opening. The effects of the toxin on stomatal opening could be reversed by the action of 2,4-dinitrophenol in both the light and the dark. These results indicated that for the normal stomatal opening process, ATP from both photosynthetic phosphorylation and oxidative phosphorylation is involved as an energy source in driving the uptake of potassium. Turner's results on potassium uptake have been confirmed by Natr (1971), who also showed that photosynthetic rates were increased in toxin-treated plants which he related to a decrease in the stomatal resistance to CO_2 transfer. Chain *et al.*, 1971 also substantiated Turner's observations on toxin-induced stomatal opening in tomato, but hastened to point out that leaves whose stomata were closed by application of abscisic acid still wilted in response to treatment with fusicoccin. These workers suggested that the most satisfactory hypothesis for the toxin action would involve a decrease in resistance of the plasmalemma to the passage of water and that the stomatal opening process is a secondary effect.

In μmolar concentrations fusicoccin induces the germination of dormant wheat seeds (Lado *et al.*, 1974). Further, it is more active than either gibberellic acid, benezyladenine, or white light in stimulating the germination of lettuce seeds. At 15 μmolar fusicoccin removes the inhibitory effect of abscisic acid on the germination of radish and lettuce seeds. It appears that fusicoccin holds many possibilities for future study both as a toxin and as a phytohormone.

VII. Some Amino Acid Derived Bacterial Toxins

Recently, Patil (1974) has published a review on toxins produced by plant pathogenic bacteria. Generally, the bacterial phytotoxins are either glycosides (glycopeptides) or are low molecular weight compounds derived from amino acids.

A. Tabtoxins

Wildfire is a highly infectious disease of tobacco plants caused by *Pseudomonas tabaci*. Leaves of plants infected by this disease are covered by circular yellow lesions about 1 cm in diameter. The causal organism can be isolated from these lesions. Cell-free culture filtrates are capable of producing the symptoms of the disease, and, in fact, symptoms can be produced on a number of plant species. Using ion-exchange chromatography in conjunction with a bioassay test, Stewart (1971) was able to isolate the pure toxin as an off-white, fluffy noncrystalline powder. Based upon extensive NMR experiments coupled with basic compositional data and other spectroscopic observations, Stewart showed what is now accepted as the correct structural formula of tabtoxin (Fig. 9). Hydrolysis of tabtoxin yields tabtoxinine and threonine. The structure of tabtoxin was recently confirmed by Taylor *et al.* (1972). Furthermore, these workers isolated the serine analog of tabtoxin and proposed the name 2-serine-tabtoxin. It, too, is capable of inducing chlorosis in plants and has the same relative specific biological activity as tabtoxin.

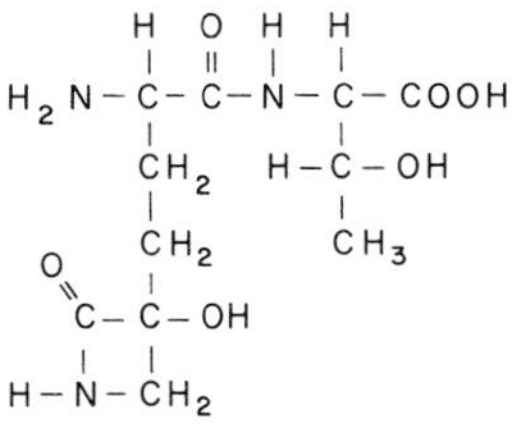

FIG. 9. Tabtoxin.

Crude toxin preparations from *P. tabaci* have been called the "wildfire toxin". Since at least two toxins have been isolated from such preparations, the name wildfire toxin no longer seems appropriate. Sinden and Durbin (1970) showed that *P. coronafaciens*, which attacks oats, also produces both tabtoxin and 2-serine-tabtoxin. In addition, *P. garcae*, a variant of *P. coronafaciens* attacking timothy, and a pseudomonad from wheat rhizosphere also produce these toxins.

B. PHASEOTOXINS

The causal agent of halo blight of bean, *Pseudomonas phaseolicola*, produces extracellular toxins that are also capable of causing chlorotic halos on treated leaves. One active component is N-phosphoglutamic acid, and the report by Patil *et al.* (1976) appears to be the first to show the occurrence of an N-phosphorylated primary amine. Suffice it to say that the toxin competitively inhibits ornithine carbamoyltransferase (Patil, 1974). Ultimately, it might be that the deficiency of arginine results from an inhibition of the carbomoyltransferase which may preclude the production of chlorophyll synthesizing enzymes resulting in chlorosis.

C. RHIZOBITOXINE

Certain strains of *Rhizobium japonicum* fix nitrogen in root nodules of soybean plants in a normal fashion and simultaneously synthesize a toxin that induces chlorosis in new leaf growth. Owens and Wright (1965) isolated a chlorosis-inducing toxin from chlorotic leaves of diseased plants, from nodules, and from bacterial cultures. The toxin, called rhizobitoxine by Owens *et al.* (1968) is not host-specific since it produces chlorosis symptoms in seedlings of many plants. Furthermore, it seems to be the sole cause of chlorosis in plants infected with disease-producing strains of *R. japonicum*. The disease is described by interactions of the host and bacterium that are not a part of the normal symbiotic relationship.

The structure of rhizobitoxine was showed by Owens and Thompson (1972) to be 2-amino-4-(2-amino-3 hydroxypropyl)-*trans*-but 3-enoic acid. Experiments with *Salmonella typhirmurium* suggested that rhizobitoxine interferes with the metabolism of cystathionine.

D. SYRINGOMYCIN

Certain isolates of *Pseudomonas syringae*, a pathogen of stone fruit trees, produce a wide spectrum antibiotic which is also phytotoxic (DeVay *et al.,* 1968). Sinden *et al.* (1971) have described the toxin as a peptide containing 8 known and 1 unknown amino acids. Backman and DeVay (1971) suggest that the toxin has its primary effect on cellular membranes.

VIII. NON HOST-SPECIFIC ALTERNARIA TOXINS

A review on the chemical and biological properties of the alternaria toxins has appeared (Templeton, 1972). One of the most interesting toxins from this genus is tentoxin, produced by *Alternaria tenuis*. This organism produces an irreversible variegated seedling chlorosis in cotton, citrus and other seedlings. The striking symptoms of this disease could be reproduced experimentally with sterile culture filtrates of the fungus. Cotton seedlings with more than 35% chlorotic areas usually die. The period of time that cucumber seeds are soaked in toxin solution directly affects the amount and pattern of chlorosis in developing cotyledons. A workable bioassay for tentoxin was developed by evaluating the amount of chlorophyll in toxin-treated seedlings that were incubated for 4 days at room temperature.

The toxin is a cyclic peptide with 4 amino acid residues, Koncewicz *et al.* (1973) proposed the structure to be *cyclo*-N-methyl-dehydrophenylalanyl-L-N-methylalanyl. Structural work by Meyer *et al.* (1974), however, shows the toxin to be *cyclo*-L-leucyl-N-methyl-*trans*-dehydrophenylanalyl-glycyl-

N-methyl-L-alanyl. Clarification of the structure will undoubtedly come when X-ray data are obtained.

Templeton *et al.* (1967) observed that cucumber seedlings were more sensitive to this toxin in darkness than under continuous light. These seedlings exposed 48 h after the initiation of germination were insensitive to the toxin. As a result of this, as well as the overall symptom expression of affected seedlings, they suggested that the activity of the toxin was associated with some step late in chloroplast development. They proposed that the toxin may: (a) interfere with chlorophyll synthesis but not plastid development; or (b) interfere with plastid development thereby affecting chlorophyll synthesis indirectly.

Relative to the effect of the toxin on photosynthesis processes, Arntzen (1972) has shown that tentoxin inhibits cyclic photophosphorylation but not reversible proton accumulation by chloroplasts of lettuce. Relatively low toxin concentrations inhibited coupled electron flow in the presence of ADP and phosphate. There was no effect on basal electron flow or on uncoupled electron transport. Arntzen suggested that tentoxin is an energy transfer inhibitor acting at the terminal steps of ATP synthesis. This explanation may ultimately account for chlorosis in the treated plants in that synthesis of ATP in the plastid would be required for normal chloroplast development.

Other phytotoxins from *Alternaria* include: (1) zinniol from *A. zinniae*, a pentasubstituted benzene (Starratt, 1968); (2) alternaric acid from *A. solani*, a semi-quinone derivative; (3) altenin, from *A. kikuchiana*, an enediol carbonyl ester (Sugiyama *et al.*, 1966, 1967). Altenin has been chemically synthesized by Sugiyama and his associates. A relationship between altenin and the host-specific toxin of *A. kikuchiana* is suspected but not known; and (4) alternariol monomethyl ether from *A. tenuis* (Pero and Main, 1969, and Tirokata *et al.*, 1969).

IX. Concluding Comments

By no means is this review intended to be comprehensive. For instance, it did not survey the excellent work on the aromatic toxins produced by the *Fusaria* so well studied by the Swiss (Kern, 1972) nor the myriad of preliminary reports on phytotoxic activities in fungal and bacterial cultures, nor any of the past work on other terpenoid toxins of the *Helminthosporia*. It is agreed that the role of any given toxin in plant disease needs to be established experimentally, i.e. recovery from infected tissues in appropriate concentrations and during the ideal time that symptoms are expressed. In some cases this has not been done because of the instability of the toxin, but nevertheless the role of the toxin in the disease seems assured, i.e. *H. victoriae* toxin.

It should be stressed that ideally, the isolation and characterization of a toxic substance should be the first order of business in any given case. From that point, biosynthetic and mode of action studies can logically begin.

Phytotoxins, especially the ones that are host-specific, have proven to be extremely useful tools in studying host–parasite interactions in as much as pathogen specificity in a given host or hosts is entirely related to the specificity of the toxin. In such cases the potential of explaining the plant disease susceptibility and disease resistance is apparent.

Toxins also appear to be tools useful in studying some of the activities of the healthy plant, not to mention their usefulness as agents in disease screening programs in plant breeding and selection.

ACKNOWLEDGEMENTS

Gratitude is expressed to the CSRS of the USDA for grant 216–15–23 and to the NSF for grant GB 43192 both of which deal with toxins in plant diseases. This report represents paper no. 588 of the Montana Agricultural Experiment Station. The author wishes to thank Mrs Teressa Jessop for her help in preparing this manuscript.

REFERENCES

Arntzen, C. J. (1972). *Biochim. biophys. Acta* **283**, 529–542.
Arntzen, C. J., Haugh, M. F. and Bobick, S. (1973). *Pl. Physiol.* **52**, 569–574.
Arntzen, C. J., Koeppe, D. E., Miller, R. J. and Peverly, J. H. (1973). *Physiol. Pl. Path.* **3**, 79–89.
Backman, P. A. and DeVay, J. E. (1971). *Physiol. Pl. Path.* **1**, 215–233.
Ballio, A., Bottalico, A., Framondino, M., Graniti, A. and Randazzo, G. (1973). *Phytopath. Med.* **12**, 22–29.
Ballio, A., Chain, E. B., deLeo, P., Erlanger, B. F., Mauri, M. and Tonolo, Al (1964). *Nature* **203**, 297.
Ballio, A., Brufani, M., Casinovi, C. G., Cerrini, S., Fedeli, W., Pellicciari, R., Santurbano, B. and Vaciago, A. (1968). *Experientia* **24**, 631–635.
Ballio, A., Pocchiari, F., Russi, S. and Silano, U. (1971). *Physiol. Pl. Path.* **1**, 95–103.
Ballio, A., Casinovi, C. G., Framondino, M., Grandolini, G., Randazzo, G. and Rossi, C. (1972). *Experientia* **28**, 1150–1151.
Ballio, A., Casinovi, C. G., d'Alessio, U., Grandolini, G., Randazzo, G. and Rossi, C. (1974). *Experientia* **30**, 844–845.
Barash, I., Karr, A. L. and Strobel, G. A. (1975). *Pl. Physiol.* **55**, 646–651.
Bjorndal, H., Hellerquist, C. G., Lindberg, B. and Svensson, S. (1967). *Angew. Chem. Inst. Ed. Engl.* **9**, 610–619.
Chain, E. B., Mantle, P. G. and Milborrow, B. V. (1971). *Physiol. Pl. Path.* **1**, 495–514.
Comstock, J. C., Martinson, C. A. and Gengenbach, B. G. (1973). *Phytopathology* **63**, 1347–1361.
DeVay, J. E., Lukezic, F. L., Sinden, S. L., English, H. and Coplin, D. L. (1968). *Phytopathology* **58**, 95–101.
Dorn, F. and Arigoni, D. (1972). *J. chem. Soc. Chem. Commun.* 1342–1343.

Eden-Green, S. (1973). Ph.D. Thesis East Malling Research Station.
Gardner, J. M. and Scheffer, R. P. (1973). *Physiol. Pl. Path.* **3**, 147–157.
Gardner, J. M., Mansour, I. S. and Scheffer, R. P. (1972). *Physiol. Pl. Path.* **2**, 197–206.
Gardner, J. M., Scheffer, R. P. and Higinbotham, N. (1974). *Pl. Physiol.* **54**, 246–249.
Goodman, R. N., Huang, J. S. and Huang, P. Y. (1974). *Science, N.Y.* **183**, 1081–1082.
Gracen, V. E., Grogan, C. O. and Foster, M. J. (1972). *Can. J. Bot.* **50**, 2167–2170.
Graniti, A. (1962). *Phytopathol. Med.* **1**, 182–185.
Hooker, A. L., Smith, D. R., Lim, S. M. and Beckett, J. B. (1970). *Plant Dis. Reptr* **54**, 708–712.
Karr, A., Karr, D. and Strobel, G. A. (1974). *Pl. Physiol.* **53**, 250–257.
Karr, D., Karr, A. and Strobel, G. A. (1975). *Pl. Physiol.* **55**, 727–730.
Keck, R. W. and Hodges, T. K. (1973). *Phytopathology* **63**, 226–230.
Kern, H. (1972). *In* "Phytotoxins in Plant Disease" (R. K. S. Wood, A. Ballio and A. Graniti, eds), Academic Press, New York and London.
Koncewicz, M., Mathiaparanam, P., Uchytil, T. F., Sparapano, L., Tam, J., Rich, P. H. and Durbin, R. D. (1973). *Biochem. biophys. Res. Commun.* **53**, 653–658.
Kuo, M. S., Yoder, O. C. and Scheffer, R. P. (1970). *Phytopathology* **60**, 365–368.
Lado, P., Pennacchioni, A., Rasi-Caldogno, F., Russi, S. and Silano, V. (1972). *Physiol. Pl. Path.* **2**, 75–85.
Lado, P., Caldogno, F. R. and Columbo, R. (1974). *Physiol. Pl.* **31**, 149–152.
Laughnan, J. R. and Gabay, S. J. (1973). *Crop Sci.* **13**, 681–684.
Lim, S. M. and Hooker, A. L. (1972). *Plant Dis. Reptr.* **56**, 805–807.
Luke, H. H. and Gracen, V. E. (1972). Helminthosporium Toxins. *In* "Microbial Toxins" (S. Kadis, A. Ciegler, S. J. Ajl, eds), Vol. VIII, pp. 139–168. Academic Press, New York and London.
Marré, E., Columbo, R., Lado, P. and Caldogno, F. R. (1974a). *Pl. Sci. Letters* **2**, 139–150.
Marré, E., Lado, P., Ferroni, A. and Denti, A. B. (1974b). *Pl. Sci. Letters* **2**, 257–265.
Marré, E., Lado, P., Rasi-Caldogno, F., Columbo, R. and DeMichelis, M. I. (1974c). *Pl. Sci. Letters* **3**, 365–374.
Mertz, S. M. and Arntzen, C. J. (1973). *Pl. Physiol.* (*Suppl.*) **51**, 16.
Meyer, W. L., Kuyper, L. F., Lewis, R. B., Templeton, G. E. and Woodhead, S. H. (1974). *Biochem. biophys. Res. Commun.* **53**, 653–658.
Miller, R. J. and Koeppe, D. E. (1971). *Science* **173**, 67–69.
Natr, L. (1971). *Photosynthetica* **5**(3), 195–199.
Okuno, T., Ishita, Y., Sawai, K. and Matsumoto, T. (1974). *Chem. Letters Chem. Soc. Japan* 635–638.
Otani, H., Nishimura, S. and Kokmoto, K. (1974). *Ann. phytopath. Soc. Japan* **40**, 59–66.
Owens, L. D. and Thompson, J. F. (1972). *J. Chem. Soc. Commun.* 714.
Owens, L. D. and Wright, D. A. (1965). *Pl. Physiol.* **49**, 927–930.
Owens, L. D., Guggenheim, S. and Hilton, J. L. (1968). *Biochim. biophys. Acta* **158**, 219–225.
Patil, S. S. (1972). *Phytopathology* **62**, 782.
Patil, S. S. (1974). *A. Rev. Phytopath.* **12**, 259–279.
Patil, S. S., Youngblood, P., Christiansen, P. and Moore, R. E. (1976). *Biochem. biophys. Res. Commun.* **69**, 1019–1027.
Pero, R. W. and Main, C. E. (1969). *Phytopathology* **60**, 1570–1573.
Pinkerton, F. and Strobel, G. A. (1976). *Proc. Natn. Acad. Sci., U.S.A.* (in press).
Pringle, R. B. (1971). *Pl. Physiol.* **48**, 756–759.
Pringle, R. B. (1973). *Pl. Physiol.* **51**, 493–404.

Pringle, R. B. and Braun, A. C. (1958). *Nature* **181**, 1205–1206.
Pringle, R. B. and Scheffer, R. P. (1963). *Phytopathology* **53**, 785–787.
Pringle, R. B. and Scheffer, R. P. (1966). *Phytopathology* **56**, 1149–1151.
Pringle, R. B. and Scheffer, R. P. (1967). *Phytopathology* **57**, 530–532.
Pringle, R. B. and Scheffer, R. P. (1967). *Phytopathology* **57**, 1169–1172.
Ries, S. M. and Strobel, G. A. (1972a). *Pl. Physiol.* **49**, 676–684.
Ries, S. M. and Strobel, G. A. (1972b). *Physiol. Pl. Path.* **2**, 133–142.
Sadowski, P. and Strobel, G. A. (1973). *J. Bact.* **115**, 668–672.
Salemink, C. A., Rebel, H., Kerline, L. C. P. and Tschernoff, V. (1965). *Science* **149**, 202–203.
Samaddar, K. R. and Scheffer, R. P. (1968). *Pl. Physiol.* **43**, 21–28.
Samaddar, K. R. and Scheffer, R. P. (1971). *Physiol. Pl. Path.* **1**, 319–328.
Scheffer, R. P. and Samaddar, K. R. (1970). *Recent Advances in Phytochem.* **3**, 123–142.
Sinden, S. L. and Durbin, R. D. (1970). *Phytopathology* **60**, 360–368.
Sinden, S. L., DeVay, J. E. and Backman, P. A. (1971). *Physiol. Pl. Path.* **1**, 199–213.
Smedgård-Petersen, V. and Nelson, R. R. (1969). *Can. J. Bot.* **47**, 951–957.
Starratt, A. N. (1968). *Can. J. Chem.* **46**, 767–770.
Steiner, G. W. and Byther, R. S. (1971). *Phytopathology* **61**, 691–696.
Steiner, G. W. and Strobel, G. A. (1971). *J. Biol. Chem.* **246**, 4350–4357.
Stewart, W. W. (1971). *Nature* **229**, 174–178.
Straley, C. S., Straley, M. and Strobel, G. A. (1974). *Phytopathology* **64**, 194–196.
Strobel, G. A. (1973a). *J. biol. Chem.* **248**, 1321–1328.
Strobel, G. A. (1973b). *Proc. Natn. Acad. Sci., U.S.A.* **70**, 1693–1696.
Strobel, G. A. (1974). *A. Rev. Pl. Phys.* **25**, 541–566.
Strobel, G. A. (1974). *Proc. Natn. Acad. Sci., U.S.A.* **71**, 4232–4236.
Strobel, G. A. and Hapner, K. (1975). *Biochem. biophys. Res. Commun.* **63**, 1151–1156.
Strobel, G. A. and Hess, W. M. (1974). *Proc. Natn. Acad. Sci., U.S.A.* **71**, 1413–1417.
Strobel, G. A. and Steiner, G. W. (1972). *Physiol. Pl. Path.* **2**, 129–132.
Strobel, G. A., Hess, W. M. and Steiner, G. W. (1972a). *Phytopathology* **62**, 339–345.
Strobel, G. A., Talmadge, K. W. and Albersheim, P. (1972b). *Biochim. biophys. Acta* **261**, 365–374.
Strobel, G. A., Steiner, G. W. and Byther, R. S. (1975). *Biochim. Genetics* **13**, 557–565.
Sugiyama, N., Kashima, C., Yamamoto, M., Sugaya, T. and Mohri, R. (1966). *Bull. Chem. Soc. Japan* **39**, 1573–1577.
Sugiyama, N., Kashima, C., Yamamoto, M. and Mohri, R. (1967). *Bull. Chem Soc. Japan* **40**, 345–346.
Tanaka, S. (1933). *Mem. Coll. Agric. Kyoto Imp. Univ.* **28**, 1–31.
Tatum, L. A. (1971). *Science* **171**, 1113–1115.
Taylor, P. A., Schones, H. K. and Durbin, R. D. (1972). *Biochim. biophys. Acta.* **286**, 107–117.
Templeton, G. E. (1972). *In* "Microbial Toxins" (S. Kadis, A. Ciegler and S. J. Ajl, eds), Vol. VIII. Academic Press, New York and London.
Templeton, G. E., Grable, C. I., Fulton, N. D. and Bollenbacher, K. (1967). *Phytopathology* **57**, 516–518.
Tipton, C. L., Mondal, M. H. and Uhlig, J. (1973). *Biochem. biophys. Res. Commun.* **51**, 725–728.
Tirokata, H., Ohkawa, M., Sassa, T., Yamada, T., Ohkawa, H.,
Tanaka, H. and Aoki, H. (1969). *Ann. phytopath. Soc. Japan* **35**, 62–66.
Turner, N. C. (1972). *Nature* **235**, 341–342.
Van Alfen, N. K. and Turner, N. C. (1975). *Pl. Physiol.* **55**, 312–316.
Van Sambeck, J. W., Novacky, A. and Karr, A. L. (1975). *Pl. Physiol.* (*Suppl.*) **56**, 53.

Watrud, L. S., Baldwin, J. K., Miller, R. J. and Koeppe, D. E. (1975). *Pl. Physiol.* **56**, 216–221.

Wheeler, H. and Black, H. S. (1963). *Am. J. Bot.* **50**, 686–693.

Wheeler, H. and Luke, H. (1954). *Phytopathology* **44**, 334 (*Abstr.*).

Wheeler, H. and Luke, H. (1963). *A. Rev. Microbiol.* **17**, 223–242.

Wood, R. K. S. (1972). *In* "Phytotoxins in Plant Diseases", (A. Ballio, and A. Graniti, eds), pp. 530. Academic Press, New York and London.

Yoder, O. C. (1973). *Phytopathology* **63**, 1361–1365.

Yoder, O. C. and Scheffer, R. P. (1973). *Pl. Physiol.* **52**, 513–517.

Yoder, O. C. and Scheffer, R. P. (1973). *Pl. Physiol.* **52**, 518–523.

CHAPTER 9

Structural Features of Resistance to Plant Diseases

D. J. ROYLE

Department of Hop Research, Wye College (University of London), near Ashford, Kent, England

I. INTRODUCTION

As long ago as 1892, Cobb proposed a "mechanical theory" of rust resistance which stated that morphological characters, such as a thick cuticle, a waxy covering, small stomata, a large number of leaf hairs and upright leaves, might be responsible for the resistance of some wheat varieties to *Puccinia graminis*. Many supported his theory in relation to both stem rust and other diseases, and during the next 30–40 years morphological features associated with the plant's defence against pathogens were investigated widely and the results included in a number of general reviews, e.g. Brown (1936) and Hart (1949).

These early studies were carried out at a time when there was also an active interest in the physiology of fungal penetration, and structural defence was largely conceived against those pathogens which were thought to enter their hosts by mechanical rather than chemical means. General explanations for resistance to infection were sought among single plant characters, mostly simple and morphological in nature. For instance, there were many claims, based largely on demonstrations of correlation, that thickness of the plant's surface layers affected pathogen entry, and there were also many others denying such a relationship. With a growing awareness that resistance operates most commonly after pathogen entry and that it cannot be explained completely by simple mechanisms, interest in structural barriers then dwindled and during the last 25 years or so systematic research on the subject has been rare, having been replaced largely by efforts to understand the chemical basis of resistance. Though aspects of structural resistance have been discussed in recent times by Dickinson (1960), Rich (1963), Martin (1964) and Martin and Juniper (1970), the entire subject has been reviewed only by Akai (1959) and Wood (1967) and briefly in textbooks on general plant pathology, e.g. Stakman and Harrar (1959), Agrios (1969), Walker (1969) and Tarr (1972). This lack of attention is understandable since nowadays many new claims for the involvement of a structural barrier in resistance are often only casually made and not substantiated by adequate experimental study. As a result, we have inherited much incomplete evidence with few well-researched examples which provide conclusions. This situation undoubtedly contributes to the present-day view, widely held, that structural elements are of little real consequence and that resistance is based largely on physiochemical interactions within the host–parasite system. While there is certainly insufficient evidence at the present time to believe other-wise, this review will try to show that, in several areas, there are good reasons to suspect that mechanisms based solely on plant structure can contribute usefully to overall disease resistance and that there is justification for renewed efforts to explore structural mechanisms, to assess their value and then perhaps to exploit them in breeding programmes.

It is first necessary to define the meaning and scope of structural resistance so that the choice and analysis of examples may be understood. "Resistance" is an attribute of the host and, though it may be detected early in the infection process, should ultimately be judged by the expression of disease. While immunity is a complete and absolute host response, resistance is partial and relative (Van der Plank, 1975). Against the few pathogens which are able to attack it, a host plant shows varying degrees of resistance or susceptibility which can relate to its genotype, to its component organs and to time, either in the long-term, as affected by age of tissue, or in the relatively short-term, as influenced by the host's diurnal activity, e.g. stomatal function-ing. A plant may be termed "resistant" to a disease only from knowledge of its response to challenge by the pathogen; it may then resist disease by preventing the pathogen from retaining contact with its surface or by restricting the pathogen's entry and subsequent development. Disease escape (klendusity) is not a part of resistance, and is therefore of no interest in this account, since it evokes an absence of challenge, i.e. "resistance" is not put to the test (Wood, 1967).

Structural features may operate in resistance passively, as preformed barriers characteristic of the healthy, unchallenged host, or actively, when they are induced by the pathogen's attempts to cause disease. Akai (1959) has discussed in detail the induced structures, e.g. cork and lignin formation, callosities, which by definition involve chemical interaction and therefore may not act purely mechanically. Of course, any form of structural resistance inevitably implicates both physical and chemical attributes between which it is often difficult to discriminate. Nevertheless this discussion will emphasize the physical qualities of host plant structure and attempt to show their relationships to some of the chemical properties. For these reasons, and because the subject of structural defence is so large, preformed features of disease resistance only are considered.

II. INOCULUM DEPOSITION

The first host barrier to invasion by a pathogen may be presented by structural characteristics of its surface which influence the arrival and retention of inoculum. To be deposited effectively, water-borne inoculum requires a surface which can be wetted with relative ease. Since plant wettability is governed largely by its superficial wax deposit or hairiness (Martin and Juniper, 1970), many workers have considered that these characters contribute to disease avoidance by affecting the ease with which pathogens can contact host surfaces. There is, however, only limited experimental evidence to support this view, probably in part because characters controlling wettability can vary widely within host genotypes, organs and ages, between different positions on the plant and with environment. The available evidence derives mostly from observations of an apparent association between increased waxiness (often recorded as "blooms") or hairiness and reduced disease, usually among plants in the field. Freeman (1911), for example, observed that water drops carrying spores of *Puccinia hordei* rolled off waxy foliage of barley. Fewer rust pustules arose on those plants on which a bloom had been induced by growing in soils containing up to 2% alkaline salts than on others without a bloom, growing in the absence of salts. Johnstone (1931) noted that waxy blooms on leaves and fruit of some apple varieties seemed to affect infection by the scab fungus, *Venturia inaequalis*. Waxy leaves tended not to become infected while wax on the fruit caused infection droplets to collect at the eye or on the lower surface where infection was enhanced. Scab resistance was also associated with leaf hairiness which seemed to prevent water from reaching the cuticle, a relationship only important when it supplemented tissue resistance. Wilson and Jarvis (1963) attributed greater infection of white blister, *Albugo candida*, on leaves of mutant forms of Brussels sprout, to the absence in the mutant of surface wax which occurred as plates and tubes on leaves of normal glaucous plants. Inheritance of the "waxless" character in wheat is known to be due to a single dominant gene and Jensen and Driscoll (1963) noticed that stem and leaf diseases generally appeared more severe on non-waxy than on waxy varieties growing in the field. This was substantiated by Troughton and Hall (1967) who recorded wheat varieties with the lowest contact angles (measures of wettability of a solid by a liquid) as being the most susceptible to leaf rust (*Puccinia recondita*) in the field. Apart from this latter work, relationships between extracuticular wax or hairs and the manner and extent of water droplet accumulation were not investigated in any of the above examples.

One of the few concerted attempts so far to assess the importance of external plant morphology to inoculum deposition was made by Jennings (1962) with the raspberry cane diseases grey mould (*Botrytis cinerea*), spur blight (*Didymella applanata*) and cane spot (*Elsinoe veneta*). In this example, cane characters which showed wide inter-varietal variation—hairiness, spininess and the thickness of the waxy bloom—were studied in relation to the liability of seedling varieties to become infected. A bloom, or bluish colouration of the plant surface, was first described by Cobb (1892) and is caused by the light scattering properties of epicuticular wax. However, it is important to note that blooms (or glaucousness) refer to the visual appearance of the plant surface and do not necessarily indicate heavy wax deposits which, conversely, may fail to cause blooms (Martin and Juniper, 1970). The bloom on raspberry canes is now known not to correspond to the amount and composition of wax (Baker *et al.*, 1964) but Jennings judged the presence, absence and degree of waxy bloom visually without relating to wax deposit so caution is needed in interpreting some of his results. Observations on disease incidence in the field were made on about 120 raspberry seedlings in each of 31 families arranged in five groups according to parentage. Except for the degree of wax, the pairs of alternative characters which segregated in the families (Table I) could be attributed to segregation of a major gene. Table I summarizes detailed records of disease severity which were fairly consistent in relation to each cane character. Spur blight was more severe on non-hairy (*h*), spiny (*S*) and waxy (*B*) canes, grey mould behaved similarly except that it was favoured by the absence of wax (*b*),

TABLE I

Cane morphological characters which allowed the greatest incidence of raspberry cane diseases[a] (from data of Jennings, 1962)

Alternative cane characters	Spur blight	Grey mould	Cane spot
Hairy (*H*) and non-hairy (*h*) canes	*h*	*h*	*H*
Spiny (*S*) and spine-free (*s*) canes	*S*	*S*	*S*
Waxy (*B*) and non-waxy (*b*) canes	*B*	(*b*)[b]	*b*
Moderately waxed (B +) and sparsely waxed (*B*) canes	*B*	(*B*)	*B*

[a] Based on data of 2 years for spur blight and cane spot, and 1 year for grey mould.
[b] Parentheses indicate only a slight difference in effect of alternative cane characters.

while cane spot increased on hairy (*H*) canes and on those with no waxy bloom. Cane spot was affected similarly to the other diseases by spine status and wax thickness. Three years' data for spur blight showed that the effectiveness of the cane characters in influencing susceptibility to infection varied between seasons. Jennings then tried to find explanation for these observations by carrying out experiments to see how wettability of the cane surfaces could be varied by the morphological characters. Lengths of cane were dipped in an aqueous dye and allowed to dry in a vertical position. The amount of staining on the nodes and internodes was then related to phenotype. Cane hairiness dispersed water at the nodes and allowed very little to adhere so it appeared that hairs could effectively remove the conditions required at the nodes for fungal infection. However, this property of the hairs was offset by the presence of wax which encouraged the accumulation of water droplets on the internodes and impeded run-off from the nodes irrespective of whether they were hairy or not. The incidence of spur blight, the only disease whose symptoms are restricted to nodes, could therefore be explained in terms of waxiness and hairiness. Reduction of all diseases from a denser wax covering (B+) was thought most likely to be due to other wax properties, possibly by acting as a mechanical barrier to penetration or by being fungistatic. The presence of spines seemed to augment the activity of wax by impeding water run-off, though this was not a consistent effect.

Three general points arise from Jennings' work that relate also to evidence for structural resistance mechanisms in other examples to be considered. First, as emphasized also by Wood (1967), evidence of correlation between disease severity and the morphological cane characters is not proof of causal relationships. Though the results of the experiment gave plausible explanations for the observations, a close inspection of the influence of interactions between the cane characters and water accumulation on the raspberry pathogens is nevertheless needed to explain the findings with complete confidence. Second, Jennings notes that although a form of structural resistance, at least to spur blight and grey mould, appears to be conferred by the cane characters, a possible though unlikely alternative explanation of the results might be that genes for tissue resistance are closely linked with those controlling cane morphology. This hypothesis could conceivably be applied to cane spot rather than the other diseases because its severity varied more between the families than within them. In addition, fungistatic properties of the cane wax may possibly be implicated. These considerations raise the practical issue, noted in the introduction, of the difficulties in distinguishing between purely physical and intrinsic physio-chemical mechanisms of resistance because of their co-existence. These difficulties can be encountered not only where physical barriers and chemical factors are spatially separated in the plant but also where they are components of the same structural barrier. Third, from his conclusions Jennings advocates that the gene for hairiness should be bred into new varieties but that since its benefits are variable according to season, it should be augmented with the gene for spinelessness plus any genes known to confer tissue resistance. Selection for heavily waxed canes was not to be recommended until the adverse effects of cane hair properties had been further researched. This emphasizes the possible value to the plant breeder of defined structural mechanisms which, providing they make a significant contribution to the total level of resistance, can help to achieve durability in resistance.

Two other examples also illustrate ways in which the water-shedding properties of surface wax appear to affect host resistance. Heather (1967a,b) investigated a disease of *Eucalyptus bicostata* leaves caused by *Phaeoseptoria eucalypti* which on susceptible old leaves produces local necrotic lesions. Young leaves are resistant and very hydrophobic however, apparently due to a double layer of wax-rods and plates over the cuticle which are probably removed by weathering as leaves age. The number of water-dispersed spores of *P. eucalypti* deposited on the leaves was inversely related to the amount of leaf wax. When the wax of young leaves was removed by wiping with cotton wool their susceptibility greatly increased. Even though a constituent of the wax inhibited germination of the spores *in vitro*, its hydrophobic property was considered to be more important in conferring substantial survival value to populations of the host plant in the field. Different angles of orientation of leaves of different age appeared to be a secondary factor in their degree of resistance.

Fisher and Corke (1971) attempted to explain why, in cider apple orchards, the red, exposed surface of the fruit of a particular variety appeared to be more susceptible to scab than the green, sheltered surface. Conidia of the imperfect stage (*Fusicladium dendriticum*) of the scab fungus were shown by experiment to be twice as likely to cause infection of the red as of the green fruit surface. This was attributed partly to differences in the extent to which water drops spread over the two surfaces which in turn related to the content and structure of surface waxes. Wettability was measured by the method of Silva Fernandes (1965) which determines the concentration of a wetting agent needed to enable a water drop to flow over the surface leaving a straight-sided trace. The red surface was easily wetted (0–0·005% wetter needed), but to wet the green surface equally well required 0·02–0·05% wetter. There was a relationship between scab infection on the fruits used in the experiment and the ease of wetting of the red surface (Table II). Analysis of the cuticles of each skin type showed the red skin to contain less surface and embedded wax and to have cuticles lighter in weight than the green skin. The wax structure also differed, for the green surface was thickly covered with overlapping wax platelets forming an irregular surface, while the red surface was smooth with fused platelets forming a continuous layer. Water repellancy appeared to be only one factor explaining differential susceptibility as the two surfaces also differed in their nitrogen content,

TABLE II

Concentration of wetter (Manoxol OT) needed to wet opposite surfaces of apple fruit, and number of red surfaces naturally infected with scab (Fisher and Corke, 1971)

	Wetter concentration (%)						
	0	0·005	0·01	0·02	0·03	0·04	0·05
% wetted on:							
green side	0	3	3	16	53	24	1
red side	51	11	10	15	9	4	0
% infected on:							
red side	70	18	10	2	0	0	0

their effects on the behaviour of conidial germ tubes, and the antifungal activity of impure cuticle fractions.

Both the capacity of the inoculum itself to be wetted and the effect of spores on the surface tension of water drops containing them, can influence the degree to which plant surface wettability contributes to disease resistance. The movement and deposition of spores on a surface is markedly affected by their wettability, for Davies (1961) demonstrated that when non-wettable spores are collected in drops of water they are deposited on inclined hydrophobic surfaces over which the drops are allowed to pass. Wettable spores, on the other hand, are carried away with water drops and are deposited only when the drops come to rest. Apart from the study of Fisher and Corke, no other examination of host surface wettability in relation to disease resistance has taken the wettability of the pathogen into account. Of interest here are some observations of Burrage (1969) who found that non-wettable uredospores of *P. graminis* adhere to the surfaces of water droplets on wheat leaves and then produce germ tubes which grow over the leaf between the droplets, not within them. Since infection occurs via stomata it is favoured by dew which condenses in many discrete droplets along the vein ridges leaving stomata dry. A continuous water film, as from rain, allows fewer spores to be retained on the leaf and depresses germ tube activity. This knowledge might be of benefit in resistance breeding if inheritable characters which favour the coalescence of dew drops can be found.

The presence of wettable spores of some pathogens in water drops lowers the surface tension of the drops which then behave as if the surfaces over which they pass were less hydrophobic (Corke, 1966). This effect contributed to the differential susceptibility of opposite apple fruit surfaces in the study of Fisher and Corke (1971) described above. A concentration of 1×10^5 conidia/ml of the apple canker pathogen, *Gloeosporium perennans*, was needed to wet the fruit surface to the same extent as a 0·01% solution of wetting agent, whereas the same concentration of scab conidia was equivalent to only 0·005% wetter; a concentration of 7×10^6 conidia/ml of the scab fungus was required to equal 0·01% wetting agent in its wetting ability. Therefore, the red surface of 62% of the fruit examined in Fisher and Corke's tests could have been wetted by drops containing 1×10^5 scab conidia/ml but only 3% of the fruit would have been wetted by similar drops on the green surface (Table II).

As Dickinson (1960) has commented, there appears to be no general evidence for assuming that hairs cause pathogens to exhaust their food supply before reaching a penetrable host surface nor that they create an unsuitable microclimate for pathogen development. However, we have already noted the findings of Jennings (1962) and there are a few other indications in the literature that plant hairs or spines may occasionally affect host resistance other than by repelling water drops. Hursh (1924) recorded fewer appressoria of *P. graminis* on varieties of wheat with a greater number of hairs and thought that when inoculum was at a low concentration the hairs might form a physical barrier to the growth of germ tubes, reducing the number reaching stomata. An alternative explanation was proposed by Burrage (1969) who suggested that an increased number of hairs might allow dew to cover the surface of leaves more efficiently thus providing an effective barrier to rust germ tubes which, as already noted, prefer dry surfaces. Liability of hairs or spines to break may occasionally increase the chances of fungal invasion as Smith (1936) has described for the infection of peach fruit by *Sclerotinia fructicola*. Pubescence on the fruit retains dirt and spray residues so brushing is required to make the fruit more attractive for marketing. This breaks many of the hairs which reduces by half the time needed for infection via the hair sockets. Rapidity of infection is thereby correlated with the relative accessibility of the fruit surface to germinating conidia.

We have already briefly mentioned instances in which chemical as well as physical properties of surface wax can affect the response of a pathogen to its host plant. Similarly, a possible chemical contribution to resistance from plant hairs cannot be disregarded since Beckman *et al.* (1972) showed that phenols are localized in some hairs and released in response to external stimuli. Some substances originating in plant hairs have been shown by Bailey *et al.* (1974) to be inhibitory to fungal growth.

Other mechanisms of disease avoidance are characteristic of host plants growing naturally in the field and can be attributed to variation of form and function in their gross habits. Plant stature and stage of maturity have been shown by Scott (1973) to influence susceptibility of wheat to *Septoria nodorum*. On about 20 cultivars in two years' field trials severity of infection was positively correlated with both short stature and earliness in maturity. The association with shortness was believed to be one of cause and effect since on a short cultivar, flag leaves and ears were nearer to the inoculum source on the ground and in the zone of higher relative humidity than on a tall cultivar. The effect of maturity on infection was less easy to explain, but probably due partly to a tendency for short cultivars to mature early. Leaf orientation is another factor which may vary sufficiently to influence inoculum deposition (Heather, 1967a; Russell, 1975) and investigation will undoubtedly reveal other examples where variation in plant form affects the efficiency of spore impaction, inoculum dispersal gradients etc. In most cases this type of "resistance" may simply determine whether the susceptibility level is achieved by large numbers of relatively small lesions or small numbers of large, highly productive lesions. Under low inoculum levels resistant individuals within plantings of multiline varieties may act effectively as a mechanical barrier to inoculum dispersal, restricting the amount of inoculum deposited on susceptible individuals and reducing the overall rate of disease increase within the population (Browning and Frey, 1969).

There is one notable demonstration of how field resistance among host genotypes can be affected primarily by differences in the extent to which particular host sites required for infection are exposed to the pathogen. Malik and Batts (1960) showed that spores of *Ustilago nuda* were prevented from reaching the ovaries of field-resistant barley varieties by closed glumes at anthesis. Over several years the reaction of varieties to loose smut had proved difficult to determine, since discrepancies had arisen in the disease levels obtained after natural infection in the field and when inoculated artifically by injecting spores, an experience not encountered with *U. tritici* on wheat varieties. Physiological resistance was known not to be implicated in the range of varieties used and a form of disease avoidance operative only in the field was suspected. Appel (1915) and Fromme (1920) had previously suggested that in some barley varieties closed glumes at anthesis prevented access of smut spores to the ovaries. In contrast, Tapke (1929) had found that glumes opened simultaneously in susceptible and resistant wheats. Malik and Batts allowed drops of dyed water or spore suspension to fall on the ears of six barley varieties at anthesis and established that the ovaries of field-resistant varieties did not become coloured nor infected whereas those of field-susceptible varieties did. They hypothesized that closed glumes prevented the pathogen from reaching the ovaries thus conferring resistance in the field. This agreed with observations that in wet years, loose smut is more common and glumes stay open for longer periods than in dry years. Artificial inoculation was therefore of value for screening physiological resistance but gave no indication of field performance in which ear structure first determined whether or not physiological factors could operate. There is some evidence that a similar mechanism might help to control infection of wheat and barley by the ergot pathogen, *Claviceps purpurea* (Puranik and Mathre, 1971). Though this structural mechanism is peculiar to these diseases, we might wonder as to the existence of mechanisms based on similar principles in other diseases and how much useful resistance, structural or otherwise, might be missed during screening procedures that are essentially artificial.

III. ENTRY OF THE PATHOGEN

A. DIRECT ENTRY

The importance of the plant's intact surface as a barrier to wound pathogens is of course not doubted, but whether a physical or chemical role exists for the outer surface in resistance to entry of directly penetrating pathogens has long been argued. In this area, the great majority of evidence for or against the participation of plant structure in resistance relates to aerial fungal pathogens which gain access to herbaceous host tissues by direct penetration of a cuticularized epidermal cell wall. Casual claims or denials of host surface structure affecting the entry of airborne pathogens abound in the literature. To refer to just a few, Ainsworth *et al.* (1938) noticed that increased size of tomato fruits was accompanied by greater resistance to *Botrytis cinerea* which in turn could be correlated with thickness of the skin, and Louis (1963) showed that cuticle thickness in bean, tomato and other hosts related to the ability of *B. cinerea* to penetrate. In contrast, Berry (1959) found no association between the thickness of the outer epidermal wall in onion and resistance to *Peronospora destructor*, and Johnstone (1931) considered that *Venturia inaequalis* is able to penetrate cuticles of all thicknesses encountered on apple fruits. Perhaps unfortunately, general judgement of the participation of structural plant features in disease resistance tends to be based mainly on evidence regarding resistance of the outer surface layers to direct entry. Much of this work has not distinguished clearly between effects of the cuticle and of the outer epidermal cell wall and quite often further confusion has arisen because of possible participation also of the entire epidermal cell layer. The role of the cuticle in plant disease has been comprehensively reviewed by Martin (1964) who concluded that it provides no serious barrier to penetration and that its contribution, whether physical or chemical, to resistance cannot be great. This view remained unchanged six years later (Martin and Juniper, 1970) and seems to be justified as it is based not only on the available evidence on cuticular penetration by pathogens but also on substantial knowledge of the organization, function and chemistry of the cuticle, described in detail by Martin in both of his excellent works. This account will therefore take a fresh look at the resistance of the surface layers to direct entry and consider the most recent evidence against a background of some of the older examples mentioned only briefly by Martin.

As pointed out in the introduction, the concept of a structural role in defence for the plant's surface layers has in the past paralleled that of the pathogen penetrating mechanically and it is difficult to imagine that physical properties of the outer cell wall operate in resistance unless entry involves mechanical force. It is inappropriate here to discuss the nature of fungal penetration, (see Wood, 1967) but, as we shall see, instances are rare in which suggestions of a physical relationship between the epidermal wall components and resistance are related to the manner by which entry is likely to occur.

It has long been recognized that fungi usually penetrate young, tender parts more easily than older, tougher parts of their hosts, so it is not surprising to find that many workers have tried to measure resistance using artificial needle punctures to represent fungal penetration. They reasoned that the ease with which various plant parts could be pierced might decrease with age and be a convenient index of the thickness of the plant's surface and therefore of resistance to infection. Measurements of the force needed to puncture the outer cell wall were obtained with a modified Joly balance fitted with a round-pointed needle, often 20–75 μm diameter, much larger than a fungal infection peg. Thus Rosenbaum and Sando (1920) observed that as tomato fruits mature, the cuticle thickness increases and infection by *Macrosporium tomato* is reduced. A correlation between age and resistance of the skin to puncture was then demonstrated (Table III). Willaman *et al.* (1925) and Willaman (1926) related the level of resistance to brown rot (*Sclerotinia cinerea*) of 11 plum varieties at 6 stages of ripeness to some chemical and physical characteristics of the fruit. Increasing ripeness was accompanied by easier needle puncturing of the fruit particularly among susceptible varieties, and this was consistent with the observed resistance to brown rot. Other factors such as firmness of flesh, the content of crude fibre and pentosan, were also related to the level of resistance. Curtis (1928)

TABLE III

Relation between age, resistance to puncture and infection of tomato fruits by *M. tomato* (Rosenbaum and Sando, 1920)

Age (days)	Pressure (g) needed to puncture skin of the fruit	% fruit infected with *M. tomato*
7	0·97	100
14	2·99-	100
21	4·21	85
28	4·90	49
35	5·08	23
41	5·96	0
48	6·74	0
55	5·56	0

later remarked that when the cuticle resists penetration, as in unripe fruit, *S. cinerea* finds entry through stomata. How skin toughness participates in plum resistance was therefore not clearly established.

The work of Melander and Craigie (1927) is often used as a model example of studies on resistance to needle puncture. It is also sometimes quoted in support of a structural resistance though it offers evidence only of correlation, as in the above examples. In this study, leaves of *Berberis* and *Odostemon* spp. were examined for their resistance to infection by basidiospores of *P. graminis*. Leaf entry was believed to be by mechanical force and occurred readily in susceptible and resistant but not in immune species. The combined thickness of the cuticle and outer epidermal wall of susceptible species was less than that of immune plants and increased more substantially with age in leaves of increasing resistance (Table IV). Species which were resistant or immune to rust were also very resistant to needle puncture. However, easily punctured leaves were not always susceptible to rust so while resistance to puncture might indicate a morphological defence mechanism, ease of puncture does not necessarily indicate susceptibility and physiological factors seemed also to be implicated. Melander and Craigie were also aware that resistance to mechanical penetration of the cuticularized cell walls cannot be judged precisely by needle puncture methods since other factors including the shape, size and possibly turgor of the cells and thickness of the radial walls all have an influence. For this reason results of needle puncture tests are perhaps better interpreted as measures of the resistance of

TABLE IV

Average combined thickness (μm) of the outer epidermal cell wall and cuticle of leaves of certain *Berberis* and *Odostemon* spp. susceptible, resistant or immune to *P. graminis* (Melander and Craigie, 1927)

Species	2 to 3-day-old leaves	5 to 6-day-old leaves	Mature leaves
Susceptible:			
B. canadensis	0·88	0·93	1·29
B. dictyophylla	0·82	1·23	1·80
B. vulgaris	1·10	1·18	1·87
Resistant:			
B. brachypoda	1·43	2·09	2·56
B. lycium	1·23	2·86	3·41
B. pruinosa	1·16	1·46	2·20
Immune:			
B. thunbergii	1·57	1·62	2·44
O. repens	1·75	—	3·01

tissues to forceful penetration, as in the work of Hawkins and Harvey (1919) to be discussed in section IV.

The approach in which correlation is sought between the level of infection and thickness of the outer wall layers has been pursued quite extensively with the powdery mildews whose ectotrophic habit makes them obvious targets for studies of resistance mechanisms at the penetration level. Compared to other diseases the powdery mildews appear to provide a unique inter-acting system, but as will become apparent, no general rules for resistance to penetration emerge in the group and interactions between the fungus and the cuticular surface can be complex. It is probably significant that where evidence exists for a structural mechanism of resistance to a powdery mildew fungus, penetration is believed to be mechanical and, conversely, where no structural involvement can be demonstrated the cuticle appears to be degraded enzymically.

Resistance of strawberry varieties to *Sphaerotheca macularis* was shown by both Peries (1962) and Jhooty and McKeen (1965) to vary with cuticle thickness. These workers found reason to believe that *S. macularis* enters the epidermis by mechanical means. Peries noted that 40–60% germinating conidia established infections on susceptible varieties while less than 1% were successful on resistant hosts. Of cuticle constituents from susceptible and resistant varieties, given amounts of cutin acids and the absorbed (not surface) wax fraction suppressed conidial germination. There was clearly insufficient evidence to assign a greater importance to physical or chemical properties of the cuticle in resistance, but of possible significance is Peries' findings that, of the various cuticle components, only the quantity of cutin acids (the hydroxy-fatty acids derived from the cutin which confers hardness to the cuticle) varied in proportion to the susceptibility of straw-berry leaves. Jhooty and McKeen made measurements of cuticle thickness after observing that, when grown in the same conditions, both surfaces of the young leaves of *Fragaria ovalis* were equally susceptible while only the lower surface of *F. chiloensis* was susceptible, its upper surface rarely being infected. The upper cuticle of *F. chiloensis* (4·0 mµ, range 2·88–5·12 mµ) was almost seven times as thick as that of *F. ovalis* (0·6 mµ, range 0·32–0·98 mµ) and was also apparently thicker than the lower cuticle, though no other measurements were offered. Cuticles of stolons of the two species dif-fered less in thickness despite wide variation in their susceptibility to *S. macularis*. Osmotic differences within epidermal cells, claimed by Schnathorst (1959) to be a basis for resistance of lettuce to powdery mildew (*Erysiphe cichoracearum*), were considered unlikely to explain differences in suscepti-bility between surfaces of the same leaf and so, like Peries, these workers pro-posed that cuticle thickness may explain resistance to *S. macularis*.

The dangers of interpreting a correlation between cell wall thickness and infection, which is found in a single set of environmental conditions, as an explanation of differential susceptibility is well illustrated in the work of

Weinhold and English (1964) with *S. pannosa* on peach. Young leaves are susceptible but they become highly resistant as they mature. The combined thickness of cuticle and epidermal cell wall correlated well with leaf age but showed no differences in leaves of comparable age between two peach varieties (Table V). However, mature leaf resistance could not be explained by the thickness of the cuticularized cell wall, first, because mature, resistant leaves became susceptible when plants were transferred to darkness even though thickness of the wall remained unchanged, and second, because rudimentary haustoria were observed in epidermal cells of resistant leaves.

TABLE V

Average combined thickness (μm) of the outer epidermal cell wall and cuticle, and susceptibility to *S. pannosa* of leaves of two peach varieties and of various age represented by leaf position (Weinhold and English, 1964)

Leaf position	Susceptible variety		Resistant variety	
	Cuticularized wall thickness	Mildew susceptibility	Cuticularized wall thickness	Mildew susceptibility
Apex	3·4[a]	+ +	3·0[a]	+ (weak)
Node 1	3·5	+ +	3·3	+ (weak)
2	4·1	+	3·6	−
3	4·5	−	4·0	−
4	4·0	−	3·9	−
5	4·4	−	—	—

[a] Mean of 20 measurements with confidence limits ± 0·2–0·4 (at 5% level).

S. pannosa is also a pathogen of rose whose epidermis it enters enzymically (Caporali, 1960). As with peach, resistance increases with leaf age and is paralleled by an increase in the combined thickness of the epidermal wall and cuticle (Mence and Hildebrandt, 1966). Haustoria were observed in the cells of resistant leaves and so these workers concluded that penetration was not prevented by the cuticle acting as a morphological barrier. Roberts *et al.* (1961) supported this conclusion since they were unable to relate quantities of waxes and cutin in the cuticles of healthy rose varieties to different levels of susceptibility.

Penetration of their hosts by *Podosphaera leucotricha, Erysiphe cichoracearum* and *E. graminis* (Woodward, 1927; Lupton, 1956; Staub *et al.*, 1974) also seems to be enzymic and thickness of the cuticularized cell wall appears not to be implicated in resistance. Though influenced by leaf age, infection of apple by *P. leucotricha* (Burchill, 1958; Roberts *et al.*, 1961) and of lettuce by *E. cichoracearum* (Schnathorst, 1959) bears no relation to thickness or composition of the outer wall layers. Both Mackie (1928) and Smith and Blair (1950) dismissed simple morphological factors as contributing to resistance of wheat and barley varieties to *E. graminis*, though a different aspect of structural defence by the cuticle against this pathogen was suggested by Ellingboe (1972) and Yang and Ellingboe (1972). The fungus can produce normal and malformed haustoria but infection only proceeds from normal appressoria whose formation is stimulated by some particular physical properties of the cuticle surface. Functional appressoria could be produced on enzymically isolated cuticles so their stimulation by chemicals diffusing from epidermal cells seemed to be a remote possibility. Furthermore, differences in the proportions of normal and malformed appressoria on cuticles isolated from upper and lower leaf surfaces suggested that possibly the superficial wax layer of the cuticle was primarily responsible. This was investigated more deeply using five barley mutants in which the components, amount and physical structure of the epicuticular wax layer were modified to a known extent. The results indicated that appressorial normality depended principally upon the spatial distribution of wax bodies; the more homogeneous their distribution the greater the stimulation of normal appressoria. Additional evidence was obtained after reconstructed cuticular surfaces failed to encourage appressorial formation to the same extent as intact leaves or isolated cuticles. Interestingly, Yang and Ellingboe foresaw that, since wax body formation was affected markedly by environmental conditions, the wax layer could conceivably play a role in field resistance to the disease. Evidence which conflicts with some of these results was recently reported by Staub *et al.* (1974) who found that functional appressoria of *E. graminis* formed on both host and non-host (cucumber) leaves, suggesting that specific wax configurations may not be necessary. However, these workers recognized a possible role for epicuticular wax structure in the immunity of cucumber to *E. graminis* and also for fungistatic and stimulatory substances in the wax. Clearly, further work is needed to clarify the precise contribution of wax configuration to infection by *E. graminis* but these studies nevertheless emphasize that mere thickness of the composite cuticular layer may be too simple a criterion to reflect a structural involvement in infection by the powdery mildews. If this is so, then none of the attempts to relate cuticle thickness to resistance in powdery mildews have come near to solving the problem.

The ways in which chemical activity of cuticle and cell wall constituents can confuse evaluation of the part physical factors play in resistance have briefly been referred to. There is ample evidence that substances both inhibitory and stimulatory to fungal activity are exuded into infection drops on the host surface (Martin and Juniper, 1970), and some cuticle components, e.g. cutin acids, absorbed and superficial waxes, are known to influence germination of powdery mildew conidia on dry surfaces (Martin *et al.*, 1957; Peries, 1962; Roberts *et al.*, 1961). These effects tend to be non-specific, cuticle constituents from susceptible and resistant varieties, old and young tissues and even host and non-host plants stimulating or suppressing activity of a pathogen in a similar manner. As factors in resistance, these substances are possibly implicated to the extent with which their concentration varies in proportion to the susceptibility level of the tissue from which they are derived. This concentration may of course be closely related to the thickness or density of the outer wall layers. Martin and Juniper review the increasing evidence that unsaturated fatty acids, which occur in leaf waxes, are involved in mechanisms of disease resistance.

There are two other examples in which both physical and chemical elements of the host's surface layers may be involved in resistance to direct entry. Nutman and Roberts (1960) considered that the ease with which the infection peg penetrates the intact cuticle of coffee berries was the sole reason for differences in susceptibility between coffee varieties to *Colletotrichum coffeanum*. This pathogen can colonize equally well berry tissues of field resistant and highly susceptible coffee varieties once it has been allowed entry by wounding. Variation in varietal susceptibility was ascribed to some unknown chemical or physical differences in the cuticle which made penetration more difficult in the resistant variety. No examination of the physical characteristics of cuticles of different coffee varieties has yet been made but chloroform extracts of the cuticular wax of berries have been shown by Lampard and Carter (1973) to be toxic to spores of *C. coffeanum* and of several other fungi. They showed a correlation between the degree of fungitoxicity of cuticular wax extracts from many coffee varieties and their field resistance to berry disease.

Resistance to Withertip disease (*Gloeosporium limetticola*) of limes in Zanzibar has been more thoroughly researched and the results suggest that this may be the most likely instance where the cuticle confers effective mechanical resistance. Wheeler (1963) observed that young leaves reacted susceptibly to artificial inoculation while older leaves remained healthy. The resistance of older leaves was shown not to be due to inhibition of spore germination and it seemed probable that the fungus was excluded from resistant leaves by tougher cuticles. The development of the cuticle of the growing lime leaf was examined by Roberts and Martin (1963) who found that cutin was deposited rapidly and in an old leaf reached the unusually high level of 250 µg/cm^2 (Table VI). More cutin was detected in the upper than in the more easily infected lower surface of leaves about to become resistant. Even though these observations do not establish cause, they at least give good evidence to suggest that the lime leaf cuticle can provide a mechanical barrier to fungal invasion. Other factors could well be involved and Martin *et al.* (1966) later found highly antifungal substances within the tissues of both young and mature lime leaves and considered that they contributed to a great extent to defence against the Withertip pathogen. Since it seems that in old, resistant leaves the fungus cannot gain access to the internal tissues, a contribution of these substances might only be imagined if they diffused into infection drops through the cuticle of mature, resistant

TABLE VI

Waxes and cutin acids ($\mu g/cm^2$) of lime leaves in relation to susceptibility to
G. limetticola (Roberts and Martin, 1963)

	Young, susceptible	Young, losing susceptibility		Old, resistant	
		Sample[a] A	Sample B	Sample A	Sample B
Internal waxes	17	84	59	94	73
Cutin { upper surface	24[b]	94	76	240	250
acids { lower surface		54	64	156	160

[a] Samples A and B of leaves collected on different occasions.
[b] Mean for both surfaces.

leaves in relatively larger amounts than in young, susceptible leaves, a possibility that seems remote. Very little activity against *G. limetticola in vitro* was found in the waxes and cutin acids derived chemically from lime cuticles suggesting no chemical protection from the cuticle. Whether cuticular penetration is in fact resisted mechanically probably depends on the ability of *G. limetticola* to degrade the cuticle enzymically, at present unknown. If it can, then as Martin *et al.* point out, the infection thread in its passage through the cuticle may activate a local concentration of cutin acids sufficient to exert a toxic effect. For the moment, however, there is clearly a good case in favour of the lime leaf cuticle offering a mechanical barrier to fungal entry and further work to confirm this or otherwise is to be welcomed.

Relationships between supposed mechanical penetration and evidence for cuticle thickness as a factor in resistance in the powdery mildews have been discussed. Marks *et al.* (1965) examined the mechanism of penetration of leaves of a species of *Populus* by *Colletotrichum gloeosporioides* which relates to cuticular resistance. They observed that infection pegs from appressoria of the fungus penetrated young but not mature leaves. Cytoplasmic factors, which were known to confer another form of resistance, could therefore be excluded and chemical stimuli from within the leaf were shown also not to influence the entry process. Resistance of older leaves to penetration appeared to exceed the mechanical strength of the weakest parts of the appressorium which seemed to collapse due to the force exerted on the infection peg by the unyielding cuticle. It was recognized that forceful penetration could be assisted by enzymes so resistance of mature leaves could not for certain be assigned solely to physical properties of the cuticle.

The structure of intact surfaces of subterranean plant parts in relation to invasion by soil-borne pathogens has been little investigated and such evidence as there is seems to relate to the structure of periderm. Even though the powdery scab organism, *Spongospora subterranea*, enters potato tubers via lenticels or eyes, Wild (1929) was able to correlate the amount of infection with the thickness (mμ and number of cell layers) of the suberized periderm in a single variety planted on various sites. If this is a genuine effect, periderm thickness could presumably influence fungal entry by controlling number, rate of development or morphology of the lenticels and eyes. Lohnis (cited by Gäumann, 1950) could find no relationship between periderm thickness and level of infection of potato tubers by *Phytophthora infestans* which enters through the eyes or wounds. Possibly the only convincing evidence yet documented that the periderm can mechanically determine resistance concerns the susceptibility of potato varieties to skin spot (*Oospora pustulans*). This fungus can enter tubers through the eyes or directly through the skin, and Nagdy and Boyd (1965) demonstrated a close correlation between the susceptibility of 30 commercial varieties to skin infection and periderm thickness, the number of cell rows in the periderm, the thickness of the suberin layer and the content of crude fibre (cellulose, hemicellulose, lignification of cell walls) in the skin. These relationships were unaffected by two different types of soil in which the varieties were grown. When the skin was ruptured, penetration of tubers of all varieties proceeded with ease. Since soil-borne diseases are in general notoriously difficult to control, more systematic study of possible resistance factors associated with the structure of underground plant parts seems particularly desirable.

B. ENTRY THROUGH NATURAL OPENINGS

1. Stomata

While the majority of pathogens of plant foliage enter through intact surfaces, the remainder use mainly stomata either solely or alternatively to direct penetration. These include several pathogens of economic importance, including most parasitic bacteria, some downy mildews and Phytophthoras, the aecidiospore and uredospore stages of rusts and several Fungi Imperfecti. Apart from brief accounts in some textbooks, e.g. Wood (1967), and a now out-of-date treatment of the general role of stomata in plant disease by Rich (1963), aspects of resistance attributable to stomata have not been reviewed before. Only the infection process is of interest here since there is no evidence of a role for stomata in restricting sporulation processes.

Several workers have specifically referred to the absence of a relation between numbers, arrangements, structure or function of stomata and resistance to pathogen entry, (e.g. Ward, 1902; Goulden *et al.*, 1930; Lepik, 1931; Radulescu, 1933; Berry, 1959), and there is little doubt that in general stomatal characteristics are not seriously considered to influence disease severity. However, there has been little systematic investigation of this subject, especially in recent times, despite a number of examples in the literature in which a possible role for stomatal regulation of susceptibility is strongly suggested.

Variation in disease can be attributable to stomata according to the manner by which they influence the pathogen at or before entry. Thus, their density and spatial arrangement may control the number of potential infection ports or their morphology or function may regulate access of a pathogen to the internal tissues. The number of stomata on herbaceous parts of plants varies greatly and their absence may confine infection to other areas, the abaxial and not the adaxial leaf surface for example, as in hop and grapevine downy mildews (Arens, 1929a,b). Quantitative differences in stomatal density are usually too slight to contribute to differential susceptibility though exceptions appear to exist. According to Hull, (cited by Brown, 1936), susceptibility of maize varieties to *Puccinia sorghi* runs parallel with the concentration of stomata on the upper leaf surface. Similarly, Nagai and Imamura (1931) suggested that the extent of infection of 76 rice varieties by *Pyricularia oryzae* might in part correspond to the density of stomata at the base of the panicle, though this could be varied by fertilizer treatment. Infection of beet leaves by *Cercospora beticola* is influenced markedly by their age which in turn may relate to stomatal density and to the length of the stomatal pore. Pool and McKay (1916) considered that the number of stomata was one of several stomatal factors influencing fungal entry but this was refuted recently by Ruppel (1972). Hop downy mildew (*Pseudoperonospora humuli*) is another disease for which there is conflicting evidence. Slabyhoudek (1969) believed there was a positive correlation between infection and varietal differences in the number of stomata on leaves and cone bracts. However, Royle and Thomas (1971a) found comparable stomatal densities on leaves of the same age from five hop varieties ranging from susceptible to highly resistant.

The exclusion of pathogens from host tissues by the structure or function of stomata seems to be a more precisely recorded aspect of resistance though stomatal structure is known to be responsible only in a single, commonly cited case. McLean (1921a) compared the characters of stomata in species of *Citrus* susceptible and resistant to bacterial canker (*Pseudomonas citri*). Leaves of all species of *Citrus* and related genera are susceptible when the bacterium is allowed to enter natural or artificially made wounds, yet in nature *C. nobilis* is found to be very resistant and *C. grandis* very susceptible. Stomata of young leaves of the two species were found to be similar in size and in their opening periodicity but strikingly different in the structure of their vestibule and cuticular ridges (Fig. 1). In the susceptible species the stomatal entrance was widely distended but in the resistant plants almost closed due apparently to occlusion by the cuticular ridges. McLean's records were necessarily made on closed stomata due to the leaf fixation process used.

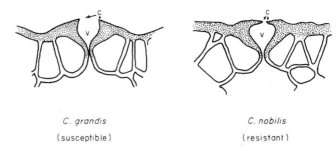

C. grandis   C. nobilis

(susceptible)   (resistant)

FIG. 1. Structure of closed stomata in two species of *Citrus* susceptible and resistant to bacterial canker (*P. citri*). v, vestibule; c, cuticular lips (redrawn from McLean, 1921a).

However, it is likely that proportional differences occurred with open stomata for he later (McLean, 1921b) clearly demonstrated that the pressures required to force water into open or closed stomata of the susceptible species were about half those needed for stomata of the resistant form. These pressures could be correlated with the average width of the stomatal aperture which in *C. grandis* was about twice that of *C. nobilis*. A similar correlation was later found with four varieties representing a range of canker resistance (McLean and Lee, 1922). When suspensions of the bacterial pathogens were substituted for water, resistant leaves became diseased only after the bacteria had been forced through the stomata by pressure whereas susceptible leaves were infected easily after merely immersing in a suspension. Convincing evidence was therefore presented that the structure of stomata in resistant *Citrus* forms prevented the natural ingress of bacteria into the leaf tissues and that only this physical characteristic was implicated in resistance. These findings suggested to McLean and Lee (1921) that it may be possible to develop periclinal chimeras in which the epidermis of the resistant type of *Citrus* could be allowed to confer resistance on other susceptible but commercially desirable species.

The functioning of stomata as a factor in resistance involves response of a pathogen to their diurnal opening and closing rhythm. In this area it is generally accepted that there are no basic structural differences between stomata when they are uniformly open or closed. Exactly what constitutes an open stoma seems to be a confused issue in the literature on both plant physiology and pathology, probably because of the variety of methods which have been used to judge the stomatal condition. Attention has been drawn to this problem by Royle and Thomas (1971a,b) who experienced difficulty in interpreting the state of hop stomata under the scanning electron microscope. As Idle (1969) also explains, stomata with open cuticular lips which reveal the vestibule beneath do not necessarily possess open pores which are delimited by the facing walls of the guard cells. Mature, functioning stomata which have closed vestibules can also be confused with immature ones with unbroken cuticular membranes. Caution may therefore be needed in interpreting observations on relationships between pathogens and "open" and "closed" stomata, especially in the older literature.

Two aspects of resistance attributable to stomatal function can be recognized. In one, the pathogen on the host surface before entry responds differentially to open and closed stomata and in the other closed stomata cannot be penetrated, or at least can be penetrated less easily than open ones. The first is rather a special aspect since only one example is known, concerning infection of hop leaves by zoospores of *P. humuli* (Royle and Thomas, 1971a,b, 1973). In this work, a correlation was found between the number of stomata open, i.e. with open vestibules, at the time of inoculation and the amount of subsequent disease, 7 days later. This occurred whether stomatal opening was varied by light, chemicals or soil moisture treatments and was independent of leaf age or intrinsic susceptibility of hop variety. Observations of the settling pattern of the zoospores on the leaf surface showed that when stomatal opening was encouraged, more than 75% (sometimes 90%) of the zoospore population settled on and penetrated stomata. Curiously, usually a single zoospore, and never more than three, settled on each stoma. By contrast, when most stomata were closed, zoospores settled randomly, only 20–25% on stomata, and as a result few germ tubes encountered and penetrated stomata. An analysis of the mechanisms concerned with the typical response of zoospores to stomata showed that two processes are involved, a shortening of their motile period on the leaf surface (from several h to < 15 min), and their selection of stomata as sites for settling and encystment. A series of experiments which allowed distinction to be made between photochemical and other factors revealed that two stimuli contribute independently to the complete response. One of these is chemical, involves photosynthesis and causes zoospores both to select stomata and to encyst and germinate quickly. The other is purely physical and is due to properties of the leaf surface which have no effect on the period of zoospore motility but help zoospores to select the open rather than closed stomata on which to settle. This response to the topography of the leaf surface was best illustrated using perspex replicas of leaves having stomata either mostly open or closed. After several hours of movement, zoospores on replicas of leaves with open stomata mostly settled on the open "stomata" while they settled at random over the replicas of leaves with closed stomata (Fig. 2). Table VII shows the results of a typical experiment with 10 pairs of leaf replicas in which open and closed stomata were induced by light and darkness respectively. It is unclear why zoospores behave in this curious fashion but probably they have an affinity for open stomata, which are topographically more prominent on the leaf surface than closed ones. As a consequence of this differential response, inherently susceptible leaves are resistant to severe infection during darkness due both to stomata being closed and no photosynthesis, and about one quarter of the level of disease achieved in light results. Epidemiologically this is highly significant

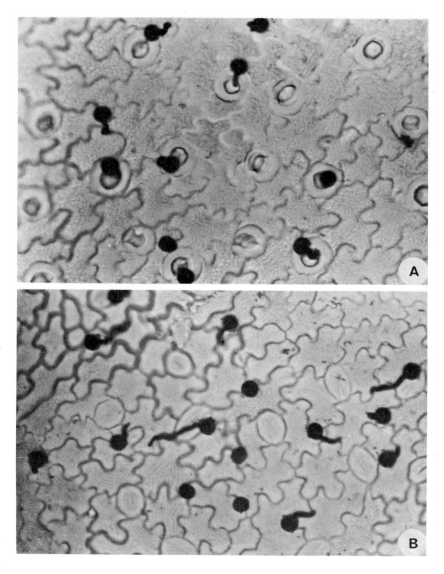

since outbreaks of infection cannot occur when dew provides the source of wetness, only when daytime rain occurs. Field experience supports these findings.

Zoospores of *Plasmopara viticola* aggregate in groups of up to 10 on open stomata of grapevine leaves in the light and in groups of up to 30 on the relatively few stomata remaining open in the dark (Royle and Thomas, 1973). Disease severity appears to be comparable in both situations, however,

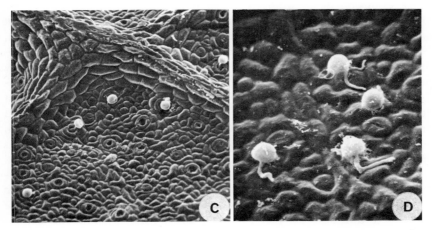

FIG. 2. Distribution of zoospores of *P. humuli* on "perspex" replicas of hop leaves with light-induced open stomata (A and C), and with dark-induced closed stomata (B and D), 6 h after inoculation. A and B are light micrographs ($\times$ 340) and C and D scanning electron micrographs ($\times$ 220 and $\times$ 470 respectively). In A 7 out of 10, and in C all 5 zoospores have settled on "stomata", while in B only 3 out of 15, and in D 1 out of 4 zoospores are on "stomata" (Royle and Thomas, 1973).

TABLE VII

Percentage zoospores of *P. humuli* which settled on "stomata" of perspex replicas prepared from leaves in the light (with mostly open stomata) and in the dark (with mostly closed stomata) (Royle and Thomas, 1973)

| | Replica number | | | | | | | | | |
	1	2	3	4	5	6	7	8	9	10
Light replicas	73 (+)	42 (+)	58 (+)	55 (+)	61 (+)	63 (+)	54 (+)	35 (−)	54 (+)	—
Dark replicas	17 (−)	20 (−)	22 (−)	31 (−)	23 (−)	11 (−)	27 (−)	21 (−)	18 (−)	21 (−)

(+) indicates significant ($P < 0.05$) departure from random zoospore settling; (−) indicates no departure.

though more work is needed on this disease and on others whose mode of stomatal infection is by zoospores.

In the above examples, zoospore germ tubes enter leaves through open and closed stomata equally well. Any basis of resistance here relies on the positioning of the spore of the leaf surface. Some other stomatal pathogens may require open stomata for entry and there are several instances where exclusion of a pathogen by stomatal function appears to confer resistance on the host plant. Thus infection of tobacco leaves by *Bacterium tabacum* was found to be less when stomata were closed than open (Diachun, 1940) and soybeans were infected most severely when inoculum of *Xanthomonas phaseoli* var. *sojae* was applied to leaves between 08.00–14.00 h, the period when most stomata were open (Allington and Feaster, 1946). Unfortunately it has often been rather too easily assumed that closed stomata deny access to fungal pathogens for initial claims have later been disproved or at least held in doubt, and this concept of resistance has broken down. Such an instance is the much quoted work of Pool and McKay (1916) who believed that penetration of sugar beet leaves by *Cercospora beticola* was effected through only open stomata and was therefore confined to the daylight hours. They also considered that the large germ tubes of this fungus were excluded from the relatively small stomata of young heart leaves which are resistant to infection. However, reappraisal by Ruppel (1972) revealed no relationship between size, density or function of stomata and the incidence of this disease in young and mature leaves of six beet cultivars.

Hursh (1924) considered that stomatal movements may have some influence on the entry into wheat of uredospore germ tubes of *P. graminis*. This was then studied in some detail by Hart (1929) who proposed that closed stomata prevented fungal entry and that the daily rhythm of stomatal movement in mature plants differed between wheat varieties. She found that the stomata of some varieties opened very shortly after sunrise and remained open during the day whereas those of other varieties were slow to open in the early morning and stayed open for only a short time. Some varieties were intermediate between these extremes. Since the critical period for rust infection is in the early morning when plants are still heavy with dew, the pathogen would enter easily if the stomata were open at this time but be excluded if they were closed. Hart was able to associate the levels of field resistance in 15 wheat varieties to their diurnal stomatal behaviour and her "functional resistance" theory became well known and stimulated others to investigate its possibilities for both black stem rust and other rust diseases. Peterson (1931) could not substantiate Hart's claims even though he used some closely related wheat varieties and was careful to apply the same criteria for judging stomatal opening. He observed some differences in stomatal periodicity between varieties but they were insufficient to account for the experienced levels of stem rust infection. Apart from doubts about differential stomatal behaviour, the other key question of whether uredospore germ tubes of *P. graminis* require open stomata for infection seems to be unresolved. Allen (1923a,b) presented cytological evidence that infection pegs of two races entered stomata of the susceptible Baart wheat but were mostly excluded from apparently narrower stomatal pores of the resistant Kanred variety. Later, Hart and Forbes (1935) could find no evidence that closed stomata are penetrated by this fungus but nevertheless showed that darkness affected infection differently according to the wheat variety and race of rust used. Their additional finding that infection of wheat leaves by *P. recondita* proceeded equally well in light and dark was confirmed by Caldwell and Stone (1936) who then discovered that in this fungus appressorial attachment actually stimulates closure of stomata which are then forcibly penetrated by infection pegs. Appressorial contact in clover rust (*Uromyces fallans*) also closed wheat and clover stomata which could then be penetrated.

The infection processes in both wheat stem and leaf rusts were examined by Yirgou and Caldwell (1963) who confirmed that *P. recondita* enters stomata with equal ease in light and dark but *P. graminis* can only do so in light. Their explanation for this behaviour was based on different sensitivity of the two pathogens to CO_2: 5% CO_2 suppressed penetration by *P. graminis* but not by *P. recondita*. Light could thus promote stomatal entry of *P. graminis* through photosynthetic reduction of CO_2 within the leaf but *P. recondita* might be expected to penetrate independently of light because of its relative insensitivity to the gas. Evidence that light also has a direct effect on *P. graminis* has been discussed by Sharp *et al.* (1958).

This issue is further complicated by the findings of Burrage (1970) who induced stomatal closure in wheat by subjecting plants in the light to various degrees of water stress and was then able to reduce the number of *P. graminis* infections to a level similar to that found in the dark when stomata would normally be closed. The weight of evidence therefore seems to favour there being some kind of stomatal regulation of penetration in *P. graminis* though whether this contributes a resistance mechanism is uncertain. Clearly, factors other than the mere physical opening and closing of stomata may be implicated. This is particularly well illustrated in the work of Romig and Caldwell (1964) who reappraised functional resistance of wheat to *P. recondita*. Differences in rust severity on peduncles, leaf sheaths and blades of susceptible varieties were ascribed to various degrees of stomatal exclusion of fungal penetration. At anthesis, the stomatal guard cells of peduncles and sheaths had thick walls and prevented germ tube entry whereas the leaf blades possessed thin-walled guard cells which allowed penetration to proceed. All stomata occupied by appressoria were closed, as Caldwell and Stone had found, so any penetration was through closed pores. Quite unexplained was an observation that when tillers were decapitated the thick-walled guard cells admitted the fungus while their ability to exclude was restored when auxin was applied to the cut stump. As senescence developed, stomatal exclusion declined so Romig and Caldwell considered that physiological as well as structural aspects of stomatal function were involved. A similar regulation of pathogen entry in wheat was found for crown rust (*P. coronata*). Reappraisal of Hart's findings and much more detailed study of stomatal control of infection by rust fungi is needed. Until then it would be dangerous to disregard a possible contribution to resistance of mechanisms based on stomatal functioning alone.

2. *Lenticels*

Lenticels occur in the periderm of nearly all plants and comprise small areas of thin-walled, loosely arranged cells with many intercellular spaces. They commonly originate from stomata and are the only means of access to the plant for a small number of pathogens which regularly infect through these breaks in the corky layer. There is a little evidence that character differences in lenticels may sometimes affect the degree of infection though it is uncertain if this due to their pre-existing condition or their subsequent responses to the pathogen's attempts at entry. Infection of plum fruit by *Sclerotinia cinerea* can proceed through the skin, and through stomata and lenticels, the structure of which differs widely between varieties (Curtis, 1928). In some resistant varieties, Valleau (1915) found that germ tube entry was impeded by the close packing and suberization of the lenticel cells, absent in susceptible plums. *Streptomyces scabies*, the cause of common scab of the potato tuber, uses lenticels as its chief avenue of infection. Darling (1937) proposed that infection was easier in a susceptible seedling whose lenticels were larger and possessed rounder and more loosely arranged cells than in a resistant seedling. Infection was also believed to be slower in the resistant seedling because cells were more rapidly suberized and suberization extended further into the tuber. Lapwood (personal communication) considers the situation is more complex than this and that the work of Darling was concerned with the structure of lenticels at a time when they would be resistant anyway. Infection of a tuber can probably only occur during a brief "susceptible" phase when the guard cells are lost from the raised stoma to expose the underlying tissue of the young lenticel (Adams, 1975). Soil moisture status then governs the success of infection which can only proceed in dry soil and before lenticel suberization which is encouraged by dry conditions. Why infection does not occur in wet soil when lenticels proliferate is not known and it is clear that until the mechanisms involved in infection are understood the manner of resistance will remain speculative. Edney (1956, 1958) has indicated that in apple fruits lenticels which have tightly packed cells, and are well suberized and impervious to gas exchange, can present an appreciable barrier to invasion by *Gloeosporium perennans*. Even partly closed lenticels might provide a level of resistance which might be important in delaying fungal invasion when fruit is in store.

IV. COLONIZATION AND SPORULATION

Though many pathogens may gain easy entry into their hosts, existing structural barriers within the tissues may contribute to resistance by limiting the pathogen's subsequent development. Such barriers may be generally distributed and reduce the pathogen's rate of colonization, be localized as particular impenetrable tissues and confine the pathogen to specific areas, or be features of the outer layers and restrict its sporulation. As might be expected, structural features akin to direct entry barriers are to be found in each of these categories.

In a few cases, the existence of general tissue barriers has been inferred from determinations of resistance to artificial needle puncture. Hawkins and Harvey (1919) found that about twice as much pressure was needed to pierce tissues of a variety of potato resistant to *Pythium debaryanum* as tissues of two susceptible varieties. Resistance to puncture was correlated also with a higher crude fibre content, thought to be due to greater secondary thickening of cell walls in the resistant variety. These workers considered that mechanical pressure of the hyphae was perhaps the most important factor in cell wall penetration and that resistance to colonization was due to the inability of *P. debaryanum* to penetrate. However, as Wood (1967) points out, mechanical factors are now believed to be unimportant to colonization by this fungus since it readily secretes cell wall-degrading enzymes which macerate potato tuber tissues (Wood and Gupta, 1958). The structural character of the cell wall in its own right may nevertheless help to limit fungal spread. Leach (1923) showed that hyphae of *Colletotrichum lindemuthianum* penetrate cell walls of resistant bean pods more slowly than those of susceptible varieties. In old, susceptible, and in resistant tissues hyphae, retarded in growth, swell and bend before cell wall penetration can be effected, suggesting that mechanical force may be a significant factor. Similar evidence for forceful penetration of the outer epidermal cell wall has been noted earlier for *C. gloeosporioides* (Marks *et al.*, 1965) and it may be that pathogenic *Colletotrichum* species in general might provide rewarding subjects for detailed investigations of mechanical interactions with plant tissues.

Fungal colonization of susceptible tissues can be localized in discrete areas by barrier tissues which are resistant to invasion. These usually comprise suberized or lignified cells and may vary in extent between varieties, plant organs or with age. Hursh (1924) noted that mycelium of *P. graminis* grew only in the chlorenchyma of wheat stems which in some varieties was very broad and confluent but in others was fragmented in small bundles by sclerenchyma, which the fungus cannot colonize. The amount of chlorenchyma was inversely proportional to the sclerenchyma and varied with variety and fertilizer treatment. Such limitation of the pathogen's growth not only reduced damage to the host but also affected the formation of uredosori which were large in varieties with little sclerenchyma and small and narrow in those with a profuse amount. Hart (1931) extended this work and found that restriction of *P. graminis* by internal barriers occurred only in adult plants; seedling tissues were all equally colonized. Within the varieties

Hart selected for observation, the size, shape and alignment of chlorenchyma in relation to the sclerenchyma were important in regulating mycelial spread and in affecting the size of rust pustules and the time they took to become erumpent. Lignified tissues as morphological barriers are also thought to become progressively more important as soybean plants age in varietal resistance to *Phytophthora megasperma* var. *sojae* (Paxton and Chamberlain, 1969). A well suberized endodermis may serve as a barrier to further progress of a pathogen for Pearson (1931) claimed that in varieties of corn semi-resistant to seedling blight (*Gibberella saubinetti*), the endodermis is a partial if not complete obstacle.

While almost all the evidence for resistance to tissue colonization by unspecialized pathogens originates from early work and is largely unsubstantiated, there is good modern evidence that colonization of xylem elements by a few specialized vascular pathogens may be restricted by vascular anatomy in resistant host varieties. The movement of spores of *Ceratocystis ulmi* within xylem vessels of susceptible species of *Ulmus* is well established and Elgersma (1970) demonstrated that spore transport in resistant elms could possibly be limited by vessel length and diameter. The time needed to draw air or water through the vascular system of elm shoots of standard length was longer for resistant than susceptible varieties and this was related to a greater proportion of shorter vessels and of those of smaller diameter in resistant elms. Using a large number of selections within the *Ulmaceae*, McNabb *et al.* (1970) demonstrated a correlation between increasing disease resistance and decreasing size of xylem vessel groups (vessel diameter × number of contiguous vessels). They suggested that the fungus would be more effectively restricted, not so much by the physical limitations of vessel size, but more by plugging of vessels by tyloses, gums and viscosity of xylary fluid which would be greater in the smaller-sized vessel groups. Because of the current importance of Dutch elm disease, these workers advocated that vessel size and arrangement be used as criteria in screening for resistance. Opportunities for modifying xylem formation by application of trichlorophenylacetic acid (Venn *et al.*, 1968) should also be explored in elms for possible effects on vessel groups.

Vascular anatomy may also be implicated in varietal resistance of alfalfa to bacterial wilt (*Corynebacterium insidiosum*). Cho *et al.* (1973) sought an explanation for a remarkable stability of resistance over the years and found that roots and stems of resistant plants had fewer vascular bundles, shorter xylem vessels and a thicker cortex than those of susceptible varieties. Each of these characteristics might, in a physical manner, impede infection and slow down multiplication and movement of the bacterium once inside the plant. Purely physical explanations for the effects of vascular characters could not, however, be offered with certainty since several other morphological parameters could not be associated with resistance, and growth of the pathogen was adversely affected by crude aqueous extracts from resistant

but not from susceptible varieties. Very recently, vascular structure, possibly
in the form of a low number of large continuous vessels through the nodes,
has been proposed as a factor in resistance of sugarcane clones to ratoon
stunting disease thought to be caused by a small bacterium (Teakle *et al.,*
1975). The evidence here is mostly from measurements of the resistance
of the vascular tracts to water flow which might reflect impedance of the
pathogen.

Though it is conceivable that a thick and tough epidermis may restrict
the fructification of a pathogen or limit its final development and effectiveness
as secondary inoculum, there has been little research to substantiate this
view. The principal evidence comes from Hart (1931) with *P. graminis* and
Sharvelle (1936) using *Melampsora lini* and relates restricted development
of uredosori to heavy lignification of epidermal cells and to resistance of the
epidermis to needle puncture, respectively.

V. Conclusions

This review has presented the main evidence for and against the participation of the plant's structural features in several aspects of resistance
to disease. "Resistance" has deliberately been interpreted loosely, for while
varietal resistance might be regarded as the most important, any other
form that can be shown to be of potential value in disease control should
not be ignored—temporal aspects, for instance, which may affect the ultimate
size or number of lesions. We have encountered little evidence to suggest
that structure alone can sustain the plant against pathogen invasion, nor
that it can substitute for physiochemical mechanisms of defence. However,
results of the most critically executed research have given strong indications
that structural characters can sometimes provide at least a background
level of resistance. The extent of this is simply not known. The common
attitude that structural elements are of small consequence in resistance and
of little practical use may be understandable on present evidence but may
prove to be premature. As we have seen, detailed research designed to
produce conclusions on the role of structural factors has been very limited
and so it would be wise to withhold general judgements based on investigations which were not carried beyond testing for a broad correlation. Much of
this work needs reappraisal but it is nevertheless valuable in pointing to areas
where structure might be the cause of some resistance and where profitable
new investigation could be carried out.

The most thoroughly studied examples also suggest that, rather than
acting in a simple, straightforward manner as has usually been envisaged,
structural characters may often behave in a relatively complex fashion.
For example, cuticle thickness is probably too simple a criterion in some
cases and success of penetration may be affected more by the pathogen's

response to the configuration of superficial cuticular wax, as claimed in cereal mildew. Similarly, pathogen entry may not simply depend on open stomata but on guard cell characteristics or on the topography of the leaf surface as governed by the stomatal condition. Complexity results also when the chemical properties of structural barriers are considered, for these may not only produce substances stimulatory or antagonistic to a pathogen, but may differentially regulate the exudation of active chemicals from internal tissues. Despite the difficulties of separating physical and chemical effects, several instances have been noted in which physical properties alone seem to account for some defence. Distinction between physical and chemical properties is of course necessary to understand how structural characters operate, but in practice the total contribution of a structural feature to resistance is of principal interest. At the present time it is not easy to judge the size of the structural contribution to resistance, partly because of the fragmentary information available, and also because standards for degrees of resistance may not be universally similar.

Finally, we have encountered several instances in which the incorporation of structural characters in breeding for resistance has been advocated. Qualities of resistance based on plant structure are of particular use to the plant breeder, for they are easy to recognize phenotypically, they can often be selected when plants are young and when diseases are not prevalent and, perhaps most important, they are probably less likely than physiochemical mechanisms to select new pathogen races. There is nowadays substantial interest in durable disease resistance thought to be based largely on polygenic control and it is quite conceivable that structural elements could contribute to a spectrum of mechanisms which make up an acceptable degree of field resistance. In future it would therefore not be surprising to find that structural features may become valuable in augmenting other resistance characters operating at both the individual plant level and within populations of the host plant. There are already signs that a new search for structural contributions to disease resistance may have begun and it is certainly to be encouraged.

REFERENCES

Adams, M. J. (1975). *Ann. appl. Biol.* **79**, 265–273.
Agrios, G. N. (1969). "Plant Pathology". Academic Press, New York and London.
Ainsworth, G. C., Oyler, E. and Read, W. H. (1938). *Ann. appl. Biol.* **25**, 308–321.
Akai, S. (1959). *In* "Plant Pathology, An Advanced Treatise" (J. G. Horsfall and A. E. Dimond, eds), Vol. I, pp. 391–434. Academic Press, New York and London.
Allen, R. F. (1923a). *J. agric. Res.* **23**, 131–151.
Allen, R. F. (1923b). *J. agric. Res.* **52**, 917–932.
Allington, W. B. and Feaster, C. V. (1946). *Phytopathology* **36**, 385–386.
Appel, O. (1915). *Science N.Y.* **41**, 773.
Arens, K. (1929a). *Jb. wiss. Bot.* **70**, 91–157.

Arens, K. (1929b). *Phytopath. Z.* **1**, 169–193.

Bailey, J. A., Vincent, G. G. and Burden, R. S. (1974). *J. gen. Microbiol.* **85**, 57–64.

Baker, E. A., Batt, R. F., Silva Fernandes, A. M. and Martin, J. T. (1964). *Ann. Rep. agric. hort. Res. Stn. Univ. Bristol for* 1963, pp. 110–118.

Beckman, C. H., Mueller, W. C. and McHardy, W. E. (1972). *Physiol. Pl. Path.* **2**, 69–74.

Berry, S. Z. (1959). *Phytopathology* **49**, 486–496.

Brown, W. (1936). *Bot. Rev.* **2**, 236–281.

Browning, J. A. and Frey, K. J. (1969). *A. Rev. Phytopath.* **7**, 355–382.

Burchill, R. T. (1958). *Rep. agric. hort. Res. Stn. Univ. Bristol for* 1957, pp. 114–123.

Burrage, S. W. (1969). *Ann. appl. Biol.* **64**, 495–501.

Burrage, S. W. (1970). *Ann. appl. Biol.* **66**, 429–440.

Caldwell, R. M. and Stone, G. M. (1936). *J. agric. Res.* **52**, 917–932.

Caporali, L. (1960). *C. R. Acad. Sci., Paris* **250**, 2822–2824.

Cho, Y. S., Wilcoxson, R. D. and Frosheiser, F. I. (1973). *Phytopathology* **63**, 760–765.

Cobb, N. A. (1892). *Agric. Gaz. N.S.W.* **3**, 181–212.

Corke, A. T. C. (1966). *Colston Pap.* **18**, 143–149.

Curtis, K. M. (1928). *Ann. Bot.* **42**, 39–68.

Darling, H. M. (1937). *J. agric. Res.* **54**, 305–317.

Davies, R. R. (1961). *Nature, Lond.* **191**, 616–617.

Diachun, S. (1940). *Phytopathology* **30**, 268–272.

Dickinson, S. (1960). *In* "Plant Pathology, An Advanced Treatise" (J. G. Horsfall and A. E. Dimond, eds), Vol. II, pp. 203–232. Academic Press, New York and London.

Edney, K. L. (1956). *Ann. appl. Biol.* **44**, 113–128.

Edney, K. L. (1958). *Ann. appl. Biol.* **46**, 622–629.

Elgersma, D. M. (1970). *Neth. J. Pl. Path.* **76**, 179–182.

Ellingboe, A. H. (1972). *Phytopathology* **62**, 401–406.

Fisher, R. W. and Corke, A. T. K. (1971). *Can. J. Pl. Sci.* **51**, 535–542.

Freeman, E. M. (1911). *Phytopathology* **1**, 109–115.

Fromme, F. D. (1920). *Phytopathology* **10**, 53.

Gäumann, E. (1950). "Principles of Plant Infection". Crosby Lockwood and Son, Ltd., London.

Goulden, C. H., Newton, M. and Brown, A. M. (1930). *Scient. Agric.* **11**, 9–25.

Hart, H. (1929). *J. agric. Res.* **39**, 929–948.

Hart, H. (1931). *Tech. Bull. U.S. Dept. Agric.* no. 266.

Hart, H. (1949). *A. Rev. Microbiol.* **3**, 289–316.

Hart, H. and Forbes, I. L. (1935). *Phytopathology* **25**, 715–725.

Hawkins, L. A. and Harvey, R. B. (1919). *J. agric. Res.* **18**, 275–298.

Heather, W. A. (1967a). *Aust. J. biol. Sci.* **20**, 769–775.

Heather, W. A. (1967b). *Aust. J. biol. Sci.* **20**, 1155–1160.

Hursh, C. R. (1924). *J. agric. Res.* **27**, 381–411.

Idle, D. B. (1969). *Ann. Bot.* **33**, 75–76.

Jennings, D. L. (1962). *Hort. Res.* **1**, 100–111.

Jensen, N. F. and Driscoll, C. J. (1963). *Crop Sci.* **3**, 504–505.

Jhooty, J. S. and McKeen, W. E. (1965). *Phytopathology* **55**, 281–285.

Johnstone, K. H. (1931). *J. Pomol.* **9**, 30–52.

Lampard, J. F. and Carter, G. A. (1973). *Ann. appl. Biol.* **73**, 31–37.

Leach, J. G. (1923). *Tech. Bull. Minn. agric. Exp. Stn.* no. 14, 41 pp.

Lepik, E. (1931). *Pflkrankh.* **41**, 228–240.

Louis, D. (1963). *Annls. Epiphyt.* **14**, 57–72.

Lupton, F. G. H. (1956). *Trans. Brit. mycol. Soc.* **39**, 51–59.

Mackie, J. R. (1928). *Phytopathology* **18**, 901–910.

Malik, M. M. S. and Batts, C. C. V. (1960). *Ann. appl. Biol.* **48**, 39–50.

Marks, G. C., Berbee, J. G. and Riker, A. J. (1965). *Phytopathology* **55**, 408–411.

Martin, J. T. (1964). *A. Rev. Phytopath.* **2**, 81–100.

Martin, J. T. and Juniper, B. E. (1970). "The Cuticles of Plants". Edward Arnold, London.

Martin, J. T., Batt, R. F. and Burchill, R. T. (1957). *Nature, Lond.* **180**, 796–797.

Martin, J. T., Baker, E. A. and Byrde, R. J. W. (1966). *Ann. appl. Biol.* **57**, 491–500.

McLean, F. T. (1921a). *Bull. Torrey bot. Club* **48**, 101–106.

McLean, F. T. (1921b). *Phillipp. J. Sci.* **19**, 115–123.

McLean, F. T. and Lee, H. A. (1921). *Phytopathology* **11**, 109–114.

McLean, F. T. and Lee, H. A. (1922). *Phillipp. J. Sci.* **20**, 309–321.

McNabb, H. S. Jr., Heybroek, H. M. and Macdonald, W. L. (1970). *Neth. J. Pl. Path.* **76**, 196–204.

Melander, L. W. and Craigie, J. H. (1927). *Phytopathology* **17**, 95–114.

Mence, M. J. and Hildebrandt, A. C. (1966). *Ann. appl. Biol.* **58**, 309–320.

Nagai, I. and Imamura, A. (1931). *Jap. J. Bot.* **5**, 101–102.

Nagdy, G. A. and Boyd, A. E. W. (1965). *Eur. Potato J.* **8**, 200–214.

Nutman, F. J. and Roberts, F. M. (1960). *Trans. Brit. mycol. Soc.* **43**, 489–505.

Paxton, J. D. and Chamberlain, D. W. (1969). *Phytopathology* **59**, 775–777.

Pearson, N. L. (1931). *J. agric. Res.* **43**, 569–596.

Peries, O. S. (1962). *Ann. appl. Biol.* **50**, 225–233.

Peterson, R. F. (1931). *Scient. Agric.* **12**, 155–173.

Pool, V. W. and McKay, M. B. (1916). *J. agric. Res.* **5**, 1011–1038.

Puranik, S. B. and Mathre, D. E. (1971). *Phytopathology* **61**, 1075–1080.

Radulescu, E. (1933). *Planta* **20**, 244–286.

Rich, S. (1963). *In* "Stomata and Water Relations in Plants" (I. Zelitch, ed.), *Bull. Conn. agric. Exp. Stn.* **644**, pp. 102–114.

Roberts, M. F. and Martin, J. T. (1963). *Ann. appl. Biol.* **51**, 411–413.

Roberts, M. F., Martin, J. T. and Peries, O. S. (1961). *Rep. agric. hort. Res. Stn. Univ. Bristol for* 1960, pp. 102–110.

Romig, R. W. and Caldwell, R. M. (1964). *Phytopathology* **54**, 214–218.

Rosenbaum, J. and Sando, C. E. (1920). *Am. J. Bot.* **7**, 78–82.

Royle, D. J. and Thomas, G. G. (1971a). *Physiol. Pl. Path.* **1**, 329–343.

Royle, D. J. and Thomas, G. G. (1971b). *Physiol. Pl. Path.* **1**, 345–349.

Royle, D. J. and Thomas, G. G. (1973). *Physiol. Pl. Path.* **3**, 405–417.

Ruppel, E. G. (1972). *Phytopathology* **62**, 1095–1096.

Russell, G. E. (1975). *Rep. Pl. Breed. Inst. for* 1974, pp. 137–139.

Schnathorst, W. C. (1959). *Phytopathology* **49**, 562–571.

Scott, P. R. (1973). *Ann. appl. Biol.* **75**, 321–329.

Sharp, E. L., Schmitt, C. G., Staley, J. M. and Kingsolver, C. H. (1958). *Phytopathology* **48**, 469–474.

Sharvelle, E. G. (1936). *J. agric. Res.* **53**, 81–127.

Silva Fernandes, A. M. (1965). *Ann. appl. Biol.* **56**, 297–304.

Slabyhoudek, K. (1969). *Chmelarstvi* **42**, 63–64.

Smith, H. C. and Blair, I. D. (1950). *Ann. appl. Biol.* **37**, 570–583.

Smith, M. A. (1936). *Phytopathology* **26**, 1056–1060.

Stakman, E. C. and Harrar, J. G. (1959). "Principles of Plant Pathology". The Ronald Press Company, New York.

Staub, T., Dahmen, H. and Schwinn, F. J. (1974). *Phytopathology* **64**, 364–372.

Tapke, V. F. (1929). *J. agric. Res.* **39**, 313–339.

Tarr, S. A. J. (1972). "Principles of Plant Pathology". Macmillan, London.

Teakle, D. S., Smith, P. M. and Steindl, D. R. L. (1975). *Phytopathology* **65**, 138–140.

Troughton, J. H. and Hall, D. M. (1967). *Aust. J. biol. Sci* **20**, 509–525.

Valleau, W. D. (1915). *J. agric. Res.* **5**, 365–395.

Van der Plank, J. E. (1975). "Principles of Plant Infection". Academic Press, New York and London.

Venn, K. O., Nair, V. M. G. and Kuntz, J. E. (1968). *Phytopathology* **58**, 1071.

Walker, J. C. (1969). "Plant Pathology" (3rd edition). McGraw-Hill, New York and London.

Ward, H. M. (1902). *Proc. Camb. phil. Soc. Biol. Sci.* **11**, 307–328.

Weinhold, A. R. and English, H. (1964). *Phytopathology* **54**, 1409–1414.

Wheeler, B. E. J. (1963). *Trans. Brit. mycol. Soc.* **46**, 193–200.

Wild, N. (1929). *Phytopath. Z.* **1**, 367–452.

Willaman, J. J., Pervier, N. C. and Triebold, H. O. (1925). *Bot. Gaz.* **80**, 121–144.

Willaman, J. J. (1926). *Proc. Soc. exp. Biol. Med.* **23**, 680–681.

Wilson, A. R. and Jarvis, W. R. (1963). *Pl. Path.* **12**, 91.

Wood, R. K. S. and Gupta, S. C. (1958). *Ann. Bot.* **22**, 309–319.

Wood, R. K. S. (1967). "Physiological Plant Pathology". Blackwell Scientific Publications, Oxford and Edinburgh.

Woodward, R. C. (1927). *Trans. Brit. mycol. Soc.* **7**, 173–204.

Yang, S. L. and Ellingboe, A. H. (1972). *Phytopathology* **62**, 708–714.

Yirgou, D. and Caldwell, R. M. (1963). *Science N.Y.* **141**, 272–273.

CHAPTER 10

Pre-existing Antimicrobial Substances in Plants and their Role in Disease Resistance

J. C. OVEREEM

Institute for Organic Chemistry, T.N.O., Utrecht, Netherlands.

I. INTRODUCTION

It is not very difficult to discuss antimicrobial substances in plants because there are so many. It is much more difficult to discuss the role that these substances may play in resistance of plants to fungal or bacterial attack because so much has been suggested and so little has been convincingly proved.

I would like to express my views on the significance (or insignificance) of preformed antimicrobials and phytoalexins in plant diseases before I discuss the chemistry.

There is an enormous number of microorganisms but only few species are able to attack living plants, although many can grow perfectly well in macerated plant tissue. All pathogens have only one or a few specific hosts. In other words, the establishment of a host–parasite relationship seems to be a very specific process. There is evidence that in order to be a successful parasite of a particular plant, the parasite has to excrete a substance which is specifically recognized or bound by a substance present in the host, for example, the specific symbiotic relationships which exist between soil bacteria of the genus Rhizobium and legume roots. Solheim and Raa (1973) from the University of Tromsö in Norway investigated the substances which are excreted by *Rhizobium trifolii*. They found two heat-labile factors, one probably a nucleic acid, the other a protein or a polysaccharide, which are able to cause curling of root hairs of clover roots. This curling phenomenon is the first visible sign of interaction between *Rhizobium* and clover roots. Both factors bind specifically to a component present in clover roots and thus, without losing their root curling activity, become heat-stable. A similar system was investigated by Bohlool and Schmidt (1974) from the University of Minnesota, U.S.A. They found that soybean roots contain a lectin or phytohemagglutinin, which is able to bind specifically to the soybean *Rhizobium, R. japonicum*. Both the Norwegians and the Americans suggest that this key-lock mechanism may explain the specificity in *Rhizobium*–legume root interactions.

It is reasonable to suppose that this is a much more general phenomenon and that in many, if not in all interactions between living plants and micro-organisms such key-lock mechanisms exist. This would explain the observed specificity. Thus, when a plant does not have the fitting lock for the specific key of a parasite no interaction occurs. The plant is then resistant to attack by that particular parasite. In this form of resistance possible antimicrobial substances in the plant do not play a role simply because the parasite does not even come into contact with them. In the recent book by Day (1974) this form of resistance is called non-host resistance, in contrast to host-resistance. Day defines host-resistance as the result of genetic modifications of the host which render it resistant to pathogens that would otherwise grow on it.

In other words, host-resistance can be considered as a special case of susceptibility. In the model I have just cited it could mean that the key-lock mechanism is destroyed; the host has been transformed into a non-host. It could also mean, in the case of hypersensitivity, that the key-lock mechanism is still operating. In the latter case resistance would then be caused by a second reaction which could be the action of an antimicrobial substance on the parasite, although here it is certainly not necessary to assume that in such a case antimicrobial substances need to be present. One can envisage several other mechanisms by which the growth of the parasite is eventually stopped.

In nature non-host resistance is the rule, host-resistance is the exception. Therefore, it would probably not be very different in nature if there were no antimicrobial substances in plants. On the other hand, for obvious reasons host-resistance is important for our agriculture and horticulture. Therefore the attention of phytopathologists is focussed on this type of resistance. The question to be examined is how much antimicrobial substances contribute to this host-resistance.

Now that I have clarified my views I shall discuss some relevant chemistry. I shall not give a long list of plant constituents with antifungal or antibacterial activity but rather take some examples from different classes of chemical compounds and discuss in detail their possible significance.

II. TULIPALIN AND TULIPOSID

The first compound is a very simple one isolated from tulip bulbs. An important disease in tulip bulbs is caused by *Fusarium oxysporum* f. *tulipae*. At a certain stage of the growth cycle young growing tulip bulbs are protected against attack by this fungus. It is known for certain that the bulbs can be attacked by this fungus and if at a certain stage of the growth this attack has not occurred it is quite reasonable to suppose that by then some internal factor must be present which prevents the disease. At the Flower Bulb Research Station in Lisse, Bergman (1966) was able to prepare a fungitoxic extract from these resistant bulbs. Later Beyersbergen and Bergman purified larger amounts of this fungitoxicant, which was named tulipalin and in Utrecht we showed that this compound has the structure α-methylenebutyrolactone (Bergman *et al.*, 1967). At about the same time

$$HO-CH_2-CH_2-\overset{\overset{\displaystyle O}{\|}}{\underset{\underset{\displaystyle CH_2}{\|}}{C}}-\overset{O}{\underset{}{C}}-O-glucosyl \longrightarrow$$

Tuliposid A   Tulipalin A

this fungicidal compound from tulips was investigated in Germany by Schönbeck (1966). In co-operation with Tschesche *et al.* (1968, 1969) the structure was elucidated. It appeared that tulipalin is an artifact and that the compound present in the tulip bulbs is the glucose-ester tuliposid. This ester is very easily converted into the lactone. When the extraction is done with phosphate buffer, pH 7·5, only tulipalin is isolated. In a recent publication Schönbeck and Schroeder (1972) discuss the role of tuliposides in tulips attacked by *Botrytis* species. Grey mould of tulips is caused by the fungus *B. tulipae*. Under artificial conditions the related *B. cinerea* is also able to attack tulips but the growth of this fungus is rapidly stopped. Schönbeck and Schroeder found that *B. cinerea* increases the permeability of cell membranes of tulips to a greater extent than *B. tulipae*, so it might be expected that the former fungus causes the release of a higher concentration of tuliposid. Furthermore, *B. cinerea* is much more sensitive to the compound which seems to be because this fungus converts the tuliposid into the toxic tulipalin while *B. tulipae* converts the tuliposid into the nontoxic free acid.

In tulips tuliposid A which I have discussed so far is accompanied by a much smaller amount of tuliposid B, which is the analogous compound in which an OH-group is attached to the β-carbon atom.

I would like to conclude my discussion of the tulip compounds with two remarks. Firstly, on the mechanism of action, SH-compounds very readily add across the double bond in tulipalin. The fungicidal action might be explained by inactivation of essential SH-enzymes in the fungus. This is also a very common mechanism of action of commercial synthetic fungicides. Secondly, the α-methylenebutyrolactone structure is often encountered in quite complicated terpenoids of plant origin, some of which have anticancer activity (Kupchan, 1974).

III. Wyerone

Probably the best known example of a polyunsaturated plant constituent with fungitoxic activity is the acetylenic compound wyerone. This compound

was first detected in healthy broad bean seedlings by Spencer at Wye College (Fawcett and Spencer, 1969). Rather than discussing whether or not this compound has any significance for resistance of broad beans to fungal attack, I would like to emphasize that a tremendous number of similar acetylenic compounds have been isolated from plants of different families. In this connection I need only mention the German Professor F. Bohlmann. The greater part of the research in this field has not been done with the aim of isolating fungicidal compounds. I am convinced, however, that several of these compounds will show antifungal activity.

IV. Phenolic Compounds

A. Pyrocatechol and Protocatechuic Acid

Much attention has been paid to the phenolic constituents of plants and their role in disease resistance. The best documented example of the protectant activity of preformed phenolics in plants is undoubtedly the presence of pyrocathecol and protocatechuic acid in dry outer scales of coloured onion varieties resistant to onion smudge (*Colletotrichum circinans*) (Walker and Stahmann, 1955).

It has been convincingly proved that these water-soluble chemicals diffuse into infection droplets on the host and there prevent spore germination. If the outer scales of the onions containing these phenolics are removed the inner fleshy scales can easily be invaded by the pathogen. It is interesting to note that from these fleshy scales allyl sulphides can be extracted, which are toxic to the smudge fungus *in vitro*. Nevertheless, these compounds cannot prevent the disease.

B. PHLORIDZIN AND PHLORETIN

At our institute in Utrecht, Raa (1968) has investigated the natural resistance of apple plants to the scab fungus *Venturia inequalis*. In apple leaves there is no mechanical defence against penetration of germinating spores of *V. inequalis*. I will discuss here the resistance which is displayed by certain apple varieties against certain physiological races of the pathogen and which is associated with a rapid necrogenic reaction and collapse of the host cells surrounding the primary hyphae of the fungus. In this so-called hypersensitive reaction the fungus stops growing without causing further damage to the surrounding tissue. In a susceptible combination the fungus develops just beneath the cuticle. Necrosis of the underlying cells occurs only after many of the leaves have been infected. Young apple leaves contain 4–7% of the glucoside phloridzin.

Raa studied the transformation reactions of phloridzin which occur under the influence of a crude enzyme preparation from apple leaves. By the hydrolytic action of β-glucosidase phloridzin is converted into the aglucone phloretin. At the same time phloridzin and, at a slower rate, phloretin are oxidized under the influence of polyphenoloxidase. Via the *o*-diphenols they are converted into the highly unstable *o*-quinones. The latter can react with nucleophilic centers such as $-NH_2$ or $-SH$ groups and they can also give rise to polymeric products. To investigate the type of products which can be formed we reacted 4-methylcatechol, which can be considered as one half of the oxyphloretin molecule, with the other half represented by phloroacetophenone. Under the influence of an oxidative agent a non-fungitoxic benzofuran derivative was obtained (Raa and Overeem, 1968; Overeem, 1969). (Incidentally, a very convenient way to prepare oxyphloretin is to homogenize apple leaves in the presence of ascorbic acid. This reducing substance stops the oxidation at the *o*-diphenol stage. The result is a homogenate from which oxyphloretin can be isolated.)

Neither phloridzin, phloretin, the corresponding *o*-diphenols nor the end products of the transformation reactions show appreciable fungitoxicity. When, however, apple leaf enzymes are added to a spore suspension of *V. inequalis* which contains a small amount of phloridzin, the spores are killed. We can conclude that the intermediate *o*-quinones are effective fungicides.

Phloridzin

Phloretin

a. Reaction with nucleophilic centers
b. Polymerization

a. Reaction with nucleophilic centers
b. Polymerization

This sequence of reactions in principle explains why a fungus penetrating apple leaf cells is killed. The disorganization of the cell will lead to mixing of the phloridzin and the enzymes and then transient fungitoxic compounds are formed. It does not explain, however, why in some resistant, apple–*Venturia* combinations these reactions occur immediately and in other susceptible combinations they do not. An attractive hypothesis is that in the resistant combinations a product of the fungus reacts specifically with a receptor molecule in the cell membrane of the host cells, thus leading to membrane rupture and subsequently mixing of phloridzin and apple leaf enzymes. It is very difficult, however, to prove conclusively that this hypothesis is true.

V. ANTIFUNGAL COMPOUNDS IN WOOD

At the beginning of this chapter I said that, because of the specificity in host–parasite interactions, I do not believe that either preformed fungitoxic compounds or phytoalexins are the key substances which determine whether or not a living plant is attacked by microorganisms. The situation is different in dead plant material, for example, wood. Wooden objects are more or less exposed nutrients. In principle a fungus which can grow on cellulose can grow on wood. Consequently there are a lot of wood-destroying organisms. Woods which are resistant to fungal decay generally contain fungitoxic compounds and it seems likely that in these cases the fungicides play a predominant role in preventing fungal attack.

Pinosylvin

in *Pinus sylvestris*

β–Thujaplicin

in *Thuja plicata*

Pinosylvin in the heartwood of Scots pine (Erdtman, 1939) and the α-, β- and γ-thujaplicins in the heartwood of Western red cedar (Gripenberg, 1948; Anderson and Gripenberg, 1948; Erdtman and Gripenberg, 1948) are well known examples of fungicides in wood.

A series of sesquiterpenoid *o*-quinones, the mansonones, occur in the heartwood of the West-African tree *Mansonia altissima* (Marini-Bettolo *et al.*, 1965; Tanaka *et al.*, 1966).

Mansonone E   Mansonone F

Mansonia wood is used for furniture making and the mansonones are held to be responsible for the skin-irritating properties of the wood, but they are also fungitoxic. I have encountered mansonones E and F as phytoalexins during an investigation of the compounds responsible for the discoloration in young elm twigs, infected with *Ceratocystis ulmi* (Overeem and Elgersma, 1970). This fungus is the causal agent of Dutch elm disease. Young healthy elm twigs do not contain mansonones or related compounds. Upon infection with *C. ulmi* the orange mansonone E and the violet mansonone F accumulate around the infection spot.

In vitro toxicity data for some non-pathogens of elm and for *C. ulmi* reveal that in this case the mansonones show characteristics of a phytoalexin.

We have not found a significant difference in the amounts of mansonones which accumulate in closely related susceptible and resistant elm varieties

TABLE I

Antifungal activity of mansonones E and F

	Botrytis allii	Penicillium italicum	Aspergillus niger	Cladosporium cucumerinum	Ceratocystis ulmi[a]
Mansonone E	10	5	50	20	> 500[b]
Mansonone F	2	2	> 50	5	> 100[b]

Medium: glucose-mineral salts agar; pH 6·3. Minimum inhibitory concentration in ppm after 8 days' incubation at 24 °C.

[a] Pyridoxal and biotin were added to the growth medium.

[b] Growth retarded in comparison with control. Due to lack of material the compounds were not tested at higher concentrations.

after infection with *C. ulmi*. Nor is there any difference in the rate of formation of the compounds (Elgersma and Overeem, 1971). According to Elgersma (Elgersma, 1970) a major factor governing susceptibility or resistance against Dutch elm disease is the length and diameter of the outer xylem vessels. In resistant varieties these vessels are smaller. The infection remains localized due to an early clogging of the infected vessel.

It is interesting that from the heartwood of different elm species several sesquiterpenes related to the mansonones Ia,b and IIa,b (Fracheboud *et al.*, 1968; Lindgren and Svahn, 1968) and also mansonone C (Krishnamoorthy and Thomson, 1971; Fa-ching Chen *et al.*, 1972; Rowe *et al.*, 1972) have been isolated.

I a: R = Me
I b: R = CHO

II a: R = Me
II b: R = CHO

Mansonone C

These compounds are also fungitoxicants.

VI. Antifungal Compounds in Gramineae

All gramineae important for agriculture contain fungicides which are regarded as significant factors in disease resistance. The longest known is 2-benzoxazolinone which was isolated in 1955 from young rye seedlings by Virtanen and Hietala (1955). This compound is slightly fungitoxic for the snow mould fungus *Fusarium nivale*. *In vitro* it inhibits the growth completely at a concentration of 500 ppm. Later it appeared that 2-benzoxazolinone is an artifact. The compound actually occurring in the

2-benzoxazolinone, R = H
6-methoxy-2-benzoxazolinone, R = OMe

plants is 2-glucosyloxy-4-hydroxy-1,4-benzoxazin-3-one. When the plants are crushed the aglucone is formed by the action of β-glucosidase and upon heating this aglucone loses the 2-carbon atom with formation of formic acid and 2-benzoxazolinone (Hietala and Virtanen, 1960; Honkanen and Virtanen, 1961). In wheat and maize plants an analogous precursor of 6-methoxy-2-benzoxazolinone is present (Wahlroos and Virtanen, 1959).

These compounds were important in a different context as they were apparently the inspiration for Klöpping to investigate benzimidazole chemistry. This finally led to the development of benomyl by Delp and Klöpping (1968). At present, benomyl is undoubtedly the most widely investigated and most widely discussed commercial systemic fungicide.

From rye, wheat and maize to barley is only a small step. Barley seedlings can resist invasion by the fungus *Helminthosporium sativum* in the first days after their emergence. Ludwig *et al.* (1960) showed than an aqueous extract of 5-day-old barley coleoptiles showed antifungal activity. However, an aqueous extract of 6-day-old coleoptiles did not. It was shown that from both aqueous solutions an antifungal factor could be extracted with butanol. The authors demonstrated that the difference between 5- and 6-day-old coleoptiles does not lie in the disappearance of the antifungal factor but in the appearance of an inhibitor of it. Furthermore, they showed that the presence of Ca ions in the extracts of the older coleoptiles can account for the inhibitory effect on the antifungal factor.

Isolation and structure determination of the antifungal factor turned out to be a quite complicated matter. Stoessl (1967) was finally able to show that the antifungal activity can be ascribed to four strongly basic compounds.

Hordatine A: R = R' = H
Hordatine B: R = OMe, R' = H
in glucosides: R' = D-glucopyranosyl

The barley seedlings contain the so-called hordatines A and B together with a larger amount of their glucosides. An interesting feature is that the available evidence (NMR) points to an α-glucosidic linkage in the glucosides. Almost all naturally occurring phenolic glucosides have the β-configuration. A study of the antifungal activity of the compounds by Stoessl and Unwin (1970) showed that both the hordatines and their glucosides have a very marked effect on spore germination of several fungi. They inhibit spore germination completely at concentrations around 10 ppm. *Helminthosporium sativum* is somewhat less sensitive than several other fungi. The compounds in which the configuration of the cinnamic acid moiety is *cis* instead of *trans* and the dihydroderivatives are less active. The activity of the compounds is antagonized by several divalent cations (Ca^{2+}, Mn^{2+}, Mg^{2+}).

An interesting example of a compound which seems to play a role in resistance of oats to a particular fungus is avenacin. Turner (1961) has shown that oat roots contain a factor which can be held responsible for the resistance of oats to *Ophiobolus graminis*, the fungus that causes the take-all disease of wheat. This fungus also penetrates into oats, but then soon dies off. *Ophiobolus graminis* var. *avenae* is a true parasite of oats. It was shown that this variety *avenae*, under the circumstances of the natural infection, produces an enzyme which can detoxify the resistance factor. Thus it is supposed that *O. graminis* var. *avenae* owes its ability to attack oats to the production of this enzyme.

The compound avenacin was isolated from oat roots by Maizel *et al.* (1964). They showed that avenacin can inhibit the growth of *O. graminis* and of several other microorganisms at levels ranging from 3–50 ppm. The structure was partly elucidated by the same authors (Burkhardt *et al.*, 1964).

Avenacin

Avenacin is a triterpenoidal saponin having a carbohydrate chain of three units, two of which are glucose. *N*-methylanthranilic acid is linked to the triterpene nucleus in an ester linkage.

The enzyme avenacinase which is produced by *O. graminis* var. *avenae* and which detoxifies avenacin is able to split off the unknown carbohydrate moiety from the rest of the molecule.

The mechanism of action of avenacin has been studied by Olsen (1971, 1972, 1973). The fungicidal activity, like that of other saponins, can be explained by accepting that the compound binds to sterols in the cytoplasmic membrane, thus changing membrane permeability.

VII. CONCLUDING REMARKS

In my introductory remarks I stressed that the significance of anti-microbials in plants should not be overemphasized. In conclusion I want to make it clear, however, that I do not consider investigations on these compounds unimportant. Probably I can best illustrate this by quoting from a recent review paper by Robinson (1974) on alkaloids. Discussing the importance of alkaloids for the plants in which they occur, Robinson mentions all the possible roles which have been suggested in the past. For example, alkaloids have been thought to protect plants against predators and they could also play a role in the competition between plants. He then writes, "but in no case is there a shred of evidence that such things really occur".

This is perhaps too extreme to apply to the fungitoxic compounds in plants. I have tried to give some examples of compounds to which one can quite confidently attribute some role in the protection of plants against microbial attack. I would stress that although production of alkaloids is probably of minor or no selective advantage for plants, one cannot deny the tremendous importance of alkaloids for chemistry, biology and medicine.

We are probably justified in expecting the same of fungicidal compounds in plants. Apart from the fact that these compounds can help in improving our understanding of the mechanism of plant–pathogen interactions, they can serve as model compounds for the synthetic chemist engaged in pesticide research. Moreover, they are important for taxonomy and it is possible that the search for a fungicidal compound in a plant will ultimately lead to the discovery of a much needed pharmaceutical.

REFERENCES

Anderson, A. B. and Gripenberg, J. (1948). *Acta chem. scand.* **2**, 644–650.
Bergman, B. H. H. (1966). *Neth. J. Pl. Pathol.* **72**, 222–230.
Bergman, B. H. H., Beyersbergen, J. C. M., Overeem, J. C. and Kaars Sijpesteijn, A. (1967). *Recl. Trav. Chim. Pays-Bas* **86**, 709–714.
Bohlool, B. B. and Schmidt, E. L. (1974). *Science* **185**, 269–271.
Burkhardt, H. J., Maizel, J. V. and Mitchell, H. K. (1964). *Biochemistry* **3**, 426–431.
Day, P. R. (1974). "Genetics of Host–Parasite Interaction". Freeman, San Francisco.
Delp, C. J. and Klöpping, H. L. (1968). *Pl. Dis. Reptr.* **52**, 95–99.
Elgersma, D. M. (1970). *Neth J. Pl. Pathol.* **76**, 179–186.
Elgersma, D. M. and Overeem, J. C. (1971). *Neth. J. Pl. Pathol.* **77**, 168–174.
Erdtman, H. (1939). *Naturwissenschaften* **27**, 130–131.
Erdtman, H. and Gripenberg, J. (1948). *Acta chem. scand.* **2**, 625–638.

Fa-ching Chen, Yuh-meei Lin and Arh-hwang Chen (1972). *Phytochemistry* **11**, 1190–1191.

Fawcett, C. H. and Spencer, D. M. (1969). *In* "Fungicides" (D. C. Torgeson, ed.), Vol. II, pp. 637–669. Academic Press, New York and London.

Fracheboud, M., Rowe, J. W., Scott, R. W., Fanega, S. M., Buhl, A. J. and Toda, J. K. (1968). *Forest Prod. J.* **18**, 37–40.

Gripenberg, J. (1948). *Acta chem. scand.* **2**, 639–643.

Hietala, P. K. and Virtanen, A. I. (1960). *Acta chem. scand.* **14**, 502–504.

Honkanen, E. and Virtanen, A. I. (1961). *Acta chem. scand.* **15**, 221–222.

Krishnamoorthy, V. and Thomson, R. H. (1971). *Phytochemistry* **10**, 1669–1670.

Kupchan, S. M. (1974). *Federation Proc.* **33**, 2288–2295.

Lindgren, B. O. and Svahn, C. M. (1968). *Phytochemistry* **7**, 1407–1408.

Ludwig, R. A., Spencer, E. Y. and Unwin, C. H. (1960). *Can. J. Bot.* **38**, 21–29.

Maizel, J. V., Burkhardt, H. J. and Mitchell, H. K. (1964). *Biochemistry* **3**, 424–426.

Marini-Bettolo, G. B., Casinovi, C. G. and Galeffi, C. (1965). *Tetrahedron Lett.* 4857–4864.

Olsen, R. A. (1971). *Physiologia Pl.* **24**, 534–543.

Olsen, R. A. (1972). *Physiologia Pl.* **25**, 204–212; **25**, 503–508; **27**, 202–208.

Olsen, R. A. (1973). *Physiologia Pl.* **28**, 507–515; **29**, 145–149.

Overeem, J. C. (1969). *Recl. Trav. Chim. Pays-Bas* **88**, 851–859.

Overeem, J. C. and Elgersma, D. M. (1970). *Phytochemistry* **9**, 1949–1952.

Raa, J. (1968). "Natural Resistance of Apple Plants to *Venturia inaequalis*". Thesis. University of Utrecht.

Raa, J. and Overeem, J. C. (1968). *Phytochemistry* **7**, 721–731.

Robinson, T. (1974). *Science* **184**, 430–435.

Rowe, J. W., Seikel, M. K., Roy, D. N. and Jorgensen, E. (1972). *Phytochemistry* **11**, 2513–2517.

Schönbeck, F. (1966). *Angew. Bot.* **39**, 173–176.

Schönbeck, F. and Schroeder, C. (1972). *Physiol. Pl. Pathol.* **2**, 91–99.

Solheim, B. and Raa, J. (1973). *J. gen. Microbiol.* **77**, 241–247.

Stoessl, A. (1967). *Can. J. Chem.* **45**, 1745–1760.

Stoessl, A. and Unwin, C. H. (1970). *Can. J. Bot.* **48**, 465–470.

Tanaka, N., Yasua, M. and Imamura, H. (1966). *Tetrahedron Lett.* 2767–2773.

Tschesche, R., Kämmerer, F. J., Wulff, G. and Schönbeck, F. (1968). *Tetrahedron Lett.* 701–706.

Tschesche, R., Kämmerer, F. J. and Wulff, G. (1969). *Chem. Ber.* **102**, 2057–2071.

Turner, E. M. C. (1961). *J. exp. Bot.* **12**, 169–175.

Virtanen, A. I. and Hietala, P. K. (1955). *Acta chem. scand.* **9**, 1543–1544.

Wahlroos, Ö. and Virtanen, A. I. (1959). *Acta chem. scand.* **13**, 1906–1908.

Walker, J. C. and Stahmann, M. A. (1955). *A. Rev. Pl. Physiol.* **6**, 351–366.

CHAPTER 11

Current Perspectives in Research on Phytoalexins

B. J. DEVERALL

Department of Plant Pathology and Agricultural Entomology, University of Sydney, N.S.W. 2006, Australia

For a Symposium of the Phytochemical Society in 1971, I explained the origin of the concept of phytoalexins as antifungal principles produced by plants in response to infection. Their existence was postulated by Müller and Börger (1941) long before any chemical entity was detected as a phytoalexin (Müller, 1958). The first compound to be isolated, characterized and termed a phytoalexin was pisatin from pea (Cruickshank and Perrin, 1960; Perrin and Bottomley, 1962). I also described research mainly in the decade to 1971 which resulted in the characterization of a number of different types of compound as phytoalexins in a range of plant families. Research on the detection and characterization of phytoalexins has proceeded rapidly in the last few years. Compounds with a wide range of chemical structures are now considered as phytoalexins. Furthermore many of the plant species studied have been shown to produce several phytoalexins rather than one as first thought. Another major development concerns the metabolism by fungi of phytoalexins to characterized products. Thus separate chapters are included in this book on the terpenoid phytoalexins which are best known in the Solanaceae, and on the metabolism of the isoflavonoids of the Leguminosae. I wish to present a general review of the different fields of development in research on phytoalexins, and to remind phytochemists and biologists of some of the doubts and questions which should be in the minds of plant pathologists considering the subject.

I. DEFINITIONS AND GENERAL PROBLEMS

Major questions can be raised about the definition and continued use of the term phytoalexin and about the ways in which we should regard many other antimicrobial compounds of plant origin. Thus Ingham (1973) has gathered examples of these compounds which can be allocated to a number of different categories. Firstly, there are compounds which are present in healthy plants. Secondly, there are compounds present in healthy plants but which undergo a further increase in concentration after infection. Thirdly, there are a number of active compounds which are released from inactive precursors after infection and are termed post-inhibitins by Ingham. These include a variety of antimicrobial substances which are released from inactive glycosides by enzymatic action following injury to cells. Fourthly, Ingham would restrict the use of the term phytoalexin to those antimicrobial compounds which are synthesized from remote precursors after infection. These are valid distinctions, although I oppose the introduction of a new set of cumbersome terms for the designation of these classes and shall later question the evidence for the assignment of some compounds to the last two classes. I particularly urge caution in giving the classes terms which imply roles in disease resistance as yet not proven. Because the term phytoalexin has been widely used in plant pathology, it is not reasonable to recommend its immediate abandonment. However I believe that the rapid development in the characterization of active compounds permits each compound to be considered under its chemical or trivial name. Thus a major emphasis of research should now be on evaluation of the roles of named compounds in host–parasite interactions.

It is useful at this stage to emphasize the nature of doubts concerning the role in disease resistance of antimicrobial substances isolated from plants (Daly, 1972). These doubts concern such questions as (1) the occurrence of the compounds at micro-sites within a plant where they might contact a parasitic bacterium or fungus, (2) the presence of the compounds in sufficient concentration in soluble form to cause cessation of growth of the parasite, (3) the accumulation of a compound at the appropriate time to cause the observed cessation of growth of the parasite and (4) the possibility that some other growth-limiting process affects the parasite before the compound accumulates. These questions are posed strongly by the demonstration by Király et al. (1972) that the growth of several parasitic fungi can be stopped in their host plants by exogenous application of an antibiotic, that host cells then die and that phytoalexins accumulate as a later event. Hence the need for critical experimentation and for caution in asserting a role for each active compound isolated from a plant.

II. Phytoalexins in the Higher Plants

A. CHARACTERIZED COMPOUNDS

The large number of compounds isolated, characterized and considered to be phytoalexins are listed in Table I for the Leguminosae and Table II for other families. These are all antifungal compounds which are thought to be synthesized as a result of the stimulation of metabolic pathways by some process which occurs during infection.

Most of the active substances in the Leguminosae are pterocarpans and related compounds, and it is therefore surprising that very different types of compound are produced in *Vicia*, where two acetylenic keto furanoid compounds have been isolated. One of these acetylenic compounds, a methyl ester named wyerone (XII), was first found in dark-grown seedlings of *V. faba* (Fawcett *et al.*, 1968) and was thus assumed to be a component of healthy tissues. However, subsequent research by Gawcett *et al.* (1971) failed to detect wyerone in healthy leaves but revealed substantial accumulation of wyerone following infection by *Botrytis*. It seems possible that wyerone accumulated in the seedlings from which it was first isolated in response to some form of stress caused by growth of the seedlings for 8 days between wet sacks. The second compound, wyerone acid (XI), also cannot be detected in healthy leaves but accumulates after infection (Mansfield and Deverall, 1974b). Although one enzymatic process presumably interconverts wyerone and wyerone acid, it seems very likely that an unknown biosynthetic pathway leading to these compounds is activated following infection. Thus these two compounds must be considered in the same category as, but present a curious phytochemical contrast with, the pterocarpanoid phytoalexins in the Leguminosae.

Most intensively studied of the legumes is bean, *Phaseolus vulgaris*, where it has been demonstrated by Bailey and Burden (1973) that at least four phytoalexins are produced by the host following infection by tobacco necrosis virus, assumed not to have a metabolic capacity itself. The related legume, *Vigna sinensis,* also produces several phytoalexins following virus infection (Bailey, 1973). The detection of a number of phytoalexins in other legumes infected by fungi probably has similar significance but some doubt must remain concerning the possible conversion of a single host product to other active compounds by fungal metabolism.

Table II shows that phytoalexins have been detected in a range of other plant families. The structures of the compounds include terpenoids in the Solanaceae, naphthaldehydes in the Malvaceae and polyacetylenes in the Compositae. Clearly many families and genera are not represented in these lists, so that it is difficult to assess the generality of phytoalexin production in the plant kingdom, but this will be discussed again below.

TABLE I
Phytoalexins of the Leguminosae

Species	Phytoalexin	Reference
Canavalia ensiformis	medicarpin (I)	Keen (1972)
Glycine max	Hydroxyphaseollin	Sims *et al.* (1972)
	(a revised structure (II): name	
	required)	Burden and Bailey (1975)

(I)

(II)

Lotus corniculatus	sativan (III)	Bonde *et al.* (1973)
	vestitol (IV)	

(III)

(IV)

Species	Phytoalexin	Reference
Phaseolus vulgaris	phaseollin (V) phaseollidin (VI) phaseollinisoflavan (VII) kievitone (VIII)	Perrin (1964) Perrin *et al.* (1972) Burden *et al.* (1972)

(V)

(VI)

(VII)

(VIII)

Vigna sinensis	phaseollidin (VI) kievitone (VIII) phaseollin (V)	Bailey (1973)
Medicago sativa	medicarpin (I) sativan (III)	Smith *et al.* (1971) Ingham and Millar (1973)
Trifolium pratense	maackiain (IX) medicarpin (I)	Higgins and Smith (1972)

(IX)

TABLE I (*continued*)

Species	Phytoalexin	Reference
Pisum sativum	pisatin (X) maackiain (inermin) (IX)	Perrin and Bottomley (1962) Stoessl (1972)

(X)

Species	Phytoalexin	Reference
Vicia faba	wyerone acid (XI) wyerone (XII)	Letcher *et al.* (1970) Fawcett *et al.* (1971)

(XI)

(XII)

B. *DE NOVO* SYNTHESIS OR RELEASE FROM PRECURSORS

It has been stated several times that the phytoalexins described are produced from remote precursors by biosynthetic pathways activated after infection. This is an assumption based on the relative slowness of the compounds to accumulate after infection and the absence of knowledge of any close precursors in the healthy plants from which they might arise by enzymatic hydrolysis or oxidation. Few detailed biosynthetic pathways for phytoalexin production have been outlined and there have been no systematic studies of changes in intermediary compounds and of the enzymes involved in the supposed syntheses. The enzyme, phenylalanine ammonia lyase, has been shown to increase in activity before phaseollin accumulates in infected bean (Rathmell, 1973) but subsequent steps in the possible pathway from cinnamic acid to phaseollin have not been examined. Experiments purporting to show activation of pathways by movement of carbon-14 from phenylalanine to pisatin and phaseollin are not readily evaluated (Hadwiger, 1967; Hadwiger *et al.*, 1970), and therefore fail to

TABLE II

Phytoalexins in plant families other than the Leguminosae

Family	Species	Phytoalexin	Reference
Chenopodiaceae	Beta vulgaris	2′5-dimethoxy-6,7-methylenedioxyflavanone (XIII) 2′-hydroxy-5-methoxy-6,7-methylenedioxyisoflavone (XIV)	Geigert et al. (1973)
Malvaceae	Gossypium barbadense	vergosin (XV) hemigossypol (XVI)	Zaki et al. (1972)
Umbelliferae	Daucus carota Pastinaca sativa	3-methyl-6-methoxy-8-hydroxy-3,4-dihydroisocoumarin (XVII) xanthotoxin (XVIII)	Condon and Kuć (1962) Johnson et al. (1973)

(XIII)

(XIV)

(XV)

(XVI)

(XVII)

(XVIII)

TABLE II (continued)

Family	Species	Phytoalexin	Reference
Convolvulaceae	*Ipomoea batatas*	ipomeamarone (XIX)	Kubota and Matsuura (1953)
		(XIX)	
Solanaceae	*Capsicum frutescens* *Lycopersicon esculentum* *Solanum* sp.	capsidiol (XX) rishitin (XXI) rishitin (XXI) lubimin (XXII) phytuberin (XXIII)	Gordon *et al.* (1973) Sato *et al.* (1968) Katsui *et al.* (1968) Metlitskii *et al.* (1971) Coxon *et al.* (1974)
		(XX)	
		(XXI)	
		(XXII)	
		(XXIII)	

Compositae   *Carthamus tinctorius*   safynol (**XXIV**)   Allen and Thomas (1971a)
dehydrosafynol (**XXV**)   Allen and Thomas (1971b)

$$CH_3 \cdot CH \overset{t}{=} CH \cdot (C \equiv C)_3 \cdot CH \overset{t}{=} CH \cdot CH(OH) \cdot CH_2OH$$

(**XXIV**)

$$CH_3 \cdot CH \overset{t}{=} CH \cdot (C \equiv C)_4 \cdot CH(OH) \cdot CH_2OH$$

(**XXV**)

Orchidaceae   *Orchis militaris*   orchinol (**XXVI**)   Hardegger *et al.* (1963)
Loroglossum hircinum   hircinol (**XXVII**)   Gäumann (1964)

(**XXVI**)

(**XXVII**)

convince. The possibility that immediate precursors of the pterocarpanoid phytoalexins are released from glycosides following infection has been suggested by Olah and Sherwood (1971, 1973), who found that glycosidases active upon numerous flavonoid glycosides in alfalfa leaves increased greatly in activity in diseased leaves. However, glycoside concentrations, as measured by fluorescence of spots on chromatograms, increased or remained the same following infection. Thus infection must promote the synthesis of the glycosides if these act as sources of flavonoids in diseased alfalfa. Rathmell (1973) failed to find any change in flavonol concentration in bean hypocotyls as phaseollin accumulated following infection, and suggested that infection stimulated specifically a pathway leading to iso-flavonoid synthesis. Much remains for investigation and clarification pertaining to the ways in which phytoalexins are formed, and a substantial revision of existing hypotheses may become necessary.

C. RESTRICTION TO SOME PLANT FAMILIES

Many plant families are not represented as sources of phytoalexins in Tables I and II, and this must be regarded with interest especially with respect to intensively studied families of great economic importance. Does this imply that phytoalexins, as conceived above, are products of rather few plants? It would be useful to know of thorough investigations which have failed to reveal antimicrobial compounds in plant tissues and in this connection I should like to refer to some of my unpublished work. After considerable experience of detecting and isolating phytoalexins in infected *Vicia* and *Phaseolus* (Mansfield and Deverall, 1974b; Bailey and Deverall, 1971), I have investigated some species in the Cucurbitaceae and wheat in recent years.

At Wye College, I detected a number of antifungal compounds in healthy leaves and fruit of *Cucurbita* and *Cucumis*. The compounds were isolated by either maceration of tissues in ethanol followed by extraction of lipophilic substances into chloroform or by steeping tissues in redistilled benzene. The active compounds were then separated by TLC on silica gel in a number of solvents, and detected by direct bioassay of TLC plates using *Cladosporium cucumerinum*. No information was obtained about the chemical nature of the compounds despite the use of many chromogenic reagents and several forms of spectroscopy. However, of significance to this book was the fact that these compounds did not increase in concentration, nor did other active compounds appear, in leaves after infection by either of the pathogens, *C. cucumerinun* or *Colletotrichum lagenarium*, even in hypersensitive reactions. Therefore use of techniques successful in the detection and assessment of phytoalexins in legumes failed to indicate phytoalexin production in the cucurbits.

In recent research at Sydney University, an antifungal compound has been detected in extracts from wheat leaves expressing an incompatible reaction to *Puccinia graminis tritici*. Leaves were homogenized in ethanol or boiling water and lipophilic compounds were extracted from the supernatant of the homogenate. These compounds were separated by silica gel TLC in chloroform/methanol and then bioassayed directly on TLC plates using *Helminthosporium sativum* or on agar gel overlays dusted with uredospores of *P. graminis tritici*. One active zone was detected. Chromogenic reactions and u.v. spectrophotometry of eluates of the active zone were consistent with the detection of methoxybenzoxazolinone (XXVIII) (Hietala and Wahlroos, 1956) or of a closely related compound. This compound was described as an antifungal product of hydrolysis of a glycoside (XXIX) present in healthy wheat (Wahlroos and Virtanen, 1959), so in our work healthy wheat leaves were homogenized in water at room temperature, and the homogenate was held at this temperature for 30 min before boiling.

(XXIX)   R = CH_3O

(XXXI)   R = H

(XXVIII) R = CH_3O

(XXXII) R = H

An active compound was isolated and crystallized. Mass and nuclear magnetic resonance spectroscopy performed and interpreted by Dr R. W. Rickards, Research School of Chemistry, A.N.U., Canberra, were consistent with the identification of methoxybenzoxazolinone (MBOA). Thus the use of techniques, comparable to many used for detection of phytoalexins, has resulted in the isolation of an antifungal compound from wheat which is probably released from an inactive glycoside present in healthy cells and hydrolyzed after infection. This compound is not a phytoalexin by the definition of Ingham (1973) although it could be argued that it is embraced within the postulates of Müller (1958). I do not wish to term MBOA a phytoalexin or a post-inhibitin, but clearly MBOA may have comparable significance to phytoalexins in other plant families and its role in resistance of wheat requires evaluation, as indicated also by Knott and Kumar (1972) in studies on the glycosidic precursor.

Possibly, different mechanisms of responding to infection have evolved in different plant families. The ability to synthesize antimicrobial compounds after infection may prove to have evolved in the Leguminosae, Solanaceae and some other families. The fact that quite different types of compound are recognized as phytoalexins, even among genera in the Leguminosae as discussed above, supports the concept that natural selection has favoured

(**XXXIII**)   A, $R_1 = R_2 = H$

           B, $R_1 = H$, $R_2 = CH_3O$

           Glucosides, $R_1 = -D-$ glucopyranosyl

(**XXXIV**)

the evolution of diverse means of defence. The ability to release antimicrobial compounds from inactive precursors in damaged cells has evolved in different chemical forms in many families (Ingham, 1972, 1973). Within the family Gramineae, it seems a phytochemical curiosity that the precursor of MBOA occurs in the widely separated genera *Zea* and *Triticum* (Wahlroos and Virtanen, 1959) whereas the precursor (XXXI) of benzoxazolinone (BOA) (XXXII) occurs in *Secale* (Virtanen and Hietala, 1959), a genus classified with *Triticum* in the tribe Hordeae. Further work is needed to reveal the distribution of this class of compound among the grasses, and to assess any post-infectional changes that might occur in concentrations of the quite different antifungal compounds, the hordatines (XXXIII) in *Hordeum* (Stoessl, 1967) and avenacin (XXXIV) in *Avena* (Burkhardt *et al.*, 1964; Maizel *et al.*, 1964). Research in the next decade should help to answer questions about the ubiquity of certain types of antifungal compound and of their modes of formation in plants, although it is probable that most of this research will be restricted to the crop plants.

III. Roles of Phytoalexins in Hypersensitivity and Lesion Limitation

Although much information exists on the occurrence of antifungal compounds in plants, few thorough studies have been made of their role in host–parasite interactions. One exception concerns phaseollin (V)

and related compounds and their accumulation in bean tissue during the hypersensitive response in cultivars resistant to *Colletotrichum lindemuthianum* and during lesion limitation in cultivars susceptible to *C. lindemuthianum* and to *Rhizoctonia solani.*

Phaseollin accumulates to high concentrations, greatly exceeding those which prevent germ tube growth *in vitro*, after cells have reacted hypersensitively to incompatible races of *C. lindemuthianum* (Bailey and Deverall, 1971; Rahe, 1973). Three other phytoalexins also form in substantial amounts in this tissue (Bailey, 1974). The phytoalexins are restricted in their distribution to the small portions of tissue which can be excised bearing the hypersensitive cells, but their location inside these cells has not been established. The phytoalexins continue to accumulate for about 2–3 days after first symptoms of hypersensitivity appear. Contrary to the implications of the work of Király *et al.* (1972) with other parasites, germ tubes continue to grow, albeit increasingly slowly, for part of this period after the death of hypersensitive cells (Skipp and Deverall, 1972). Therefore it is very likely that phytoalexin accumulation stops fungal growth in necrotic cells, but it is possible that the phytoalexins never contact the germ tubes or that they do so too late to bring about growth limitation. Further investigations are needed to reveal the micro-sites at which the phytoalexins accumulate. Research on the virulence of phytoalexin-tolerant mutants should provide a test of the hypothesis that the phytoalexins are responsible for the final expression of resistance.

High concentrations of some of the bean phytoalexins are also present in limited lesions in susceptible hypocotyls. Little or no phaseollin could be detected in susceptible hypocotyls as compatible races of *C. lindemuthianum* made profuse hyphal growth in cells for several days after inoculation (Bailey and Deverall, 1971; Rahe, 1973). Phaseollin accumulated rapidly when brown lesions appeared and became limited, but not in some lesions which continued to spread in etiolated hypocotyls (Rahe, 1973). Bailey (1974) found that at 17 °C very little phytoalexin accumulated in green hypocotyls and the fungus spread and rotted the entire tissue, but at 25 °C high concentrations of phaseollin and phaseollinisoflavan (VII) were present in the limited lesions which formed. *Rhizoctonia solani* also makes substantial hyphal growth in bean hypocotyls for a period after inoculation, but ceases growth after lesions appear as water-soaked areas then become firm and brown. Kievitone (VIII) was found in young lesions, and it accumulated to high concentrations together with phaseollin as the lesions matured (Smith *et al.*, 1975). Thus it seems very likely that phytoalexin accumulation creates an antifungal environment in limited lesions and that this possibly with other factors prevents further fungal growth in susceptible hypocotyls.

Another host–parasite interaction in which changes in phytoalexin concentration have received close attention involves a different process of lesion formation from that described above. Soon after germ tubes of *Botrytis* spp. penetrate cells in broad bean (*Vicia faba*) leaves, the penetrated cells and their immediate neighbours become necrotic and brown so that small lesions may appear within a day of inoculation. *Botrytis cinerea* usually remains limited to these initial lesions, but *Botrytis fabae* is necrotrophic in its parasitism, growing in these dead cells, killing more cells in advance and then growing on into the dead tissue (Mansfield and Deverall, 1974a).

The phytoalexins wyerone (XII) and wyerone acid (XI) accumulate to high concentrations in the epidermal tissue infected by *B. cinerea* two days after inoculation, and fluorescence microspectrography has been used to attempt to reveal their localization within this tissue (Mansfield *et al.*, 1974). Excitation at 405nm caused samples of both phytoalexins to fluoresce between 450 and 650nm and most intensely near 500nm. Excitation of vacuolar contents of some living cells adjacent to necrotic cells caused similar emission spectra, suggesting that the phytoalexins were being formed in living cells at this stage of infection. The presence of the phytoalexins in the dead brown cells could not be confirmed because the contents absorbed the excitation wavelength. It is likely that the phytoalexins were also produced in these cells before or during their death. As an indication of this, wyerone acid has been isolated in high yield from completely necrotic segments of infected leaves (Mansfield and Deverall, 1974b). The hypothesis that these phytoalexins prevent growth of *B. cinerea* in necrotic cells of broad bean should be tested with the aid of phytoalexin-tolerant mutants if these can be obtained. Wyerone acid also begins to accumulate at sites of infection by *B. fabae* but then disappears as the fungus grows (Mansfield and Deverall, 1974b). The disappearance of wyerone acid is almost certainly caused by its reduction by the fungus to reduced wyerone acid, which is found in increasing concentration in the lesions, and is also rapidly produced from wyerone acid by *B. fabae in vitro* (Mansfield and Widdowson, 1973). If the capacity to reduce wyerone acid is an essential attribute for the virulence of this fungus, then mutants lacking this capacity should be incapable of spreading from infection sites. This experiment remains to be attempted.

IV. THE INDUCTION OF PHYTOALEXIN FORMATION

Important problems concern the processes of induction of phytoalexin formation during pathogenesis and the related one of the location of the cellular sites at which phytoalexins are formed.

Firstly, the process of induction will be considered. An early discovery was that heavy metal ions induced the formation of pisatin (X) in pea pods (Cruickshank and Perrin, 1963). This was followed by demonstrations that some metabolic inhibitors and antimetabolites did the same thing (Perrin and Cruickshank, 1965; Bailey, 1969). Adequate explanations in molecular terms of how such diverse compounds cause pisatin accumulation are not available, but it can be suggested that the active molecules impose a biochemical stress on the plant cells which respond by producing a number of substances amongst which are those recognized as phytoalexins. Of greater immediate relevance to pathogenesis was the detection of a fungal metabolite, termed monilicolin A, in cultures of *Monilinia fructicola,* capable of inducing phaseollin (V) formation in bean pods (Cruickshank and Perrin, 1968) without causing visible damage to bean cells (Paxton *et al.,* 1974). However, observations of cytological changes during host–parasite interactions coupled with chemical analyses show that cellular necrosis induced by the parasite precedes accumulation of phytoalexins. This has been recorded for the interactions between bean and *C. lindemuthianum* (Bailey and Deverall, 1971; Rahe, 1973), bean and tobacco necrosis virus (Bailey and Ingham, 1971) and broad bean and *Botrytis* (Mansfield and Deverall, 1974b). Unpublished experiments with bean and the rust *Uromyces appendiculatus* and soybean and the rust *Phakopsora pachyrhizi* at Sydney University corroborate these findings. This suggests that phytoalexin formation during pathogenesis is induced by processes accompanying death of those host cells adversely affected by the presence of the parasite. Phytoalexin formation accompanies cellular death caused by certain physical stimuli such as moderate bruising of broad bean leaves (Deverall and Vessey, 1969) and local freezing applied to bean hypocotyls (Rahe and Arnold, 1975). However, it remains possible that phytoalexin formation, which is easily induced by a variety of stimuli, can be caused by emanations from fungal hyphae during infection and also by substances released during host necrosis. Evidence that phytoalexins can accumulate in infected plants before or in the absence of cell death would be of obvious relevance to this problem, but attempts to obtain this evidence have been negative to date.

The problem of the cellular sites of formation of phytoalexins is closely related to the above discussion. The close association in space and time between cellular death following infection and detection of phytoalexins has been noted several times in this chapter. Phytoalexins might be formed in dead cells, but this seems unlikely if extensive syntheses are involved although Rathmell and Bendall (1972) have suggested that final conversions of precursors to isoflavonoid phytoalexins might occur under the influence of peroxidases in dead cells. Phytoalexins might be formed in cells which are dying slowly, and they might be formed in live cells adjacent to dying cells. Mansfield *et al.* (1974) have detected an unusual fluorescence in response to u.v. irradiation emanating from vacuoles of some live cells adjacent to necrotic cells infected by *B. cinerea* in broad bean leaves. The emission spectrum of this fluorescence was similar to that of solutions of wyerone acid (XI). The possibility that the wyerone acid had diffused from the dead cells was rendered unlikely by the absence of a zone of fluorescent cells around the dead cells. The occurrence of isolated fluorescent cells suggested that these cells were synthesizing wyerone acid in response to metabolites diffusing from either dead cells or the fungal hyphae. This demonstration of phytoalexin synthesis by live cells near sites of cell death should be followed by further investigations in this and other host–parasite interactions to reveal the sequence of events following infection.

ACKNOWLEDGEMENTS

I wish to acknowledge the research assistance of Stella McLeod and the support of the University of Sydney Research Grant and the Australian Research Grants Committee in part of this work.

REFERENCES

Allen, E. H. and Thomas, C. A. (1971a). *Phytochemistry* **10**, 1579–1582.
Allen, E. H. and Thomas, C. A. (1971b). *Phytopathology* **61**, 1107–1109.
Bailey, J. A. (1969). *Phytochemistry* **8**, 1393–1395.
Bailey, J. A. (1973). *J. gen. Microbiol.* **75**, 119–123.
Bailey, J. A. (1974). *Physiol. Pl. Path.* **4**, 477–488.
Bailey, J. A. and Burden, R. S. (1973). *Physiol. Pl. Path.* **3**, 171–177.
Bailey, J. A. and Deverall, B. J. (1971). *Physiol. Pl. Path.* **1**, 435–449.
Bailey, J. A. and Ingham, J. L. (1971). *Physiol. Pl. Path.* **1**, 451–456.
Bonde, M. R., Millar, R. L. and Ingham, J. L. (1973). *Phytochemistry* **12**, 2957–2959.
Burden, R. S. and Bailey, J. A. (1975). *Phytochemistry* **14**, 1389–1390.
Burden, R. S., Bailey, J. A. and Dawson, G. W. (1972). *Tetrahedron Lett.*, 4175–4178.
Burkhardt, H. J., Maizel, J. V. and Mitchell, H. K. (1964). *Biochemistry* **3**, 426–431.
Condon, P. and Kuć, J. (1962). *Phytopathology* **52**, 182–183.
Coxon, D. T., Curtis, R. F., Price, K. R. and Howard, B. (1974). *Tetrahedron Lett.*, 2363–2366.
Cruickshank, I. A. M. and Perrin, D. R. (1960). *Nature, Lond.* **187**, 799–800.
Cruickshank, I. A. M. and Perrin, D. R. (1963). *Aust. J. biol. Sci.* **16**, 111–128.
Cruickshank, I. A. M. and Perrin, D. R. (1968). *Life Sci.* **7**, 449–458.
Daly, J. M. (1972). *Phytopathology* **62**, 392–400.
Deverall, B. J. (1972). *In* "Phytochemical Ecology" (J. B. Harborne, ed.), pp. 217–233. Academic Press, London and New York.
Deverall, B. J. and Vessey, J. C. (1969). *Ann. appl. Biol.* **63**, 449–458.
Fawcett, C. H., Spencer, D. M., Wain, R. L., Fallis, A. G., Jones, E. R. H., LeQuan, M., Page, C. B., Thaller, V., Shubrook, D. C. and Whitham, P. M. (1968). *J. chem. Soc.* (C) **1968**, 2455–2462.
Fawcett, C. H., Firn, R. D. and Spencer, D. M. (1971). *Physiol. Pl. Path.* **1**, 163–166.
Gäumann, E. (1964). *Phytopath. Z.* **49**, 211–232.
Geigert, J., Stermitz, F. R., Johnson, G., Maag, D. D. and Johnson, D. K. (1973). *Tetrahedron* **29**, 2703–2706.
Gordon, M., Stoessl, A. and Stothers, J. B. (1973). *Can. J. Chem.* **51**, 748–752.
Hadwiger, L. A. (1967). *Phytopathology* **57**, 1258–1259.
Hadwiger, L. A., Hess, S. L. and Von Broembsen, S. (1970). *Phytopathology* **60**, 332–336.
Hardegger, E., Biland, H. R. and Corrodi, H. (1963). *Helv. chim. Acta* **46**, 1354.
Hietala, P. K. and Wahlroos, O. (1956). *Acta chem. scand.* **10**, 1196–1197.

Higgins, V. J. and Smith, D. G. (1972). *Phytopathology* **62**, 235–238.
Ingham, J. L. (1972). *Bot. Rev.* **38**, 343–424.
Ingham, J. L. (1973). *Phytopath. Z.* **78**, 314–335.
Ingham, J. L. and Millar, R. L. (1973). *Nature, Lond.* **242**, 125.
Johnson, C., Brannon, D. R. and Kuć, J. (1973). *Phytochemistry* **12**, 2961–2962.
Katsui, N., Murai, A., Takasugi, M., Imaizumi, K., Masamune, T. and Tomiyama, K. (1968). *Chem. Commun. 1968*, 43–44.
Keen, N. T. (1972). *Phytopathology* **62**, 1365.
Király, Z., Barna, B. and Érsek, T. (1972). *Nature, Lond.* **239**, 456–458.
Knott, D. R. and Kumar, J. (1972). *Physiol. Pl. Path.* **2**, 393–399.
Kubota, T. and Matsuura, T. (1953). *J. chem. Soc. Japan, Pure Chem. Sect.* **74**, 248–251.
Letcher, R. M., Widdowson, D. A., Deverall, B. J. and Mansfield, J. W. (1970). *Phytochemistry* **9**, 249–252.
Maizel, J. V., Burkhardt, H. J. and Mitchell, H. K. (1964). *Biochemistry* **3**, 424–426.
Mansfield, J. W. and Deverall, B. J. (1974a). *Ann. appl. Biol.* **76**, 77–89.
Mansfield, J. W. and Deverall, B. J. (1974b). *Ann. appl. Biol.* **77**, 223–235.
Mansfield, J. W. and Widdowson, D. A. (1973). *Physiol. Pl. Path.* **3**, 393–404.
Mansfield, J. W., Hargreaves, J. A. and Boyle, F. C. (1974). *Nature, Lond.* **252**, 316–317.
Metlitskii, L. V., Ozeretskovskaya, O. L., Vul'fson, N. S. and Chalova, L. I. (1971). *Dokl. Akad. Nauk S.S.S.R.* **200**, 1470–1472.
Müller, K. O. (1958). *Aust. J. biol. Sci.* **11**, 275–300.
Müller, K. O. and Borger, H. (1941). *Arb. biol. Anst. (Reichsanst) Berl.* **23**, 189–231.
Olah, A. F. and Sherwood, R. T. (1971). *Phytopathology* **61**, 65–69.
Olah, A. F. and Sherwood, R. T. (1973). *Phytopathology* **63**, 739–742.
Paxton, J., Goodchild, D. J. and Cruickshank, I. A. M. (1974). *Physiol. Pl. Path.* **4**, 167–171.
Perrin, D. R. (1964). *Tetrahedron Lett.*, 29–35.
Perrin, D. R. and Bottomley, W. (1962). *J. Am. chem. Soc.* **84**, 1919–1922.
Perrin, D. R. and Cruickshank, I. A. M. (1965). *Aust. J. biol. Sci.* **18**, 803–816.
Perrin, D. R., Whittle, C. P. and Batterham, T. J. (1972). *Tetrahedron Lett.*, 1673–1676.
Rahe, J. E. (1973). *Can. J. Bot.* **51**, 2423–2430.
Rahe, J. E. and Arnold, R. M. (1975). *Can. J. Bot.* **53**, 921–928.
Rathmell, W. G. (1973). *Physiol. Pl. Path.* **3**, 259–267.
Rathmell, W. G. and Bendall, D. S. (1972). *Biochem. J.* **127**, 125–132.
Sato, N., Tomiyama, K., Katani, N. and Masamune, T. (1968). *Ann. phytopath. Soc. Japan* **34**, 344–345.
Sims, J. J., Keen, N. T. and Honwad, V. K. (1972). *Phytochemistry* **11**, 827–828.
Skipp, R. A. and Deverall, B. J. (1972). *Physiol. Pl. Path.* **2**, 357–374.
Smith, D. G., McInnes, A. G., Higgins, V. J. and Millar, R. L. (1971). *Physiol. Pl. Path.* **1**, 41–44.
Smith, D. A., VanEtten, H. D. and Bateman, D. F. (1975). *Physiol. Pl. Path.* **5**, 51–64.
Stoessl, A. (1967). *Can. J. Chem.* **45**, 1745–1760.
Stoessl, A. (1972). *Can. J. Biochem.* **50**, 107–108.
Virtanen, A. I. and Hietala, P. K. (1959). *Suomen Kemist.* **32B**, 252.
Wahlroos, O. and Virtanen, A. I. (1959). *Acta chem. scand.* **13**, 1906–1908.
Zaki, A. I., Keen, N. T., Sims, J. J. and Erwin, D. C. (1972). *Phytopathology* **62**, 1398–1401.

CHAPTER 12

Terpenoid Phytoalexins

J. KUĆ, W. W. CURRIER AND M. J. SHIH

Department of Plant Pathology, University of Kentucky, Lexington, Kentucky, U.S.A.,
Department of Plant Pathology, Montana State University, Bozeman, Montana, U.S.A. and
Department of Biological Sciences, Simon Fraser University, Burnaby 2, B.C., Canada

I. INTRODUCTION

The biochemical nature of major or "R" gene resistance of potato to *Phytophthora infestans* has been intensively studied for more than 30 years. Müller and his colleagues (1940, 1949, 1957) found that potato tubers inoculated with an incompatible race of *P. infestans* developed localized resistance to a compatible race. This observation lead to the development of the "Phytoalexin Theory". The basic tenet of this theory was that a chemical compound, a phytoalexin, was produced by plant cells as a result of metabolic interaction between host and infectious agent. This compound was not detected in the plant before infection and it inhibited non-pathogens of the plant. It was proposed that pathogens were either insensitive to the concentrations of phytoalexin which accumulated or were unable to cause its accumulation. In recent years, phytoalexins have been detected in trace quantities in some healthy tissues and their accumulation is not dependent upon infection. Fungi and fungal metabolites, viruses, bacteria, mechanical injury, u.v. radiation and numerous chemicals, including some pesticides, can elicit the accumulation of phytoalexins. It also appears unlikely that a single phytoalexin is responsible for the resistance of a plant to all of its non-pathogens, and a number of phytoalexins have been demonstrated to accumulate after infection in many interactions. Questions have also been raised as to whether phytoalexin accumulation is the principal phenomenon responsible for the containment of an infectious agent in a resistant or immune plant (Ersek et al., 1973; Király et al., 1972).

Two classes of phytoalexins in potato, the phenolics and terpenoids, have received a great deal of attention. I will largely limit this presentation to a consideration of the terpenoids, although there is evidence to support the role of phenolic compounds in the disease resistance mechanism of potato (Clarke, 1973; Metlitskii and Ozeretskovskaya, 1968; Sato et al., 1971; Kuć, 1972). Evidence is accumulating that resistance is a multi-response phenomenon and it may be impossible to choose one component out of the complex and assign to it the role of "primary determinant" of resistance or susceptibility.

I will also limit this presentation to potato, though terpenoid phytoalexins have been reported in pepper, other plants in the Solanaceae, and plants in other families.

II. TERPENOIDS PRODUCED IN INFECTED POTATO

Rishitin (Fig. 1a), a bicyclic norsesquiterpene alcohol was isolated by Tomiyama et al. (1968a) from potato tubers inoculated with incompatible races of P. infestans, and its structure was established by Katsui et al. (1968). A comparison of the development of P. infestans in susceptible and resistant tuber or leaf petiole tissue with the onset and rate of rishitin accumulation, strongly suggests that rishitin has a role in disease resistance (Sato et al., 1971; Sato and Tomiyama, 1969). Rishitin is first detected 7–8 h after infection and reaches a concentration of at least 100 µg/g fresh wt. at the site of infection after 24 h. This concentration is attained when inhibition of fungal growth is apparent in incompatible interactions and is sufficient to completely stop the growth of zoospore germ tubes in vitro. There also is a close association between the onset of necrosis and the onset of rishitin accumulation in infected tissue, though Varns et al. (1971a) demonstrated that cell death of the host per se does not cause accumulation and boiled cell-free sonicates of the fungus cause rishitin accumulation with little necrosis.

Rishitin and other isoprenoid derivatives also accumulate in response to inoculation with non-pathogens of potato (Tomiyama et al., 1968a; Varns et al., 1971a). Sato et al. (1968) demonstrated that rishitin accumulated in incompatible interactions of four cultivars, and a cultivar susceptible to all known races of the fungus accumulated rishitin when dipped into a cell-free homogenate of the fungus. Using eleven cultivars and three races of P.

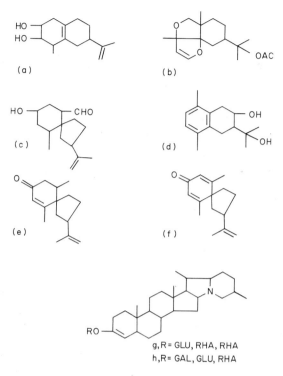

FIG. 1. Terpenoids isolated from potato infected with *P. infestans*: (a) rishitin (b) phytuberin (c) lubimin (d) rishitinol (e) Spirovetiva-1 (10), 11 dien-2-one (f) Spirovetiva-1 (10), 3, 11-trien-2-one (g) α-chaconine (h) α-solanine

infestans, Varns *et al.* (1971b) demonstrated that incompatible interactions were associated with rapid necrosis of host tissue and the accumulation of 16–18 terpenoids which included rishitin and phytuberin (Fig. 1a,b). The latter compound, an unsaturated, sesquiterpene acetate, was isolated by Varns (1970) and recently characterized by Hughes and Coxon (1974). Cell-free sonicates of all races of the fungus studied elicited the accumulation of rishitin, phytuberin, and other terpenoids equally well in tubers of differentially resistant cultivars and cultivars susceptible to all known races (Varns *et al.*, 1971a,b; Varns and Kuć, 1971). The sonicates were at least as effective, and often more effective, in eliciting the accumulation of terpenoids as were incompatible races of *P. infestans*. Rishitinol (Fig. 1d) (Katsui *et al.*, 1971) and lubimin (Fig. 1c) (Metlitskii *et al.*, 1971; Stoessl *et al.*, 1974; Masamune, 1974) are phytoalexins which also accumulate in incompatible interactions, and lubimin may be as abundant under certain conditions as rishitin. Two new spirovetiva compounds (Fig. 1e,f) (Coxon *et al.*, 1974) have been isolated from tubers infected with *P. infestans*. It is not yet clear, however, whether these compounds are antifungal and whether they accumulate principally in incompatible interactions. Since cell-free sonicates of *P. infestans* elicit accumulation of rishitin, phytuberin, derivatives of phytuberin and lubimin, it would appear that enzymes of the host are responsible for their synthesis. At least these compounds are not solely degradative products resulting from the action of the pathogen.

III. RELATIONSHIP OF TERPENOID METABOLISM TO DISEASE RESISTANCE

It is obvious that the potential for resistance exists in differentially resistant cultivars. It is also apparent that this potential exists in completely susceptible cultivars, but it is not expressed in the presence of *P. infestans*. Evidence for the above includes the protection of tubers inoculated with incompatible races of *P. infestans* from damage caused by subsequent inoculations with compatible races, and the accumulation of terpenoids and protection elicited by cell-free sonicates of the fungus in cultivars of potato reported susceptible to all races of *P. infestans*. Even if terpenoid accumulation is not the initial or primary factor resulting in the containment of the fungus in the tuber, the protection implies the presence of a potential for an effective resistance mechanism. Evidence for this potential is found in other studies of phytoalexins (Kuć, 1972) and plant–parasite interactions and is consistent with mechanisms for disease resistance in animals.

Further work by Tomiyama (1966) indicated that cells in leaf petioles invaded by a compatible race had delayed hypersensitive response to subsequent infection by an incompatible race. Varns and Kuć (1971, 1972) demonstrated that a compatible race suppressed necrosis and the accumulation of rishitin and phytuberin in tubers subsequently inoculated with an incompatible race or treated with a cell-free sonicate of *P. infestans* (Table I). The suppression of the response to the incompatible race, or inoculation with the compatible race, was accompanied by the accumulation of non-fungitoxic terpenoids which were not detected in incompatible reactions (Subramanian *et al.*, 1970; Varns and Kuć, 1972). It appears that an alteration of cellular response in the host suppresses the ability of the host to respond normally to a subsequent infection by an incompatible race. Once the host is inoculated with an incompatible race, suppression from a subsequent inoculation with a compatible race does not occur. The compatible race prevents the "metabolic switch" controlling resistance from being turned on but cannot turn it off. An incubation of 12 h is sufficient either to establish suppression or elicit the hypersensitive response. This time period is in agreement with that reported by Tomiyama (1966) in work with leaf petioles. Since a compatible interaction suppresses the hypersensitive host response to cell-free sonicates of the fungus, suppression is not due to competition or antagonism between races of the fungus. Antagonism is not evident

TABLE I

Regulation of terpenoid accumulation in Kennebec potato slices by *P. infestans*

Treatment[b]	µg/g Dry wt.[a]	
	Rishitin	Phytuberin
R 24 h S	334	23
S 24 h R	56	2
Aged 24 h R	238	4
Aged 24 h S	—[c]	—[c]
S 24 h sonicate	52	2
Aged 24 h sonicate	415	3
R 24 h sonicate	400	20

[a] Determined 72 h after second inoculation or treatment.

[b] R 24 h S = Inoculated with an incompatible race, and after 24 h, inoculated with a compatible race. Aged 24 R = Aged for 24 h after slicing and inoculated with an incompatible race.

S 24 h sonicate = Inoculated with a compatible race, and after 24 h, treated with a cell-free sonicate of the fungus.

[c] None detected.

between different races of *P. infestans* grown on lima bean agar. In at least the potato–*P. infestans* interaction, susceptibility may be determined by the suppression of a general resistance mechanism. The work of Clarke (1973) supports this hypothesis.

At least two other isoprenoids, the steroid glycoalkaloids, α-solanine and α-chaconine (Fig. 1h,g), may also be associated with resistance. They have been reported in potato tubers and foliage and appear localized around sites of injury in tubers (McKee, 1955). The steroid glycoalkaloids are largely restricted to the peel of whole tubers (Allen and Kuć, 1968) and they are major antifungal compounds in potato peel. Tomiyama *et al.* (1968b), Shih *et al.* (1973) and Shih and Kuć (1973) reported that the accumulation of α-solanine and α-chaconine at the surface of cut tissue slices is markedly suppressed by inoculation with *P. infestans* (Tables II, III). The suppression is most marked after inoculation with incompatible races of the fungus, with *Helminthosporium carbonum*, a pathogen of corn, or treatment of the slice surface with a cell-free sonicate of a compatible or incompatible race of *P. infestans* (Tables II, III). Aging tubers for 72 h before inoculation or treatment with sonicate also reduced rishitin accumulation and this effect did not appear to be caused by the steroid glycoalkaloids *per se* which accumulated during aging. Thus, tubers harvested in 1971 had a steroid glycoalkaloid content of 139 µg solanine/g fresh wt., and these tubers accumulated 72 µg/g fresh wt. rishitin in the top mm of slices (Table III). The use of cell-free sonicates of the fungus as elicitors eliminates effects

TABLE II

Steroid glycoalkaloid accumulation[a] in the top and second mm of Kennebec potato tuber slices after cutting or cutting followed by inoculation with *P. infestans*

Treatment[b]	Expt 1	Expt 2
Fresh (unaged)	15	14
Aged	251 (215)	265 (217)
Sprayed with water and aged	169 (166)	190 (177)
P. infestans incompatible race 4	30 (60)	28 (61)
P. infestans compatible race 1.2.4	90 (137)	91 (149)

[a] Expressed as µg α-solanine/g fresh wt. Figures in parentheses are the contents in the second mm.

[b] Except for fresh (unaged) tissue, all determinations were made 72 h after cutting, treatment with water or inoculation. Potatoes were grown in 1970 and tested in 1970–1971.

TABLE III

Rishitin and steroid glycoalkaloid accumulation[a] in Kennebec potato tuber slices after cutting or cutting followed by inoculation or treatment with sonicates

Treatment[b]	µg/g Fresh wt. in top mm	
	Rishitin	Steroid glycoalkaloid
Fresh (unaged)	—[c]	139
Sprayed with water and aged	4	754
H. carbonum	53	264
P. infestans incompatible race 4	72	130
P. infestans compatible race 1.3.4	15	349
Sonicate of race 4	80	122
Sonicate of race 1.3.4	71	118

[a] Expressed as α-solanine.

[b] Potatoes were grown in 1971 and tested in 1971–1972. Except for fresh (unaged) tissue, all determinations were made 72 h after treatment of the slice surface with water or sonicates or inoculation with *P. infestans*.

[c] None detected.

of the steroid glycoalkaloids on the germination of zoospores or growth of the fungus.

Tomiyama (1960) and Sato *et al.* (1971) demonstrated that a 5–24 h period between cutting tubers and inoculating the surface decreased the time required for cell death and rishitin accumulation in incompatible inter-actions. In our studies (Shih *et al.*, 1973) aging for more than 24 h prior to treatment with cell-free sonicates of the fungus decreased the time required for the detection of rishitin from 24–30 h to 6–12 h. The peak of rishitin accumulation in aged tubers occurred approximately 20 h after treatment with sonicate as opposed to 60–72 h for unaged tissue. Aging for more than 24 h markedly reduced the total amount of rishitin accumulated. Thus, slices aged for 72 h and then treated with sonicate accumulated 25% as much rishitin as slices treated with sonicate immediately after cutting. A lag period of 18–30 h is necessary for the detection of rishitin and steroid glycoalkaloids (Currier, 1974). Rishitin and steroid glycoalkaloid accumulation may be controlled by the availability of acetate and the activation or synthesis of enzymes in their specific biosynthetic pathways. A suggested partial explanation for the suppression of steroid glycoalkaloid accumulation during rapid rishitin accumulation is based on the likely assumption that biosynthesis of rishitin proceeds from the acetate-mevalonate pathway at a branch point before the steroid glycoalkaloids. Inoculation with an incompatible race of *P. infestans* or treatment with sonicate either blocks the pathway to steroid glycoalkaloid synthesis, and thereby diverts substrate to rishitin, or activates enzymes in the rishitin pathway which compete for substrate. This explanation, however, does not explain why active steroid glycoalkaloid biosynthesis, once established, reduces rishitin accumulation.

Rishitin is not a component of *P. infestans* or uninoculated freshly cut tuber slices. It is also not found in culture filtrates of the fungus (Tomiyama *et al.*, 1968a). It appears to be synthesized *de novo* via the acetate-mevalonate pathway as is shown by the rapid incorporation of ^{14}C acetate and mevalonate into rishitin (Table IV). Suppression of steroid glycoalkaloid accumulation

TABLE IV

Incorporation of isotope from acetate-1-^{14}C and mevalonate-2-^{14}C into rishitin fractions in Kennebec potato tuber slices treated with water or inoculated with *P. infestans*

Isotope added[a]	h after isotope added	Incorporation into rishitin fraction[b] (%)		
		Water	Race 1.2.4[c]	Race 4[c]
Acetate-1-[14]C	7	0·5	0·5	2·3
	14	0·4	1·2	2·9
	21	0·3	1·0	2·2[d]
Mevalonate-2-[14]C	7	0·2	0·7	2·1
	14	0·4	1·5	8·5
	21	0·4	2·2	7·4[e]

[a] Isotope added 40 h after cutting or cutting and inoculation.
[b] Based on total isotope applied.
[c] Slices inoculated with *P. infestans* race 1.2.4 (compatible) or race 4 (incompatible) immediately after cutting.
[d] Accumulated 55 µg rishitin/g fresh wt. of top mm.
[e] Accumulated 50 µg rishitin/g fresh wt. of top mm.

TABLE V
Incorporation of isotope from acetate-1-^{14}C and mevalonate-2-^{14}C into steroid glycoalkaloid fraction of Kennebec potato tuber slices treated with water or inoculated with *P. infestans*

| | | Incorporation into steroid glycoalkaloid fraction (%)[b] | | |
Isotope added[a]	h after isotope added	Water	Race 1.2.4[c]	Race 4[c]
Acetate-1-^{14}C	7	0·7	0·1	0·2
	14	1·3	0·5	0·1
	21	1·4	0·5	0·1
Mevalonate-2-^{14}C	7	2·7	0·5	0·5
	14	3·8	1·0	0·5
	21	4·5	2·0	0·5

[a] Isotope added to tissue 40 h after cutting or cutting and inoculation.
[b] Based on total isotope applied.
[c] Slices inoculated with *P. infestans* race 1.2.4 (compatible) or race 4 (incompatible) immediately after cutting.

by an incompatible race is reflected in the reduced incorporation of ^{14}C acetate and mevalonate into steroid glycoalkaloids (Table V) (Shih and Kuć, 1973).

The optimum temperature for rishitin and phytuberin accumulation in response to incompatible races of *P. infestans* or sonicates of *P. infestans* is *ca* 19 °C (Currier and Kuć, 1975). Little or no rishitin and phytuberin accumulate at 14 or 37 °C and accumulation is markedly reduced at 25 and 30 °C. Steroid glycoalkaloid accumulation increases with increased temperature from 14 to 30 °C with a marked reduction at 37 °C. The accumulation of steroid glycoalkaloids is suppressed by compatible and incompatible races of the fungus even at temperatures at which rishitin is not accumulating. Suppression is most effective with incompatible races and fungal sonicates. The optimum temperature for disease and growth of the fungus is 18–21 °C. Thus, one resistance mechanism of the host appears most effective at a temperature which is optimal for disease. It also appears that the suppression of the wound response by *P. infestans* (accumulation of steroid glycoalkaloids and phenolics) is most effective at temperatures which are not optimal for the response. Thus at 19 °C the level of steroid glycoalkaloids accumulated in cut slices after 72 h is approximately 50% that at 25 or 30 °C. A hypothetical series of events leading to pathogenicity of potato by *P. infestans* might include selection for a fungus that grew well at temperatures which were not optimal for the wound response in potato and the ability to suppress the wound response. A similar series of events leading to resistance in the potato would be the diversion of compounds from steroid glycoalkaloid biosynthesis to the biosynthesis of fungitoxic terpenoids, and this mechanism would be most effective at the optimal temperature for development of *P. infestans*.

IV. NATURE OF THE INITIATOR OF RISHITIN BIOSYNTHESIS (CURRIER, 1974)

Sonicates of *P. infestans* retain their rishitin-inducing activity (RIA) after autoclaving. RIA is completely pelleted from sonicates (pH 4·5, acetate buffer) by centrifugation at 30 000 × g and considerable RIA is pelleted at 10 000 × g. Most of the RIA is recovered on the filter when the sonicate is filtered through a 0·22 μ membrane. RIA is solubilized in pH 8·8 borate buffer and remains in the supernatant when centrifuged at 30 000 × g. RIA in the supernatant passes through a 0·22 μ filter but not through an Amicon PM 30 or XM 300 filter. RIA is quantitatively extracted into ethyl ether from pH 8·8 supernatant. The ether fraction appears as a thick gel which can be freeze-dried without loss of RIA. Resuspension of the freeze-dried powder in pH 8·8 buffer followed by ether extraction retains RIA in the ether layer which contains 29% protein and 28% carbohydrate (mainly glucose with traces of rhamnose after hydrolysis). Boiling for 6 h in N-NaOH does not effect RIA, whereas similar treatment with N-H_2SO_4 completely destroys RIA. Cell wall fractions of the fungus prepared by mechanical disruption (Bartnicki-Garcia, 1966) or repeated washing with boiling methanolic KOH (Aronson *et al.*, 1967) retain RIA. Electron microscopic examination of the fractions indicated they were largely cell wall in nature but did contain cytoplasmic and membrane components. RIA is unaffected by pronase, trypsin, chymotrypsin, amylase, cellulase, DNase or RNase. A rishitin-inducing compound (RIT) was isolated from a methanolic-KOH extract of purified freeze-dried ether extract. Isolation was accomplished by successive separation of fractions containing RIA on 3 silica gel columns eluted with cyclohexane : ethyl acetate (1 : 1 v/v), benzene, and cyclohexane : ethyl acetate (4 : 1 v/v), respectively. RIT appeared as a single component when charred after separation on silica gel plates developed with 95% ethanol, pro-panol : ammonia 0·88 sp. gr. : water (6 : 3 : 1 v/v), cyclohexane : ethyl acetate (1 : 1 v/v) or cyclohexane : ethyl acetate (4 : 1 v/v). The compound reacted positively when sprayed with antimony trichloride and vanillin-sulfuric acid reagents for terpenoids as well as anisaldehyde-sulfuric acid and phenol-sulfuric acid reagents for sugars. Rishitin was detected in the upper mm of potato slices 72 h after treatment with RIT (1–7 μg/g fresh wt. of the top mm of potato slices). This level is considerably less than the RIA in the crude soni-cate from which RIT was isolated (30–80 μg/g fresh wt.). The following hypothesis is presented to explain specificity in the *P. infestans*—potato interaction. It is based on the presence of RIA in all races of *P. infestans* (possibly as a saponin linked to cell walls) and the production of a labile blocker of RIA.

1. Incompatible interaction (R_1 cultivar inoculated with race 4): RIA shifts acetate-mevalonate pathway from steroid glycoalkaloid to rishitin synthesis. Specific site S_1, which occurs prior to rishitin in the synthetic pathway, is not blocked by blocker B_4 from race 4 and rishitin accumulates (Fig. 2a).

2. Compatible interaction (R_1 cultivar inoculated with race 1): RIA shifts acetate-mevalonate pathway from steroid glycoalkaloid to rishitin synthesis. Blocker B_1 from race 1 interacts with site S_1. Rishitin synthesis is suppressed and non-fungitoxic terpenoids accumulate.

3. Compatible interaction ($R_1 R_4$ cultivar inoculated with race 1:4): RIA shifts acetate-mevalonate pathway from steroid glycoalkaloid to rishitin synthesis. Blockers B_1 and B_4 interact with specific sites S_1 and S_4. Both sites occur at a branch prior to rishitin in the synthetic pathway. Rishitin synthesis is suppressed and non-fungitoxic terpenoids accumulate (Fig. 2b).

4. Incompatible interaction ($R_1 R_4$ cultivar inoculated with race 1): RIA shifts acetate-mevalonate pathway from steroid glycoalkaloid to rishitin

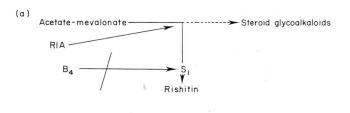

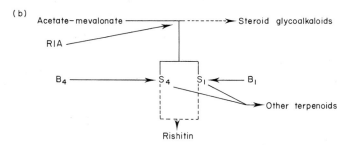

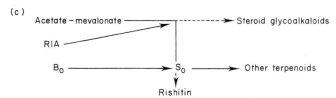

FIG. 2. Three interactions of the elicitor-blocker hypothesis (a) incompatible interaction (R_1 cultivar inoculated with race 4) (b) compatible interaction ($R_1 R_4$ cultivar inoculated with race 1:4) (c) compatible interaction (r cultivar inoculated with race 0).

synthesis. Blocker B_1 produced by race 1 interacts with site S_1. Site S_4 is not blocked and rishitin accumulates.

5. Incompatible interaction (R_1 cultivar inoculated with race 0): RIA shifts acetate-mevalonate pathway from steroid glycoalkaloid to rishitin synthesis. Race 0 has the more specialized blocker B_0 which is effective only with r cultivars containing relatively non-specific site S_0. Site S^1 is not blocked by B_0 and rishitin accumulates.

6. Compatible interaction (r cultivar inoculated with race 0): RIA shifts acetate-mevalonate pathway from steroid glycoalkaloid to rishitin synthesis. Blocker B_0 reacts with relatively non-specific site S_0 (S_0 will react with B_0, B_1, B_4, etc.). Rishitin synthesis is suppressed and non-fungitoxic terpenoids accumulate (Fig. 2c).

The blocker has not been isolated but its existence is suggested by the ability of compatible races of the fungus to suppress cell death, browning and the accumulation of rishitin and phytuberin in potato slices inoculated with incompatible races of the fungus or treated with cell-free fungal sonicates.

V. STABILITY OF TERPENOIDS PRODUCED IN INFECTED POTATO SLICES

How stable are the terpenoids in potato slices? Are they end products of biosynthesis or does the extent of accumulation reflect rates of biosynthesis and degradation? Degradation could be a function of the host and/or the pathogen.

Rishitin accumulation reaches a peak in tuber slices *ca* 96 h after inoculation with an incompatible race or treatment with sonicate (Kuć, 1975; Varns, 1970). The concentration of rishitin then drops markedly and reaches levels which are barely detectable after 7–10 days. Phytuberin levels follow a similar course. Rishitin is stable for more than a year in methanol, ethanol, chloroform or carbon tetrachloride. Phytuberin, however, is rapidly modified in chloroform and carbon tetrachloride to yield 4–6 products. The modifications occur more slowly in methanol or ethanol, however, concentration with heat in methanol or ethanol markedly hastens modification. It is apparent that at least 4–6 of the terpenoids which are apparent after TLC of extracts are the result of modifications of phytuberin. The structure of lubimin also suggests instability. Modifications of the terpenoids do not arise only during extraction, but they appear to occur in the affected tubers.

VI. CONCLUSION

The primary event determining a compatible or incompatible interaction of potato to *P. infestans* depends upon an interaction (recognition) which occurs within hours and perhaps seconds after penetration. All the profound metabolic alterations, including the accumulation of phytoalexins, are the result of this initial interaction.

ACKNOWLEDGEMENTS

Journal paper No. 75–11–64 of the Kentucky Agricultural Experiment Station, Lexington, Kentucky, 40506.

The authors' research reported in this manuscript was supported in part by grant GB-13994 A No. 1 of the National Science Foundation; grant 316-15-51, P.L. 89-106 of the Cooperative Research Service USDA; and a grant from the Herman Frasch Foundation.

REFERENCES

Allen, E. and Kuć, J. (1968). *Phytopathology* **58**, 776–781.
Aronson, J., Cooper, B. and Fuller, M. (1967). *Science* **155**, 332–333.
Bartnicki-Garcia, S. (1966). *J. gen. Microbiol.* **42**, 57–69.
Clarke, D. (1973). *Physiol. Pl. Path.* **3**, 347–358.
Coxon, D., Price, K., Howard, B., Osman, S., Kalan, E. and Zaccharius, M. (1974). *Tetrahedron Lett.* **34**, 2921–2924.
Currier, W. (1974). *In* "Characterization of the Induction and Suppression of Terpenoid Accumulation in The Potato–*Phytophthora Infestans* Interaction". pp. 13–85. Ph.D. Thesis, Purdue University.
Currier, W. and Kuć, J. (1975). *Phytopathology* **65**, 1194–1197.
Ersek, T., Barna, B. and Király, Z. (1973). *Acta phytopath. Acad. Scientiarum Hungaricae* **8**, 3–12.
Hughes, D. and Coxon, D. (1974). *Chem. Comm.* 822–823.
Katsui, N., Murai, A., Takasugi, M., Imaizumi, K. and Masamune, T. (1968). *Chem. Comm.* 43–44.
Katsui, N., Matsunaga, A., Imaizumi, K., Masamune, T. and Tomiyama, K. (1971). *Tetrahedron Lett.* **2**, 83–86.
Király, Z., Barna, B. and Ersek, T. (1972). *Nature* **239**, 456–458.
Kuć, J. (1972). *A. Rev. phytopath.* **10**, 297–232.
Kuć, J. (1975). *Rec. Adv. Phytochem.* **9**, 139–150.
Masamune, A. (1974). *Tetrahedron Lett.* **51/52**, 4483–4486.
McKee, R. (1955). *Ann. appl. Biol.* **43**, 147–148.
Metlitskii, L. and Ozeretskovskaya, O. (1968). *In* "Plant Immunity", pp. 67–79. Plenum Press, New York.
Metlitskii, L., Ozeretskovskaya, O., Vulfson, N. and Chalova, L. (1971). *Mikologiya i Fitopatologiya* **5**, 439–443.
Müller, K. (1957). *Aust. J. biol. Sci.* **10**, 189–196.
Müller, K. and Behr, L. (1949). *Nature, Lond.* **171**, 781–783.
Müller, K. and Borger, H. (1940). *Arb. biol. Reichsanst. Landw. Forstw. Berlin* **23**, 189–231.
Sato, N. and Tomiyama, K. (1969). *Ann. phytopath. Soc. Japan* **35**, 202–207.
Sato, N., Tomiyama, K., Katsui, N. and Masamune, T. (1968). *Ann. phytopath. Soc. Japan* **34**, 140–142.
Sato, N., Kitazawa, K. and Tomiyama, K. (1971). *Physiol. Pl. Path.* **1**, 289–295.
Shih, M. and Kuć, J. (1973). *Phytopathology* **63**, 826–829.
Shih, M., Kuć, J. and Williams, E. (1973). *Phytopathology* **63**, 821–826.

Stoessl, A., Stothers, J. and Ward, E. (1974). *Chem. Comm.* 709–710.

Subramanian, S., Varns, J. and Kuć, J. (1970). *Proc. Indiana Acad. Sci.* **80**, 367.

Tomiyama, K. (1960). *Phytopath. Z.*, **39**, 134–138.

Tomiyama, K. (1966). *Ann. phytopath. Soc. Japan* **4**, 181–185.

Tomiyama, K., Sakuma, T., Ishizaka, N., Sato, N., Katsui, N., Takasugi, M. and Masamune, T. (1968a). *Phytopathology* **58**, 115–116.

Tomiyama, K., Ishizaka, N., Sato, N., Masamune, T. and Katsui, N. (1968b). *In* "Biochemical Regulation in Diseased Plants and Injury", pp. 287–292, Phytopathol. Soc. Japan, Tokyo.

Varns, J. (1970). *In* "Biochemical Response and its Control in the Irish Potato Tuber (*Solanum tuberosum* L.)–*Phytophthora Infestans* Interaction", pp. 47–71. Ph.D. Thesis, Purdue University.

Varns, J. and Kuć, J. (1971). *Phytopathology* **61**, 178–181.

Varns, J. and Kuć, J. (1972). *In* "Phytotoxins in Plant Diseases". (R. K. S. Wood and A. Graniti, eds), pp. 465–468. Academic Press, New York and London.

Varns, J., Currier, W. and Kuć, J. (1971a). *Phytopathology* **61**, 968–971.

Varns, J., Kuć, J. and Williams, E. (1971b). *Phytopathology* **61**, 174–177.

CHAPTER 13

Isoflavonoid Phytoalexins

H. D. VANETTEN AND S. G. PUEPPKE*

Department of Plant Pathology, Cornell University, Ithaca, N.Y., U.S.A.

I. INTRODUCTION

Müller, who originated the phytoalexin concept (Müller and Börger, 1940), defined phytoalexins as:

* Present address: Dept. of Biology, Univ. of Missouri, St. Louis, Missouri, U.S.A.

"antibiotics which are the result of an interaction of two different metabolic systems, the host and the parasite, and which inhibit the growth of micro-organisms pathogenic to plants" (Müller, 1956).

Although revisions of Müller's interpretation have been proposed (Deverall, 1972b; Fawcett and Spencer, 1969; Fuchs and Andel, 1968; Hunter et al., 1968; Ingham, 1973; Wood, 1967), a generally acceptable, explicit definition of phytoalexin is lacking. We choose to apply the term phytoalexin to antimicrobial plant metabolites which undergo enhanced or *de novo* synthesis, i.e. induction, in response to microbial infection. This construction is, in fact, already the commonly accepted working definition. Antimicrobial isoflavonoids, for which exists tentative evidence of enhanced or *de novo* synthesis in response to microbial infection, will be considered as phytoalexins in this review.

A distinction between induced and constitutive isoflavonoids is artificial from a phytochemical standpoint, because a given isoflavonoid may be produced constitutively or inducibly, sometimes even by the same plant. For example, in the roots of *Sophora japonica* L. (Shibata and Nishikawa, 1963) and *Trifolium pratense* L. (Bredenberg and Hietala, 1961a,b), 3-hydroxy-8,9-methylenedioxypterocarpan is the aglycone of a constitutive glucoside. In addition, this pterocarpan is elaborated as a phytoalexin in the leaves of *T. pratense* (Duczek, 1974; Higgins and Smith, 1972). Nonetheless, in evaluating mechanisms of disease resistance, the distinction is fundamental: isoflavonoid phytoalexin accumulation is an active response of the plant to the pathogen. In contrast the plant is passive when pre-existing isoflavonoids are considered as the basis for disease resistance.

II. TYPES AND SOURCES OF ISOFLAVONOID PHYTOALEXINS

We have adopted the Ring Index numbering system (Patterson *et al.*, 1960) for isoflavonoids possessing the basic fused four-ring molecular skeleton, i.e. pterocarpans, 6a,11a-dehydropterocarpans, and coumestans. The traditional numbering system is retained for isoflavonoid derivatives

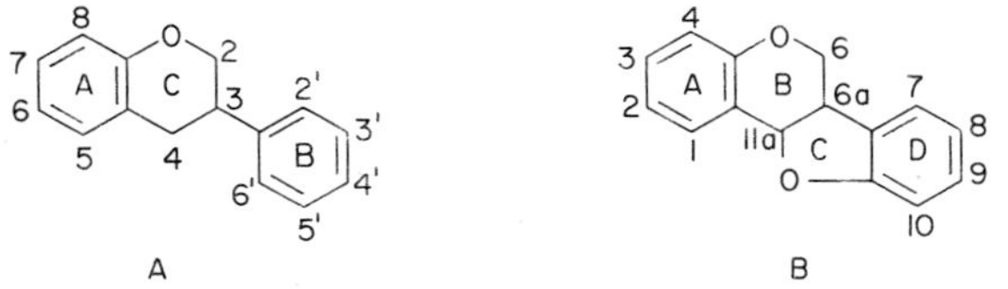

Fig. 1. A) Numbering scheme for isoflavones, isoflavanones, and isoflavans. B) Numbering scheme for pterocarpans, 6a,11a-dehydropterocarpans, and coumestans. The phytoalexins phaseollin and glyceollin are numbered like the pterocarpans, as in B, rather than according to the Ring Index numbering system.

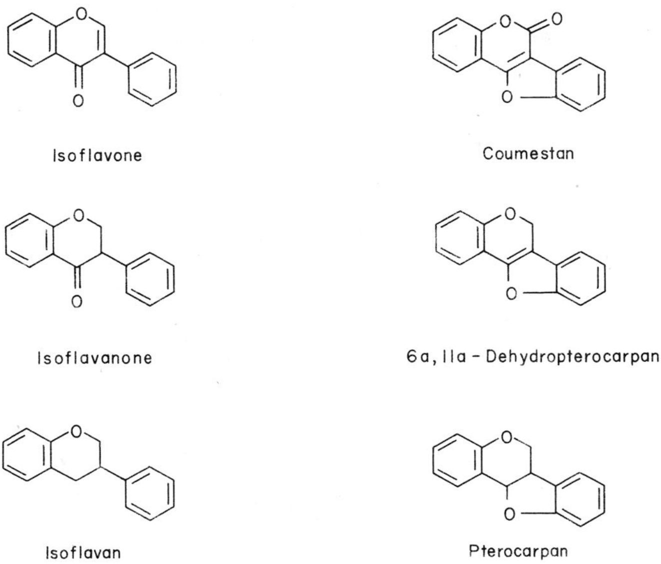

FIG. 2. Ring structures of some classes of the isoflavonoids.

of 3-phenyl-3,4-dihydro[1]benzopyran. The isoflavonoid numbering systems are illustrated in Fig. 1. Figure 2 contains the ring structures of isoflavonoid classes discussed in this review. Isoflavonoids which have been designated as phytoalexins are listed in Table I. Some of the compounds have been recovered only from microbially infected tissue. That these compounds are entirely of host origin is thus unproved; they may be microbially altered metabolites of other isoflavonoids, which are produced during infection.

Isoflavonoids are produced primarily by genera of the Leguminosae, subfamily Lotoideae (Harborne *et al.*, 1971). Genera of the Lotoideae produce all of the known isoflavonoid phytoalexins except 2′-hydroxy-5-methoxy-6,7-methylenedioxyisoflavone (Table I). Yet not all phytoalexins from legumes are isoflavonoids; *Vicia faba* L. synthesizes the acetylenic keto-acid phytoalexin, wyerone acid (Letcher *et al.*, 1970).

The pterocarpan pisatin, initially isolated from *Pisum sativum* L., holds the distinction of being the first chemically identified compound to be called a phytoalexin (Cruickshank and Perrin, 1960; Perrin and Bottomley, 1962; Perrin and Perrin, 1962). Soon after the identification of pisatin the pterocarpan phaseollin (Fig. 3) was isolated as a phytoalexin from *Phaseolus vulgaris* L. (Cruickshank and Perrin, 1963a; Perrin, 1964). Although pterocarpans remain the largest, most studied group of isoflavonoid phytoalexins (Table I), several isoflavans, one isoflavanone, and one isoflavone are classified as phytoalexins.

TABLE I

Isoflavonoid phytoalexins

Isoflavonoid, synonym(s) and reference(s) for structure determination	Sources	Phytoalexin references
Pterocarpans		
3-Hydroxy-9-methoxypterocarpan (demethylhomopterocarpin, medicarpin) Harper *et al.*, 1969	*Canavalia ensiformis* (L.) DC. (Jack bean)	(I) Keen, 1972
	Cicer arietinum L. (Chick pea, garbanzo)	(A) Lampard, 1974
		(I) Ingham, 1975 personal communication
		(I) Keen, 1975b
	Medicago sativa L. (Alfalfa, lucerne)	(iI, A) Higgins and Millar, 1968
		(I) Smith, D. G., *et al.*, 1971
	Melilotus alba Desr. (White sweetclover)	(I) Ingham, 1973
	M. officinalis (L.) Lam. (Yellow sweetclover)	(I) Ingham and Millar, 1973
		(I) Ingham, 1973
	Trifolium pratense L. (Red clover)	(I) Higgins and Smith, 1972
	T. repens L. (White clover)	(I) Cruickshank *et al.*, 1974b
	T. subterraneum L. (Subterranean clover)	(I) Cruickshank *et al.*, 1974b
	Trigonella foenum-graecum L. (Fenugreek)	(I) Ingham, 1973
	Vigna unguiculata Walp.	(I) Lampard, 1974
6a-Hydroxy-3-methoxy-8,9-methylenedioxypterocarpan (pisatin) Perrin and Bottomley, 1962; Perrin and Perrin, 1962	*Pisum abyssinicum* A. Br. *P. arvense* L. *P. elatius* Ster. *P. fulvum* Sibth. and Sm.	(I) Cruickshank and Perrin, 1965
	P. sativum L. (Garden pea)	(iI) Cruickshank and Perrin, 1960, 1961
		(A) Cruickshank and Perrin, 1963b

Compound (references)	Plant source		References
3-Hydroxy-8,9-methylenedioxy-pterocarpan (demethylpterocarpin, inermin, maackiain) Bredenberg and Hietala, 1961a,b; Cocker et al., 1962; Suginome, 1962	Cicer arietinum L.	(I)	Ingham, 1975 personal communication
		(I)	Keen, 1975b
	Pisum sativum L.	(I)	Stoessl, 1972
	Trifolium pratense L.	(I)	Duczek, 1974
		(I)	Higgins and Smith, 1972
	Trigonella foenum-graecum L.	(I)	Ingham, 1973; Ingham and Millar, 1973
3-Hydroxy-2,9-dimethoxypterocarpan 4-Hydroxy-2,3,9-trimethoxyptero-carpan, 2,3,9-Trimethoxypterocarpan Pueppke and VanEtten, 1975	Pisum sativum L.	(I)	Pueppke and VanEtten, 1975
3,9-Dihydroxy-10-(3-methyl-2-butenyl)-pterocarpan (phaseollidin) Burden et al., 1972; Perrin et al., 1972	Phaseolus vulgaris L. (French bean)	(A,I)	Bailey and Burden, 1973
	Vigna sinensis (Torner) Savi (Cowpea)	(I,A)	Bailey, 1973
Phaseollin (Structure in Fig. 3) Perrin, 1964	Phaseolus leucanthus Pip. P. lunatus L. P. radiatus L. P. vulgaris L.	(I)	Cruickshank and Perrin, 1971
		(iI)	Cruickshank and Perrin, 1963a
		(A)	Perrin and Cruickshank, 1965
	Vigna sinensis (Torner) Savi	(I,A)	Bailey, 1973
Glyceollin[a] (Structure in Fig. 3) Burden and Bailey, 1975	Glycine max (L.) Merr. (Soybean)	(iI,A)	Klarman and Sanford, 1968

[a]This compound was originally incorrectly identified as 6a-hydroxyphaseollin (Sims et al., 1972). It has now been assigned the trivial name glyceollin (Burden, Bailey and Keen, personal communications).
(iI) A phytoalexin was isolated, but the structure was not determined.
(I) Identified as a phytoalexin.
(A) Phytoalexin abiotically induced and thus of host origin.

continued

TABLE I—continued

Isoflavonoid, synonym(s) and reference(s) for structure determination	Sources	Phytoalexin references
Isoflavans		
2',7-Dihydroxy-4'-methoxyisoflavan (vestitol) Kurosawa et al., 1968	*Lotus corniculatus* L. (Birdsfoot-trefoil)	(I) Bonde et al., 1973
7-Hydroxy-2',4'-dimethoxyisoflavan (sativan, sativin[b]) Ingham and Millar, 1973	*L. corniculatus* L. *Medicago sativa* L.	(I) Bonde et al., 1973 (I) Ingham and Millar, 1973
Phaseollinisoflavan (Structure in Fig. 3) Burden et al., 1972	*Phaseolus vulgaris* L.	(I,A) Bailey and Burden, 1973
2'-Methoxyphaseollinisoflavan (Structure in Fig. 3) VanEtten, 1973b	*P. vulgaris* L.	(I) VanEtten, 1973b
Isoflavanones		
2',4',5,7-Tetrahydroxy-8-(3-methyl-2-butenyl) isoflavanone (kievitone) Burden et al., 1972; Smith, D.A., et al., 1973b	*P. vulgaris* L. *Vigna sinensis* (Torner) Savi	(ii,A) Pierre and Bateman, 1967 (I,A) Bailey and Burden, 1973; Smith, D. A., et al., 1973a (I,A) Bailey, 1973
Isoflavones		
2'-Hydroxy-5-methoxy-6,7 methylenedioxyisoflavone Geigert et al., 1973	*Beta vulgaris* L.	(I) Geigert et al., 1973

[b]The trivial name sativan now replaces sativin, because sativin was first assigned to another non-isoflavonoid compound (Bonde et al., 1973).

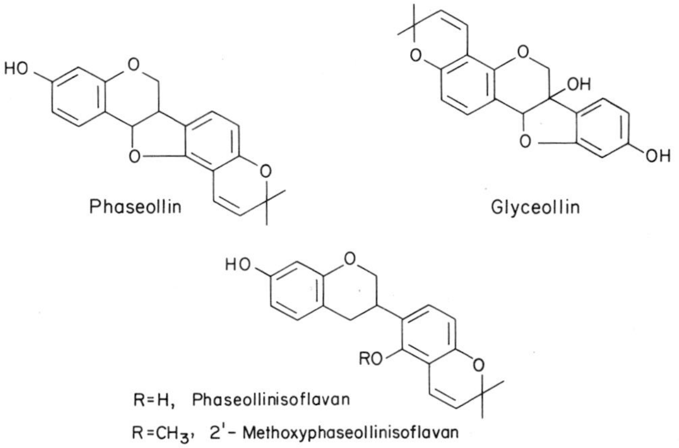

FIG. 3. Structures of some isoflavonoid phytoalexins.

Two additional isoflavones which may eventually warrant designation as phytoalexins are 7-hydroxy-4′-methoxyisoflavone (formononetin) and 4′,7-dihydroxyisoflavone (daidzein). Both are antifungal (Naim *et al.*, 1974; Virtanen and Hietala, 1958) and both accumulate in *Medicago sativa* L. in response to fungal infection (Olah and Sherwood, 1971). However, accumulation may in part reflect hydrolysis of preformed isoflavone glycosides, rather than enhanced synthesis (Olah and Sherwood, 1973). Daidzein concentration also increases in bacteria- and fungus-infected *Glycine max* L. (Merr.) (Keen and Kennedy, 1974; Keen *et al.*, 1972b).

It has been known for some time that members of another class of isoflavonoids, the coumestans, accumulate in several plant species as a response to microbial infection (Bickoff *et al.*, 1967, 1969; Keen and Kennedy, 1974; Keen *et al.*, 1972a; Loper and Hanson, 1964; Olah and Sherwood, 1971; Sherwood *et al.*, 1970; Wong and Latch, 1971). They apparently lack antifungal activity (Bickoff *et al.*, 1969; Perrin and Cruickshank, 1969), but the inducible compound, 3,9-dihydroxycoumestan (coumestrol), was recently reported to possess weak antibacterial activity (Keen and Kennedy, 1974). Thus it may merit classification as a phytoalexin. If coumestrol is accepted as a phytoalexin, all the "common" isoflavonoid classes have phytoalexin representatives.

Among the flavonoids, antimicrobial activity is not restricted to the isoflavonoids. Flavones, flavanones, and chalcones reportedly exhibit antimicrobial activity (Ben-Aziz, 1967; Chowdhury *et al.*, 1974; Gasha *et al.*, 1972; Hergert, 1962; MacDonald and Bishop, 1952; Pinkas *et al.*, 1968). In some cases fungi simultaneously activate flavone as well as isoflavonoid biosynthesis (Olah and Sherwood, 1971, 1973). It thus seems reasonable that flavonoids other than isoflavonoids will be discovered and designated as phytoalexins. Indeed, a flavanone from *Beta vulgaris* L. may be a phytoalexin (Geigert *et al.*, 1973).

Two characteristics of isoflavonoid phytoalexin production justify greater consideration. First, a given isoflavonoid phytoalexin is often produced by several plant species (Table I). For example, 3-hydroxy-9-methoxypterocarpan is reported as a phytoalexin of at least ten plant species and may be a common phytoalexin of the tribe Trifolieae (John Ingham, personal communication). Because of the limited skeleton and substituent diversity possible in isoflavonoid molecules and the substantial number of species in the Lotoideae, other examples of the production of the same isoflavonoid phytoalexin by different plant species will likely be described. Second, a plant species may produce several different isoflavonoid phytoalexins (Table I). The hypothetical advantage to a plant species of multiple phytoalexin production was briefly mentioned in several early reviews (Cruickshank, 1963, 1965b). Although the phenomenon was supported by early experimental findings (Ende, 1965; Pierre and Bateman, 1967), multiple phytoalexin systems have been only recently verified. Future scientific scrutiny will probably reveal more examples of multiple phytoalexin systems.

III. BIOSYNTHESIS OF ISOFLAVONOIDS

The biosynthetic pathways which culminate with the specific formation of isoflavonoid phytoalexins have been little studied. For the most part experimental results are in agreement with data from the more detailed research on the biosynthesis of isoflavonoids which are not phytoalexins. We will therefore discuss isoflavonoid biosynthesis in general, rather than concentrate specifically on the biosynthesis of the isoflavonoid phytoalexins. Where it is pertinent we will emphasize those phenomena which are particularly relevant to phytoalexin biosynthesis.

The biosynthetic pathway of isoflavonoids can be divided conveniently into three sequential portions: (1) early pathways which are shared with other secondary metabolites, (2) reactions mutual to flavonoid and isoflavonoid biosynthesis, and (3) biosynthetic steps unique to isoflavonoids. Only the first, general portion of the biosynthetic route is well understood. Although researchers have made extensive use of radioactive labelling techniques to identify paths of carbon flow, many of the later intermediates in isoflavonoid biosynthesis remain hypothetical. Enzymes from higher plants which act on isoflavonoid substrates generally are uncharacterized.

A. FORMATION OF THE C_{15} SKELETON

The keystone of our knowledge about flavonoid and isoflavonoid biosynthesis is that the basic C_{15} molecular skeleton results from the convergence of two unrelated metabolic routes, i.e. the acetate-malonate and shikimic acid pathways. This is supported by incorporation of radioactive label from acetate into the A-ring of flavonoids and label from shikimic acid precursors into the B- and heterocyclic rings (Grisebach, 1965). Both pathways are ubiquitous in higher plants. We will not further consider the role of the acetate-malonate pathway in isoflavonoid biosynthesis, although the germane portions of the pathway have been appraised elsewhere (Ribéreau-Gayon, 1971).

One function of the shikimic acid pathway in primary metabolism is synthesis of phenylpropanoid units such as phenylalanine and tyrosine. Phenylalanine is the only good phenylpropanoid precursor of flavonoids in the Leguminosae (Towers, 1974). Therefore, the enzyme phenylalanine ammonia-lyase (PAL), which reductively deaminates phenylalanine to *trans*-cinnamic acid (Koukol and Conn, 1962), occupies an ideal position to control flow of carbon into flavonoids and other phenolic secondary metabolites. Enhanced PAL activity is often observed prior or concurrent to increased phenol synthesis (Camm and Towers, 1973; Rathmell, 1973). The metabolic role of this enzyme is the subject of recent reviews (Camm and Towers, 1973; Creasy and Zucker, 1974).

The enzymes which hydroxylate and activate *trans*-cinnamic acid prior to its condensation with acetate units have been intensively studied in parsley (*Petroselinum hortense* Hoffm.) (Grisebach and Hahlbrock, 1974; Hahlbrock *et al.*, 1971). The enzymes are subject to light-mediated control and may function together in phenylpropanoid metabolism. Condensation of the activated phenylpropanoid unit with three 2-carbon acetate-derived units is the first unique step of flavonoid biosynthesis. Although it has been suggested that the C_{15} intermediate is a polyketide (Kreuzaler and Hahlbrock, 1972), the initial stable C_{15} structure is a chalcone or its isomeric flavanone (Grisebach, 1965; Grisebach and Barz, 1969). Distinction between a chalcone or flavanone C_{15} product is not yet possible, because the chalcone-flavanone synthesizing enzyme is contaminated with chalcone-flavanone isomerase (Grisebach and Hahlbrock, 1974). The events which culminate in chalcone-flavanone biosynthesis are schematically represented in Fig. 4.

B. THE POINT OF DIVERGENCE BETWEEN FLAVONOID AND ISOFLAVONOID BIO-SYNTHESIS

Grisebach and associates (Grisebach, 1965) unambiguously demonstrated that the isoflavonoid carbon skeleton arises from flavonoid precursors by a 1,2-phenyl shift of the B-ring. Whether the immediate flavonoid precursor is a chalcone or its isomeric flavanone is subject to debate. Moreover, it is questionable that the rearrangement proceeds straightforwardly to an iso-flavanone. Since isoflavanones are a relatively rare class of isoflavonoids,

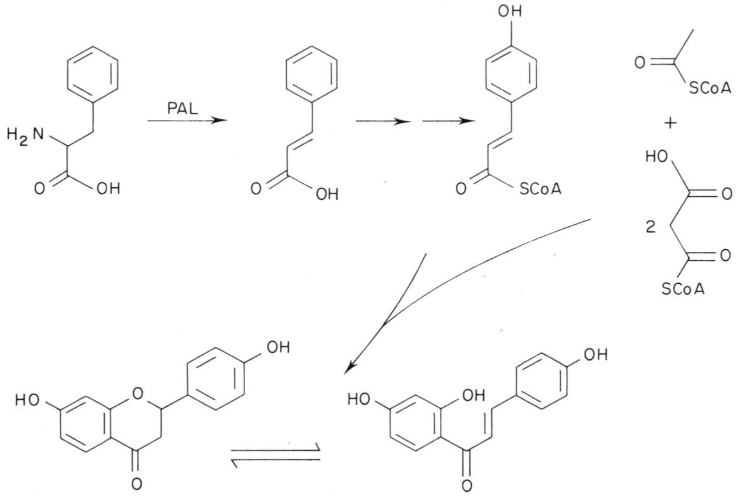

FIG. 4. Formation of the C_{15} intermediate.

it has been postulated that the rearrangement is oxidative and leads to an isoflavone (Wong, 1970).

Seedlings of *Cicer arietinum* L. stereospecifically incorporate the natural ($-$)-enantiomer of 4′,5,7-trihydroxyflavanone into 5,7-dihydroxy-4′-methoxyisoflavone (Patschke *et al.*, 1966). The stereospecificity of the conversion could be interpreted to mean that an optically inactive compound, such as a chalcone, is an unlikely possibility as an intervening structure. Recently, however, chalcone-flavanone isomerases, which are all specific for ($-$)-flavanones, have been purified from a number of sources (Hahlbrock *et al.*, 1970; Moustafa and Wong, 1967). Thus Wong (1968) suggested that conversion of ($-$)-flavanones to isoflavones may indeed proceed via a chalcone intermediate and that incorporation of flavanones into isoflavones may be indirect. To test the hypothesis, parallel competitive feeding experiments were performed with subterranean clover (*Trifolium subterraneum* L.) seeds (Wong, 1968). The seeds were fed either ^{14}C-isoliquiritigenin (2′,4,4′-trihydroxychalcone) diluted with an equal amount of ($-$)-liquiritigenin (4′,7-dihydroxyflavanone) or ($-$)-^{14}C-liquiritigenin diluted similarly with isoliquiritigenin. Isotopic dilution patterns of isolated isoflavones indicated that the chalcone was the most immediate precursor of the isoflavone. The experimental findings were supported by later double-labelling experiments (Wong and Grisebach, 1969). Furthermore, no isoflavanones with the same substitution pattern were isolated from *C. arietinum*, and thus the conversion indeed appears to be oxidative (Grisebach and Zilg, 1968). Verification that ($-$)-flavanones are only indirectly involved in isoflavonoid biosynthesis awaits confirmation that the initial C_{15} product

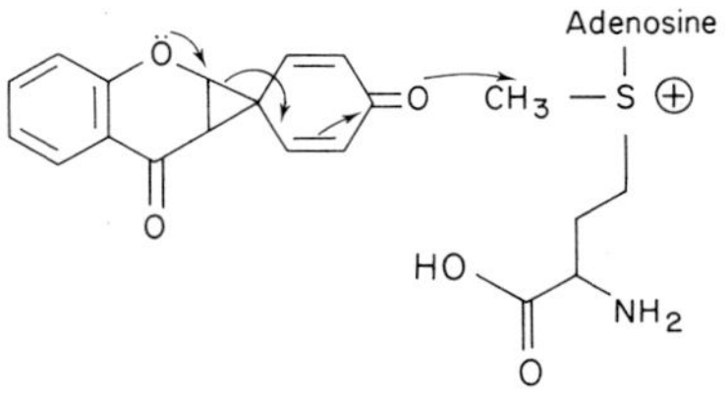

FIG. 5. Hypothetical chalcone to isoflavone biosynthetic intermediate.

from the acetate-malonate and shikimic acid pathways is a chalcone. Pelter and associates (Pelter, 1968; Pelter *et al.*, 1971), on the basis of a chemical analogy, have suggested that the mechanism of conversion of chalcones to isoflavones may involve a spirodienone intermediate (Fig. 5).

C. BIOSYNTHETIC RELATIONSHIPS AMONG ISOFLAVONOIDS

A plausible scheme for the interconversion of isoflavonoid classes pertinent to this review is shown in Fig. 6. A more complete scheme is presented by Wong (1970). The pathway may more resemble a metabolic grid than a sequential precursor to endpoint relationship. Evidence for the existence of an isoflavonoid grid is less documented than that for a flavonoid grid (Harborne, 1973; Ribéreau-Gayon, 1971), and evidently separate flavonoid and isoflavonoid pools exist (Rathmell and Bendall, 1971).

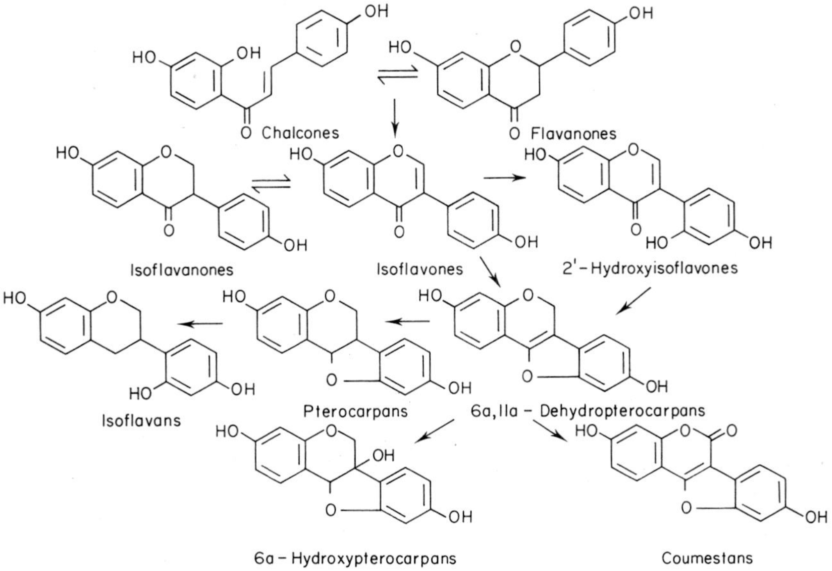

FIG. 6. A plausible biosynthetic scheme for the isoflavonoids.

Unfortunately, at the present time much of the isoflavonoid biosynthetic scheme is based, not on experimental verification, but on logical deductions and chemical analogies. The most informative experimental studies are those of Grisebach and associates (Berlin *et al.*, 1972; Dewick *et al.*, 1970). Mung bean (*Phaseolus aureus* Roxb.) seedlings or cell suspension cultures were fed various radioactive-labelled precursors, and isotope dilution into 3,9-dihydroxycoumestan was determined (Fig. 7). 4′,7-Dihydroxyisoflavone (I) and 4′,7-dihydroxyisoflavanone (II) were excellent precursors; however, label from 2′,4′,7-trihydroxyisoflavone (III) was also incorporated into the coumestan. Thus carbon can flow via several pathways, i.e. in a grid, although pathway A is of more consequence than pathway B.

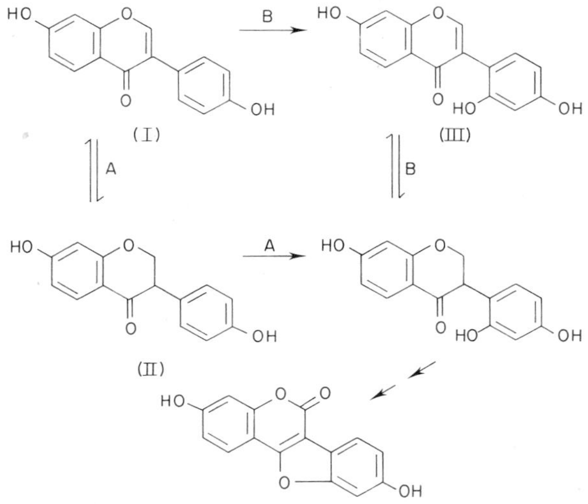

FIG. 7. Metabolic grid for biosynthesis of coumestans from isoflavones. Scheme from Berlin *et al.*, 1972.

Dehydropterocarpans are favoured candidates for the central position in pterocarpanoid biosynthesis (Berlin *et al.*, 1972; Dewick *et al.*, 1970; Hijwegen, 1973; Keen *et al.*, 1972b; Wong, 1970). Oxidation of dehydropterocarpans could produce coumestans, and reduction would yield pterocarpans. Similarly, 6a-hydroxypterocarpans could be derived from addition of water across the 6a-11a double bond. Chemical analogies to all but the latter reaction have been summarized by Wong (1970). More satisfying is the fact that 3,4,8,9-tetramethoxy-6a,11a-dehydropterocarpan spontaneously oxidizes to 3,4,8,9-tetramethoxycoumestan (Ferreira *et al.*, 1971). Such a reaction could explain the difficulty in isolating 6a,11a-dehydropterocarpan intermediates from plant tissue. However, 6a-hydroxypterocarpans themselves may undergo dehydration reactions to 6a,11a-dehydropterocarpans

(Joshi and Kamat, 1973; Perrin and Bottomley, 1962; Sims *et al.*, 1972; VanEtten *et al.*, 1975).

Catalytic hydrogenolysis of a pterocarpan under mild conditions yields the corresponding 2'-hydroxyisoflavan. Two lines of evidence summarized by Wong (1970) support his position that an analogous reaction probably is responsible for isoflavan biosynthesis by higher plants. First, all known natural isoflavans are 2'-oxygenated, as would be expected if they arise from cleavage of the benzylphenyl ether linkage of pterocarpans. Second, although enantiomeric pterocarpans and isoflavans occur naturally, in plants where pterocarpans and isoflavans co-occur, their stereochemistry corresponds. Furthermore, it has been demonstrated that some fungi can metabolize pterocarpans to the corresponding 2'-hydroxyisoflavans. (See section VIII.) Analogous enzymes may also occur in higher plants.

D. HYDROXYLATION, METHYLATION, AND PRENYLATION OF ISOFLAVONOIDS

Since the aromatic rings of isoflavonoids originate from different biosynthetic pathways, it is not surprising that hydroxylation patterns are determined by different series of reactions. Grisebach (1965) fed labelled chalcone precursors to plants and demonstrated that in isoflavone products the hydroxylation pattern from the chalcone A-ring was retained, although it was modified in the B-ring. Thus hydroxylation patterns of the A-ring seem to be fixed prior to the chalcone stage. Cyclization of the acetate units at the appropriate stages of reduction can accommodate the A-ring oxygenation patterns at carbons 5 and 7 (traditional numbering system) of isoflavonoids (Ribéreau-Gayon, 1971). Yet some compounds, such as the phytoalexins 3-hydroxy-2,9-dimethoxypterocarpan, 2,3,9-trimethoxypterocarpan, and 4-hydroxy-2,3,9-trimethoxypterocarpan, clearly possess more A-ring oxygenation than could have been supplied by acetate precursors. Whether some of the oxygen substitution in induced compounds is from microbial alteration is moot; however, even constitutive pterocarpans such as 2-hydroxy-3-methoxy-8,9-methylenedixoypterocarpan from *Neorautanenia edulis* C.A. Sm. possess additional A-ring oxygenation (Rall *et al.*, 1970).

Almost all isoflavonoid natural products are oxygenated at the 4'-position (9-position in the Ring Index numbering scheme). Such oxygenation, which takes place at the phenylpropanoid stage, is catalyzed by *trans*-cinnamic acid-4-hydroxylase (Hahlbrock *et al.*, 1971; Russell and Conn, 1967; Sutter and Grisebach, 1969). Additional B-ring hydroxylation probably occurs at or subsequent to the chalcone stage (Crombie *et al.*, 1973; Grisebach, 1967).

Methionine is a methyl donor for isoflavonoids, although the nature of the immediate reactant (*S*-adenosylmethionine?) is not known (Barz and Roth-Lauterbach, 1969; Hadwiger, 1966). Methylation of isoflavonoids apparently proceeds by several sorts of reactions and can occur with various intermediates serving as substrates. Although an O-methyltransferase with specificity for flavonoids has been purified, direct aromatic methoxylation may occur (Ebel et al., 1972; Swain, 1962). Phenylpropanoid compounds which are 4-O-methylated are good precursors for 4'-O-methylated iso-flavonoids, but they are demethylated prior to incorporation (Barz and Grisebach, 1967; Crombie et al., 1973; Ebel et al., 1970). The requirement for a free 4'-hydroxy-group may reflect the intervention of B-ring quinoid compounds in isoflavonoid biosynthesis (Pelter, 1968). Crombie and associates (Crombie et al., 1973) have recently proposed that methylation of the 4'-position proceeds concurrently with transformation of chalcones to isoflavones (Fig. 5).

Several isoflavonoid phytoalexins, e.g. phaseollin, kievitone, glyceollin, and phaseollidin, possess isoprenoid substituents from the mevalonic acid pathway. These substituents, which may survive as dimethylallyl groups or be modified to fused 2,2-dimethylpyran rings, can occupy various positions on both aromatic rings. Prenylation is a post-chalcone phenomenon, since unprenylated chalcones serve as efficient precursors for prenylated iso-flavonoids (Hess et al., 1971; Keen et al., 1972b). An enzyme which couples prenyl units to an aromatic nucleus has been described recently (Ellis and Brown, 1974).

E. TURNOVER OF ISOFLAVONOIDS IN HIGHER PLANTS

Although isoflavans, coumestans, and 6a-hydroxypterocarpans are logical choices for end products of isoflavonoid biosynthesis (Fig. 2), other isoflavonoids commonly accumulate in plant tissue—often when the above mentioned compounds are absent. This begs the frequently overlooked question of turnover of isoflavonoids in higher plants. Several questions, which are especially pertinent to the biosynthesis of isoflavonoid phytoalexins, can be posed about turnover of isoflavonoids: (1) Does turnover of iso-flavonoids to non-flavonoid compounds occur in higher plants? (2) Are any of the isoflavonoids bona fide metabolic end products? (3) By what mechanism are the levels of different isoflavonoids maintained in balance in higher plants?

Although conclusive answers to the latter two inquiries are not available, the first can be answered in the affirmative (Barz and Adamek, 1970; Barz and Hosel, 1971; Barz and Roth-Lauterbach, 1969; Barz et al., 1970; Berlin and Barz, 1971; Berlin et al., 1974). However, the rate of turnover is a function of the compound in consideration (Barz, 1969). Thus although the biological half-lives of 7-hydroxy-4'-methoxyisoflavone, 4',7-dihydroxyiso-flavone, and 3,9-dihydroxycoumestan are about 50 h, turnover of 5,7-dihydroxy-4'-methoxyisoflavone is very slow (Barz, 1969). Label from

^{14}C-4′,7-dihydroxyisoflavone is assimilated into lipid material, carbo-hydrates, phenols, amino and organic acids (Barz *et al.*, 1970).

Interestingly, turnover of exogenously-supplied pisatin in *Pisum sativum* is nil, or nearly so (Hadwiger, 1967). However, the pathway for pisatin synthesis (and turnover?) was not operative during the experiments. It is tempting to speculate that the enormous accumulation of some isoflavonoid phytoalexins in perturbed plants may be a reflection of their status as metabolic end products. Yet this phenomenon may be misleading; after high concentrations of such compounds accumulate in perturbed cells, the cellular metabolic system may be inactivated by death of the cells. An alternate interpretation of phytoalexin accumulation is that in healthy plants rates of synthesis and turnover are high and in equilibrium. Build-up of phytoalexins could thus result from a block of the turnover system.

Isoflavonoid phytoalexins offer themselves as lucrative tools for the elucidation of biogenetic pathways. For instance, the *Phaseolus vulgaris* phytoalexins phaseollidin, phaseollin, phaseollinisoflavan, and 2′-methoxy-phaseollinisoflavan are in a logical biosynthetic sequence. The steps can be envisaged as pterocarpan prenylation (phaseollidin), side chain cyclization (phaseollin), benzylphenyl ether cleavage (phaseollinisoflavan), and 2′-*O*-methylation (2′-methoxyphaseollinisoflavan). Study of this and other plausible isoflavonoid phytoalexin pathways may provide information to our as yet unanswered queries into control of isoflavonoid metabolism in healthy as well as diseased tissue.

IV. INDUCTION

The mechanism of induction of phytoalexin biosynthesis has long intrigued phytopathologists, both in its potential application to the differentiation of susceptible and resistant host–pathogen interactions, and its implications for molecular biology. While the investigations have opened a Pandora's box of inducers and have permitted induction hypotheses to accumulate, a unifying mechanistic concept of induction consistent with the available data is lacking. The absence of such an all encompassing explanation may indicate the uniqueness of each individual host–pathogen system. Neverthe-less, we shall briefly turn our attention to inducers and induction systems, from which specificity, localization, and mechanism of induction emerge as research frontiers.

A. INDUCERS

Müller (1956, 1958), although he did not know the chemical identity of the compounds he studied, was undoubtedly the first to induce experimentally an isoflavonoid phytoalexin. His method, which is commonly called the drop-diffusate technique, consists of incubating droplets of a spore suspension on pod endocarp tissue and then assaying for inhibitory compounds which diffuse into the droplets. Although the procedure has been labelled more recently as biologically unnatural (Keen and Horsch, 1972; Kuć, 1972), its redeeming virtues are simplicity and the purity of phytoalexin preparations obtained from the diffusates. The technique has been modified to utilize leaves (Bonde *et al.*, 1973; Heath and Wood, 1971; Higgins and Millar, 1968) and stems (Rathmell, 1973) as challenged tissue. The drop-diffusate technique is equally useful when abiotic agents are tested.

Tissue from all parts of the plant have the ability to produce phytoalexins in response to infection by a variety of biotic agents. Fungi are particularly efficient as inducing organisms. For example, in consequence to infection of *Glycine max* by *Phytophthora megasperma* Drechs. var. *sojae*, A. A. Hildeb., glyceollin may increase from an undetectable level to a concentration greater than 10% of the dry weight of the infected tissue in 24–48 h (Keen and Horsch, 1972). Phytoalexins are also induced in nematode- (Abawi *et al.*, 1971), bacteria- (Cruickshank and Perrin, 1971; Keen and Kennedy, 1974; Stholasuta *et al.*, 1971), and virus-infected tissue (Bailey, 1973; Bailey and Burden, 1973; Klarman and Hammerschlag, 1972). Rotting, naturally infected germinating seed is an excellent source of phytoalexins (Keen, 1975b; Keen and Sims, 1973; VanEtten *et al.*, 1975). Recently it has been shown that roots of pea and French bean exude phytoalexins into bathing solutions of non-sterile water (Burden *et al.*, 1974b).

An intriguing array of chemicals induce phytoalexins. The compounds, which are classed by chemical groups in Table II, are structurally diverse. For instance, scorpion venom, the respiratory pigment cytochrome *c*, mercuric salt solutions, and the gaseous hormone, ethylene, are all inducers. Thus phytoalexin inducers probably do not share a common structural feature responsible for their induction capacity. It is more likely that the compounds injure plant cells by diverse mechanisms. Such perturbations culminate in initiation of the biochemical lesion for phytoalexin induction. Whether a compound is an inducer may thus simply reflect its ability to injure plant cells (Day, 1974; Deverall, 1972a). For instance, ophiobolin, a toxin of *Cochliobolus miyabeanus* (Ito and Kuribay) Drechs. ex Dast., and ascochytine, a toxin of *Ascochyta pisi* Lib. and *A. fabae* Speg., induce pisatin in pea pod endocarp tissue (Land *et al.*, 1975b; Oku *et al.*, 1973). Nevertheless, much research directed at the mechanism of induction has dealt with identification of the biochemical lesion by analysis of the physiological activities of chemical inducers.

In addition to the non-specific, abiotic inducers of phytoalexins, several highly potent inducers of fungal origin have been identified. Monilicolin A, a polypeptide of molecular weight *ca* 8 000, has been separated from the mycelium of the fruit pathogen *Monilinia fructicola* (Wint.) Honey (Cruickshank and Perrin, 1968). It is an active inducer at nanomolar concentrations,

TABLE II

Chemical inducers of isoflavonoid phytoalexins

Inducer class	Phytoalexin	Reference
Antibiotics and metabolic inhibitors	Phaseollidin	Cruickshank et al., 1974a
	Phaseollin	Cruickshank and Perrin, 1971
		Hess and Hadwiger, 1971
		Perrin and Cruickshank, 1965
	Pisatin	Bailey, 1969a
		Hadwiger, 1972a,b
		Hadwiger and Martin, 1971
		Hadwiger and Schwochau, 1971a
		Land et al., 1975b
		Oku et al., 1973
		Perrin and Cruickshank, 1965
		Schwochau and Hadwiger, 1968, 1969
Coconut milk and cyto-kinins	Pisatin	Bailey, 1969b, 1970
Cyclic AMP	Glyceollin	Keen and Kennedy, 1974
Ethylene	Pisatin	Chalutz and Stahmann, 1969
Heavy metal salts	Glyceollin	Klarman and Sanford, 1968
	3-Hydroxy-9-meth-oxypterocarpan	Higgins and Millar, 1968
	Kievitone	Smith et al., 1973a
	Phaseollidin	Cruickshank et al., 1974a
	Phaseollin	Perrin and Cruickshank, 1965
	Pisatin	Cruickshank and Perrin, 1963b
		Perrin and Cruickshank, 1965
		Uehara, 1963
Organic acids	Pisatin	Bailey, 1969a
		Perrin and Cruickshank, 1965
Organic fungicides	Glyceollin	Reilly and Klarman, 1972
	Pisatin	Oku et al., 1973
Enzymes	Pisatin	Hadwiger et al., 1974
		Hadwiger and Schwochau, 1970
		Schwochau and Hadwiger, 1968

but is neither phytotoxic nor fungitoxic (Cruickshank and Perrin, 1968; Paxton et al., 1974). The polypeptide is an inducer ostensibly only of the French bean phytoalexins, phaseollin and phaseollidin (Cruickshank and Perrin, 1968; Cruickshank et al., 1974a), yet germinating spores and culture filtrates of M. fructicola cause pisatin to accumulate in challenged pea pod endocarp (Cruickshank and Perrin, 1963b). Although monilicolin A is a promising research tool, the role in nature of a specific French bean phytoalexin inducer from a fruit pathogen is difficult to rationalize.

Albersheim and associates (Anderson and Albersheim, 1974; Ayers *et al.*, 1974) have purified and partially characterized phytoalexin-inducing macromolecules from culture fluids of *Colletotrichum lindemuthianum* (Sacc. and Magn.) Briosi and Cav., a French bean pathogen and *Phytophthora megasperma* var. *sojae*, a soybean pathogen. The compounds are polysaccharides, active in submicrogram concentrations, and may be components of the fungal cell wall. Cultures of *Fusarium solani* (Mart.) Sacc. f. sp. *phaseoli* (Burk.) Snyd. and Hans. and *F. solani* (Mart.) Sacc. f. sp. *pisi* (F. R. Jones) Snyd. and Hans. have also been examined for the presence of phytoalexin inducers (Teasdale *et al.*, 1974). Although from both fungi multiple inducers of various molecular weights were detected, the most active preparations were of high molecular weight.

TABLE III

Glyceollin induction in near-isogenic Harosoy and Harosoy 63 soybean hypocotyls by culture fluids from three races of *P. megasperma* var. *sojae* and a partially purified specific elicitor from race 1. Reaction of soybean cultivars to challenge by the intact pathogen: S = susceptible, R = resistant. Data from Keen, 1975a.

Treatment	Cultivar Harosoy		Cultivar Harosoy 63	
	Reaction	Glyceollin (ppm)	Reaction	Glyceollin (ppm)
Race 1 fluids	S	16 ± 10	R	81 ± 27
Race 2 fluids	S	35 ± 18	R	75 ± 22
Race 3 fluids	S	50 ± 5	S	50 ± 14
Race 1 specific elicitor	S	22 ± 12	R	105 ± 24

Keen and associates (Keen *et al.*, 1972a; Keen, 1975a) reported that races of *P. megasperma* var. *sojae* produce phytoalexin inducers that may function as determinants of varietal resistance to pathogen races, i.e. race specificity (section IX). They propose that such compounds be called specific elicitors, and that they may function by de-repressing phytoalexin production in resistant interactions. Fungus races 1, 2, and 3 were assayed for elicitor activity by placing drops of culture fluids on wounded hypocotyls of two soybean varieties. In addition, a specific elicitor from race 1 was partially purified and analyzed. Glyceollin accumulation and host reaction for each combination are presented in Table III. More glyceollin accumulates in response to preparations obtained from races which give the resistant interaction.

B. LOCALIZATION OF PHYTOALEXIN ACCUMULATION

An essential feature of the Phytoalexin Theory *sensu stricto* is the localization of phytoalexin accumulation at the site of induction (Müller and Börger, 1940). Phytoalexins are produced by cells invaded by and juxtaposed to challenging organisms, but are absent or in greatly reduced quantity in nearby tissues (Bailey and Deverall, 1971; Cruickshank and Perrin, 1965; Keen, 1971; Pierre and Bateman, 1967; Pueppke and VanEtten, 1974; Rahe, 1973; Rathmell, 1973; Smith et al., 1975). An evaluation of the response at a cellular or subcellular level is more difficult. Potent fungal inducers of phytoalexins may be extracellular or intimately associated with cultured mycelium (Anderson and Albersheim, 1974; Ayers et al., 1974; Cruickshank and Perrin, 1968; Keen 1975a; Teasdale et al., 1974). Therefore, in the plant, phytoalexins may be elicited either at the hyphal surfaces or in adjacent areas where extracellular inducers have penetrated. Such localization of induction fixes the concentration of phytoalexins to which a potential pathogen is exposed and may be critical in distinguishing resistance from susceptibility.

C. MECHANISM OF INDUCTION

Metabolic control of phytoalexin biosynthesis may be exerted at the level of nucleic acids, e.g. transcription and translation, or by control of enzyme activity, e.g. co-factor activation, removal of inhibitors, or allosteric modification. In any case, inducers alter metabolic control such that phytoalexin-synthesizing machinery is functional. A basic unresolved question is whether RNA synthesis is a prerequisite for phytoalexin accumulation. Increased synthesis of protein and certain RNA fractions accompanies pisatin production in pea pod endocarp (Hadwiger, 1968, 1972a,b; Hadwiger and Martin, 1971; Hadwiger and Schwochau, 1970, 1971a,b; Hadwiger et al., 1973, 1974; Schwochau and Hadwiger, 1968, 1969; Teasdale et al., 1974). Inhibitors of protein and RNA synthesis, such as cycloheximide and actinomycin D, block pisatin induction, but at low concentrations these compounds themselves serve as inducers. Under certain circumstances, pisatin induction is accompanied by a decrease in total incorporation of precursors into RNA (Hadwiger, 1972b; Schwochau and Hadwiger, 1968) or protein (Hadwiger, 1972a). Nevertheless, Hadwiger and associates interpret their data to indicate that pisatin inducers promote transcription of normally inaccessible regions of DNA. Consequently, proteins necessary for pisatin production can be synthesized. Their evidence is: (1) inhibitors of RNA synthesis, when applied with the inducer or shortly thereafter, reduce or prevent pisatin production (Hadwiger et al., 1973; Schwochau and Hadwiger, 1969; Teasdale et al., 1974); (2) treatments which are known to alter the conformation of DNA induce pisatin (Hadwiger, 1972b; Hadwiger and Martin, 1971; Hadwiger and Schwochau, 1970, 1971a,b; Hadwiger et al., 1973, 1974; Schwochau and Hadwiger, 1969). These treatments include application of planar DNA-intercalating compounds, irradiation with u.v. light, exposure to light subsequent to application of photosensitive psoralen compounds, and incubation with polypeptides or even heavy metals. The most recent statement of Hadwiger's transcription induction hypothesis is:

> "that the diverse groups of phytoalexin inducers affect multiple segments of nuclear DNA and that the changes in DNA conformation which occur in the globular regions improve the accessibility or 'melting in' of polymerase to regions previously inaccessible" (Hadwiger et al., 1974).

A similar phytoalexin induction hypothesis, based on the Jacob-Monod model of gene regulation, has been proposed (Hadwiger and Schwochau, 1969; Keen, 1975a). Its viability, however, rests on the assumption that a Jacob-Monod regulatory system of phytoalexin biosynthesis is operative.

The induction mechanism for phaseollin may be similar to that of pisatin. The DNA-intercalating compound, 9-aminoacridine, induces phaseollin in French bean pod endocarp (Hess and Hadwiger, 1971). Yet in some experiments phaseollin-inducing concentrations of 9-aminoacridine depressed incorporation of radioactive-labelled RNA percursors. When monilicolin A serves as inducer, inhibition of RNA synthesis by actinomycin D does not prevent production of phaseollin (Biggs, 1972). In some experiments utilizing actinomycin D, incorporation of ^{3}H-uridine into RNA was reduced by 70% prior to and 95% subsequent to monilicolin A application. Yet phaseollin production was equivalent to that of control tissue not treated with the antibiotic. Biggs (1972) presents evidence that in short-term studies incorporation of radioactive precursors into RNA is affected by choice of precursor and the order of addition of precursor and actinomycin D to the tissue. Thus inducer-inhibitor experiments should be interpreted cautiously, and unequivocal proof that *de novo* RNA synthesis is necessary for phaseollin production is lacking.

Cruickshank and associates (Cruickshank et al., 1971) provide a comprehensive critique of postulated induction mechanisms involving enzyme activation. None has been verified experimentally. Increased phenylalanine ammonia-lyase activity is temporally associated with phytoalexin induction, but its significance is debatable (Cruickshank et al., 1971; Hadwiger et al., 1970; Rathmell, 1973). In some plant species, induction of phytoalexins specifically stimulates the isoflavonoid biosynthetic pathway, since levels of flavonoids and endogenous phenols are unchanged (Rathmell, 1973; Rathmell and Bendall, 1971). In others flavonoid biosynthesis *per se* is activated (Olah and Sherwood, 1971, 1973). Unfortunately, little else is known about inducer-mediated changes in enzyme activity. An understanding of enzyme activation mechanisms is dependent on a thorough knowledge of the isoflavonoid biosynthetic pathway and such knowledge is incomplete.

Past work on the specificity of phytoalexin induction has dealt with the biosynthetic pathway of a single isoflavonoid phytoalexin. We now know that a plant species may present different pathogens with distinctive arrays of phytoalexins. This may indicate that additional controls within the isoflavonoid biosynthetic pathway are operative. A case in point is presented in Table IV. For four pathogen–*Phaseolus vulgaris* interactions, the amounts of four isoflavonoid phytoalexins relative to phaseollin are compared. Although hypocotyl tissues were analysed in each interaction, experimental techniques varied. In addition, possible differential rates of metabolism of the phytoalexins by the fungal pathogens were not determined. Nevertheless, the substantial differences between interactions strongly imply that biosynthetic control of kind and quantity of isoflavonoid phytoalexins is operative.

V. Biological Action Spectrum

A. ACTIVITY AGAINST FUNGI

The most intensively studied biological property of isoflavonoid phytoalexins is antifungal activity. The commonly used bioassays assess the effects of the compounds on spore germination, germ tube growth, radial growth of mycelium on solid medium, or accumulation of fungal mycelium in liquid culture. Isoflavonoid phytoalexins are active generally at concentrations from 10^{-5} M to 3×10^{-4} M, i.e. about 1–100 µg/ml (Bailey and Burden, 1973; Cruickshank, 1962; Cruickshank and Perrin, 1971; Perrin and Cruickshank, 1969; VanEtten, 1976). However, *in vitro* antifungal activity is a function of both the growth parameter measured in the bioassay and the conditions under which the bioassay is performed (Bailey, 1974; Cruickshank and Perrin, 1971; Cruickshank *et al.*, 1971; Heath and Higgins, 1973; VanEtten, 1973a). Bioassay conditions are rarely standardized among researchers, and therefore it is difficult and possibly risky to draw general conclusions about the antifungal activity of these compounds. In this review the terms tolerance and insensitivity are used frequently and interchangeably. These terms are, of course, relative. We will consider an organism to be tolerant, in a given type of bioassay, if it can grow at a concentration (generally $1–3 \times 10^{-4}$ M) which substantially inhibits or prevents the growth of most fungi. The upper extremity of this concentration range corresponds roughly to the limit of water solubility for many of the isoflavonoid phytoalexins.

The most comprehensive screens for antifungal activity of the isoflavonoid phytoalexins have been performed by Cruickshank and associates (Cruickshank, 1962; Cruickshank and Perrin, 1971; Perrin *et al.*, 1974). In a pioneering study, Cruickshank (1962) assayed the effect of pisatin on fifty isolates of "filamentous" fungi. Only five isolates were insensitive to the compound; all five were pathogens of pea. Similar results were obtained

TABLE IV

Differential accumulation of isoflavonoid phytoalexins in infected *Phaseolus vulgaris* L. hypocotyls

Pathogens	Days after inoculation	Molar ratio of *P. vulgaris* phytoalexins compared to phaseollin			
		Phaseollidin: Phaseollin	Phaseollinisoflavan: Phaseollin	Kievitone: Phaseollin	2'-Methoxyphaseollinisoflavan: Phaseollin
Rhizoctonia solani	2[a]	<0·01	<0·01	8·39	0·00
	3·6[b]	0·06	0·02	0·74	0·00
Fusarium solani f. sp. *phaseoli*	2[c]	<0·01	0·21	0·00	<0·01
	6[d]	0·02	0·62	0·00	0·04
Colletotrichum lindemuthianum (Race β)	6[e]	0·19	0·65	0·26	
Tobacco necrosis virus	5[f]	0·12	0·92	0·37	

[a] Tissue contained 33 µg phaseollin/g fresh wt. (Smith et al., 1975).
[b] Tissue contained 782 µg phaseollin/g fresh wt. (Smith et al., 1975).
[c] Tissue contained 352 µg phaseollin/g dry wt. (VanEtten and Smith, 1975).
[d] Tissue contained 13 600 µg phaseollin/g dry wt. (VanEtten and Smith, 1975).
[e] Tissue contained 8·3 µg phaseollin/g fresh wt. (Bailey, 1974).
[f] Tissue contained 540 µg phaseollin/g fresh wt. (Bailey and Burden, 1973).

when twenty-seven fungal isolates, pathogenic and non-pathogenic to French bean, were bioassayed against phaseollin (Cruickshank and Perrin, 1971). The five fungi less sensitive to phaseollin were pathogens of French bean. In these two studies they observed only one pathogen which was sensitive to its host's phytoalexin. The pattern was for pathogens to be tolerant of their host's phytoalexin and non-pathogens to be sensitive (Cruickshank, 1962; Cruickshank and Perrin, 1971). Although other studies with these and different phytoalexins (phaseollidin and 3-hydroxy-9-methoxypterocarpan) have revealed additional exceptions to this pattern, in general, the results have confirmed the type of differential sensitivity first observed by Cruickshank (Higgins, 1972; Nonaka, 1967; Perrin et al., 1974; Pueppke and VanEtten, 1974; Uehara, 1964; VanEtten, 1973a). In each study radial growth of mycelium on a solid medium was the bioassay. Results typical of differential sensitivity studies are illustrated in Tables V and VI, section VIII.

Some fungal species which are inhibited differentially by a phytoalexin in radial growth bioassays respond identically in spore germination or germ tube growth bioassays (Cruickshank and Perrin, 1971; Higgins, 1972; Higgins and Millar, 1968; Uehara, 1960). For example, Cruickshank and Perrin (1971) observed that the dosage–response curves (spore germination versus phaseollin concentration) of *Colletotrichum lindemuthianum*, a French bean pathogen, and *Monilinia fructicola* a non-pathogen, were identical. However, in radial growth bioassays, *M. fructicola* was much more sensitive. Cruickshank and associates (Cruickshank and Perrin, 1971; Cruickshank et al., 1971) have suggested that radial growth bioassays are more relevant to *in situ* events, because phytoalexin production is a post-infection phenomenon, and therefore occurs after spore germination (Cruickshank and Perrin, 1971; Cruickshank et al., 1971).

Nevertheless, it is axiomatic that all *in situ* events are not duplicated in *in vitro* bioassays. Therefore, conclusions about how *in vitro* antifungal activity relates to pathogenesis should rely on data from as many types of bioassays as possible. Bailey and Burden (1973) found that fungi are differentially sensitive to phaseollin and three other French bean phytoalexins in spore germination bioassays. Interestingly, the most sensitive fungus was the only French bean pathogen tested, *C. lindemuthianum*. Heuvel and Glazener (1975) assayed liquid cultures of twelve fungi against phaseollin. The fungi were differentially inhibited, but the response was not correlated with pathogenicity to bean. Twelve fungi exhibited differential sensitivity to kievitone in the radial growth bioassay, but again the effects were not correlated with pathogenicity to bean (Smith et al., 1975).

If a phytoalexin is produced by several plant species, correlation of pathogenicity to one species and tolerance to the phytoalexin of that species need not be absolute. That is, a fungus not pathogenic to species A, but tolerant of phytoalexin A, may be a pathogen of species B, which also produces phytoalexin A. For example, *Medicago sativa* and *Trifolium repens* L. each produce 3-hydroxy-9-methoxypterocarpan. *Stemphylium botryosum* Wallr. is tolerant of this phytoalexin, but is a pathogen only of *M. sativa*.

In several well documented instances, pathogens of a given plant are sensitive in a variety of bioassays to the isoflavonoid phytoalexin(s) produced by the plant: *Aphanomyces euteiches* Drechs. versus pisatin, *Rhizoctonia solani* Kühn versus bean phytoalexins, and *Phytophthora megasperma* var. *sojae* versus glyceollin (Klarman and Sanford, 1968; Keen *et al.*, 1971; Pueppke and VanEtten, 1974; Smith *et al.*, 1975; VanEtten, 1973a; VanEtten and Bateman, 1971). Thus it is clear that a prerequisite for pathogenicity by a fungus on a given plant species is not tolerance by the pathogen (as judged by *in vitro* bioassays) of that species' phytoalexin(s).

Since some plant species can produce several isoflavonoid phytoalexins, it is possible that *in situ* the compounds act synergistically to produce a more highly antifungal environment than one based solely on the additive effects of the compounds. In the only study to date that addressed this question the effects of phaseollin and kievitone *in vitro* on the bean pathogen, *R. solani*, were additive only (Smith *et al.*, 1975).

Yeasts, possibly because they can reproduce by budding, are often considered separately from other fungi. Since most of the isoflavonoid phytoalexins exhibit a broad spectrum of antifungal activity, it would be expected that they also inhibit the growth of yeasts. Indeed, 3-hydroxy-8,9-methylenedioxypterocarpan reportedly inhibits the growth of two yeast species (Suginome, 1966). When pisatin was assayed against nine yeast species, it was inhibitory to most (Cruickshank, 1962) but, surprisingly, when phaseollin was assayed against eight of the same species, it was not inhibitory to any of them (Cruickshank and Perrin, 1971).

B. ACTIVITY AGAINST OTHER ORGANISMS

Little is known about the activity of the isoflavonoid phytoalexins against lower organisms other than fungi. Although eleven out of twenty-four bacterial species tested showed some sensitivity to pisatin (Cruickshank, 1962), only three out of sixteen species assayed against phaseollin exhibited any sensitivity to it (Cruickshank and Perrin, 1971; Stholasuta *et al.*, 1971). Glyceollin is antibacterial (Keen and Kennedy, 1974), but 3-hydroxy-8,9-methylenedioxypterocarpan apparently is not (Suginome, 1966). The nematode, *Pratylenchus penetrans* Cobb, is insensitive to phaseollin (Abawi *et al.*, 1971), but pisatin, phaseollin, and phaseollinisoflavan were all toxic to brine shrimp, *Artemia salina* L. (Whalen and VanEtten, unpublished results).

Although isoflavonoid phytoalexins are usually produced in association with necrosis of plant tissue, surprisingly little is known about their phytotoxicity. Pisatin retards growth of primary roots of wheat (Cruickshank and Perrin, 1961), and it is inhibitory to the growth of pea callus cultures (Bailey, 1970). Trifolirhizin [(−)-3-hydroxy-8,9-methylenedioxypterocarpan-β-D-glucoside] represses root and hypocotyl growth and seed germination of several clover species (Chang et al., 1969). Like many other flavonoids (Stenlid, 1970), phaseollin (but not pisatin) inhibited the formation of ATP in mitochondria from cucumber, Cucumis sativus L., hypocotyls (Stenlid, personal communications).

Substantial quantities of isoflavonoid phytoalexins can accumulate in leguminous plant species. Many are important food sources for man and livestock. Yet little is known about the effects of the phytoalexins on mammals. Some coumestans and isoflavones are mildly oestrogenic in mammals (Bickoff et al., 1969; Stob, 1973). 4',7-Dihydroxyisoflavan (equol) is oestrogenic in mice (Shutt and Braden, 1968); both 2',7-dihydroxy-4'-methoxy- and 7-hydroxy-2',4'-dimethoxyisoflavan are phytoalexins. Trifolirhizin may also be oestrogenic to mice (Oung-Boran et al., 1969). However, in our preliminary studies with mice (Schwark and VanEtten, unpublished results) phaseollin was not oestrogenic. Phaseollin, 3-hydroxy-9-methoxypterocarpan, and glyceollin lyse red blood cells in vitro (VanEtten and Bateman, 1971; VanEtten, 1972).

Most of the above types of activity are expressed at phytoalexin concentrations which are also antifungal. Higher concentrations are often present in infected plants. Therefore, in nature, it might be expected that isoflavonoid phytoalexins could affect organisms other than fungal plant pathogens.

VI. Cytological and Physiological Effects on Fungi

Müller (1956, 1958) most likely made the first cytological observation of the effects of an isoflavonoid phytoalexin on a fungal cell. When he placed zoospores of Phytophthora infestans (Mont.) Dby. into antifungal diffusates obtained from bean pods in response to germinating spores, the zoospores usually swelled and burst within 60 seconds. Occasionally the zoospores shrank instead. We have observed a similar reaction by P. infestans zoospores, when they are placed in solutions containing phaseollin (VanEtten, unpublished results). Zoospores and encysted spores of Aphanomyces euteiches responded to 10^{-4} M phaseollin, phaseollinisoflavin, and 2'-methoxyphaseollinisoflavan in a similar manner (Fig. 8) (VanEtten, 1976). Within a minute or less the protoplasm became granular, and the spore assumed an irregular shape. Long-term studies indicate that treated spores were not viable (VanEtten, 1976). Surprisingly, 3×10^{-4} M pisatin does not elicit a similar effect on encysted spores, even though at this concentration pisatin almost completely inhibited mycelial growth of A. euteiches (Table V) (Pueppke and VanEtten, 1974; VanEtten, 1973a).

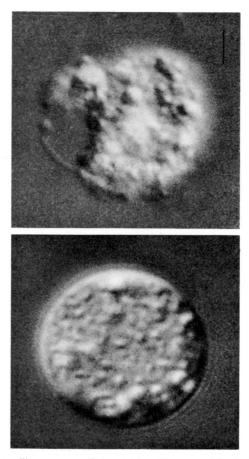

FIG. 8. Effect of phaseollin on spores of *A. euteiches*. The spore at the bottom is an untreated control. The spore at the top has been exposed to phaseollin. Experimental procedure as described by VanEtten, 1976. Interference contrast micrographs. The bar represents 2 μm. (Pueppke and VanEtten, unpublished data.)

When suddenly exposed to phaseollin, hyphae of both *Rhizoctonia solani* and *Neurospora crassa* Sher and Dodge also reacted drastically (Slayman and VanEtten, 1974; VanEtten and Bateman, 1971). Streaming of *R. solani* protoplasm ceased immediately, the hyphae shrank, and the protoplasm often contracted from the wall apex (Fig. 9). Furthermore, the protoplasm became granular, and it was difficult to recognize organelles. As with spores, the hyphal response may occur within a minute after exposure to the phaseollin solution. The response is more extreme and easier to observe in tips of young growing hyphae. Exposure of germ tubes of

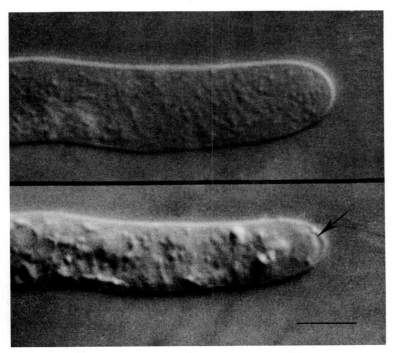

FIG. 9. Effect of phaseollin on a hyphal tip of *R. solani*. The tip prior to exposure is above. Below is the same tip immediately after exposure to phaseollin. Note the shrunken protoplasm (indicated by the arrow) and reduced hyphal diameter of the treated tip. Experimental procedure as described by VanEtten and Bateman, 1971. The bar represents 5 μm. Interference contrast micrographs. (Pueppke and VanEtten, unpublished data.)

Colletotrichum lindemuthianum to phaseollin results in a decrease in length, concurrent with swelling and distortion (Bailey and Deverall, 1971).

At concentrations which elicit the above cytological abnormalities, phaseollin alters many of the physiological processes of *R. solani* and *N. crassa*. In *R. solani* it causes a loss in dry weight, leakage of fungal metabolites, reduction of nutrient uptake, reduction of exogenous respiration, and a stimulation of endogenous respiration (VanEtten and Bateman, 1971). With *N. crassa*, whole-cell oxygen uptake is reduced 70–80% (Fig. 10), potassium uptake by potassium-starved cells is blocked (Fig. 11) and membrane potential is decreased (Slayman and VanEtten, 1974). Concurrently, intracellular ATP is only slightly decreased.

Müller (1958) suggested that the effects he observed on zoospores of *P. infestans* were attributable to disruption of the plasmalemma. Although most of the responses to phaseollin by *R. solani* and *N. crassa* could be explained by such an effect, it is difficult to separate the primary event from the spectrum of physiological responses observed. The two earliest known

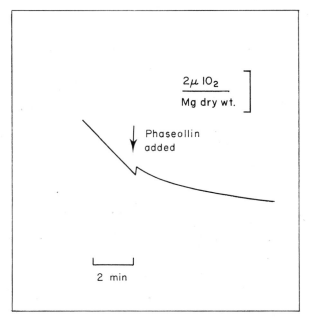

FIG. 10. Effect of phaseollin on cellular respiration of *N. crassa*. A conventional polarized platinum cathode was used to measure oxygen tension in a stirred suspension of 15–16 h old cells of *N. crassa* (0·97 mg dry wt./ml). A control rate was established and phaseollin was added (indicated by the arrow) to give a final concentration of $1·2 \times 10^{-4}$ M. (Slayman and VanEtten, 1974 and unpublished data.)

responses (*N. crassa*) are the reduction of whole-cell respiration and the destruction of the ability of potassium-depleted cells to take up potassium (Figs 10 and 11). Both occur as quickly as the cytological response, and serve to illustrate the rapidity with which at least one isoflavonoid phytoalexin can affect cells of a sensitive fungus. The actual concentration of a phyto-alexin(s) encountered by a parasite in plant tissue is unknown. A fungal parasite in plant tissue could suddenly encounter a high concentration of phytoalexin if the phytoalexin is swiftly decompartmentalized. Thus rapid responses should be considered by researchers attempting to correlate the timing of the restriction of fungal growth with phytoalexin accumulation. Rapid responses might be especially important in studies on resistance of the hypersensitive type, where it is known that the events which discriminate between a resistant and a susceptible reaction occur very quickly.

VII. STRUCTURAL REQUIREMENTS FOR ANTIFUNGAL ACTIVITY

Each of the isoflavonoid classes exhibiting antifungal activity possesses a characteristic three-dimensional shape. At one extreme are the almost

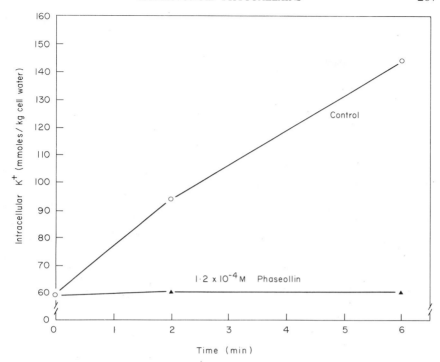

FIG. 11. Effect of phaseollin on potassium uptake by potassium-starved cells of *N. crassa*. *N. crassa* was grown in the absence of potassium for 15 h prior to treatment with phaseollin for 5 min. Potassium chloride was added at time 0 to give a final concentration of 30 mM. (Slayman and VanEtten, 1974 and unpublished data.)

planar isoflavones and isoflavanones; at the other are the skewed, aplanar isoflavans and pterocarpans (Fig. 12). Thus it is unlikely that all isoflavonoids must assume a common conformation before they can express antifungal activity. However, as a result of an interesting study of eight pterocarpans and related compounds for antifungal activity against *Monilinia fructicola*, Perrin and Cruickshank (1969) suggested that a specific molecular shape is responsible for the antifungal activity associated with the pterocarpans.

Pterocarpans possess asymmetric carbons at 6a and 11a. Although diastereomeric structures are possible, the 6a-11a junction has been established as a *cis*-fusion of the two heterocyclic rings (Suginome, 1962; Suginome and Iwadare, 1962). Therefore, only a pair of enantiomers exist. Pachler and Underwood (1967) have concluded, from nmr evidence, that the dihydrofuran ring (Ring C) is planar and the dihydropyran ring (Ring B) is in a staggered half-chair form. Thus the overall conformation of the molecule is such that the two aromatic rings are almost perpendicular, as illustrated in Fig. 12. Perrin and Cruickshank (1969) suggested that the antifungal activity of pterocarpans is dependent on this conformation and have speculated on the

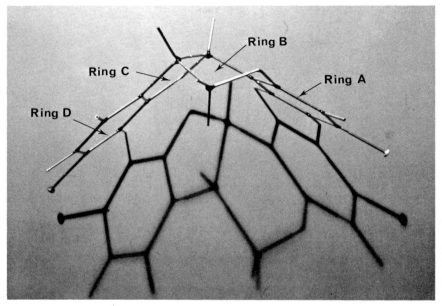

FIG. 12. Dreiding stereomodel of the pterocarpan ring structure.

existence of a receptor site in the cells of sensitive fungi. Only molecules with the bent type of conformation would fit in the site. This hypothesis is based primarily on two lines of experimental evidence. First, antifungal activity was lacking in all the coumestans which they tested. Second, the only 6a,11a-dehydropterocarpan tested, 3-methoxy-8,9-methylenedioxy-6a, 11a-dehydropterocarpan (anhydropisatin), also lacked antifungal activity. Although the ring structure of coumestans and 6a,11a-dehydropterocarpans is analogous to that of the pterocarpans (Fig. 2), they possess the 6a-11a double bond, and thus are essentially planar molecules.

In an attempt to evaluate further the hypothesis, we tested 10^{-4} M concentrations of six pterocarpans and eleven related isoflavonoids for their antifungal activity against *Fusarium solani* (Mart.) Sacc. f. sp. *cucurbitae* Snyd. and Hans. and *Aphanomyces euteiches* (VanEtten, 1976). Like Perrin and Cruickshank, we found that coumestans and anhydropisatin lack antifungal activity. Yet the three other 6a,11a-dehydropterocarpans we tested all exhibited antifungal activity (Fig. 13). The comparison between 3-hydroxy-8,9-methylenedioxy-6a,11a-dehydropterocarpan (anhydrosophorol) and the analogous 6a-hydroxypterocarpan is particularly interesting. The bent pterocarpan lacks antifungal activity, but the planar 6a,11a-dehydropterocarpan possesses antifungal activity by all criteria used. This is exactly opposite of what is expected from the proposed structure–activity hypothesis. (At higher concentrations this 6a-hydroxypterocarpan is in-

Isoflavonoid	% Inhibition of radial growth		
	Aphanomyces euteiches	*Fusarium solani* f. sp. *cucurbitae*	% Survival of *A. euteiches*
R = CH₃; Pisatin	52	20	10-100
6a,11a-dehydro derivative (anhydropisatin)	1	-2	10-100
R = H; 3,6a-Dihydroxy-8,9-methylenedioxy-pterocarpan	3	6	10-100
6a,11a-dehydro derivative (anhydrosophorol)	38	46	0
Tuberosin	55	97	10-100
6a,11a-dehydro derivative (anhydrotuberosin)	37	46	0
Glyceollin	75	77	1-10
6a,11a-dehydro derivative (anhydroglyceollin)	35	18	·001-·01
Control treatment	0	0	10-100

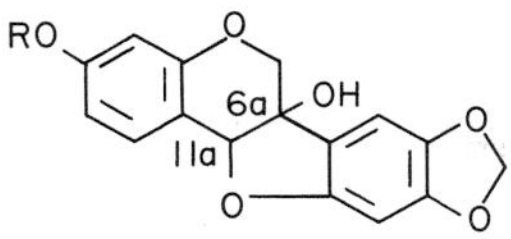

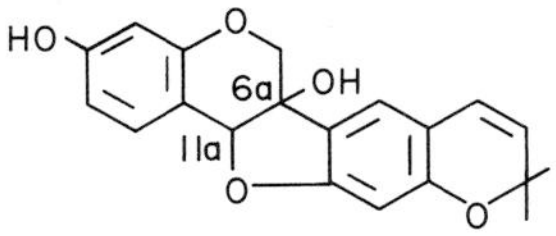

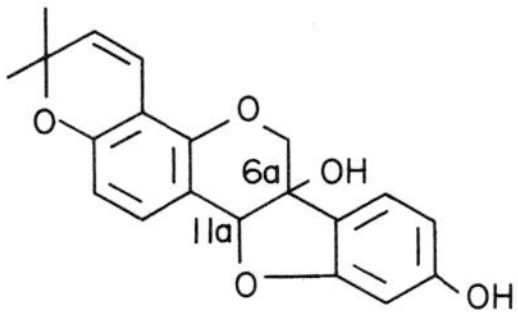

FIG. 13. Antifungal activity of pterocarpans and 6a,11a-dehydro-pterocarpans against *A. euteiches* and *F. solani* f. sp. *cucurbitae*. All compounds were bioassayed at 10^{-4} M (6a,11a-dehydroglyceollin was tested at $\leq 10^{-4}$ M). Radial growth measurements were recorded when net growth in controls was 24 ± 3 mm. Effect of the compounds on viability of *A. euteiches* spores was estimated by exposing the spores ($1-2 \times 10^{5}$/ml) to the compounds for 5 min and then serially diluting with a liquid growth medium to a 10^{6} dilution. Presence or absence of growth was recorded after 3 weeks and compared to controls (VanEtten, 1976).

hibitory to the radial growth of *F. solani* f. sp. *cucurbitae* and *A. euteiches*. See Table V.) It has also been observed that some enantiomeric pterocarpans possess equal antifungal activity (Perrin and Cruickshank, 1969; VanEtten, 1976). A rationalization of this fact has been advanced (Perrin and Cruickshank, 1969). Yet consideration of the evidence *in toto* leads us to the conclusion that the available data do not support the proposed structure-activity hypothesis for antifungal activity.

The properties that determine whether an isoflavonoid is antifungal remain elusive. Perhaps the critical factor is some physicochemical property other than three-dimensional shape. Alternatively, it may be that different classes of isoflavonoids or individual members within a class are antifungal because they have different modes of action. Therefore, different receptor sites may exist in the fungal cell, and a common physicochemical basis for activity of antifungal isoflavonoids may not exist. A hypothesis on different modes of action is, of course, speculative. Nevertheless, fragmentary experimental results are consistent with this possibility. For example, the 6a-hydroxypterocarpans were more inhibitory to fungi in the mycelial stage of growth than to spores (Fig. 13). The reverse is true for the 6a,11a-dehydro-derivatives. Perhaps the two classes of compounds affect two different metabolic processes, one of which is more critical to spores and one of which is more necessary for mycelial growth. In addition, we have observed that some of the physiological and cytological responses of fungi to the pterocarpan pisatin (VanEtten, 1972, 1976 and unpublished results) are not the same as those caused by the pterocarpan phaseollin (section VI). These observations also could be interpreted to indicate that the compounds have different modes of action. At any rate, there is at present little evidence of a common chemical or physiological basis for the activity of the antifungal isoflavonoids.

VIII. Fungal Metabolism of the Isoflavonoid Phytoalexins

Müller (1958) was the first to suggest that phytoalexins may be metabolically altered by fungi. He offered this possibility as one explanation for the adaptation of several fungi to the inhibitory compound(s) present in diffusates obtained from bean pods. Once the chemical properties of some of the isoflavonoid phytoalexins were determined it was possible to specifically monitor the concentrations of these compounds after their addition to fungal cultures. Christenson (1969), Nonaka (1967), Uehara (1964), and Wit-Elshove (1968) reported loss of pisatin; Pierre and Bateman (1967) loss of phaseollin; and Higgins and Millar (1969b) loss of 3-hydroxy-9-methoxypterocarpan from phytoalexin-amended cultures of growing fungi. Some of these researchers and others (Christenson and Hadwiger, 1973; Cruickshank et al., 1974a; Heath and Higgins, 1973; Higgins, 1972; Higgins and Millar, 1970; Sakuma and Millar, 1972; Wit-Elshove, 1969; Wit-Elshove and Fuchs, 1971) documented that new compounds appeared in such cultures. These compounds were ostensibly fungal metabolites of the phytoalexins, but it is only recently that the structures of some of the fungal metabolites of phytoalexins have been determined. The nature of the metabolites and their antifungal activities, relative to the parent phytoalexins, provides fundamental information pertinent to the role of isoflavonoid phytoalexins in plant disease.

A. IDENTITY OF FUNGAL METABOLITES OF PHYTOALEXINS

Pterocarpans are the only isoflavonoid phytoalexins for which structures of fungal metabolites are known. *Stemphylium botryosum, Fusarium solani*

f. sp. *pisi*, and *Ascochyta pisi* each initiate metabolism of pisatin by *O*-demethylation at C-3 (Fig. 14) (Land *et al.*, 1975a; Van Etten *et al.*, 1975).

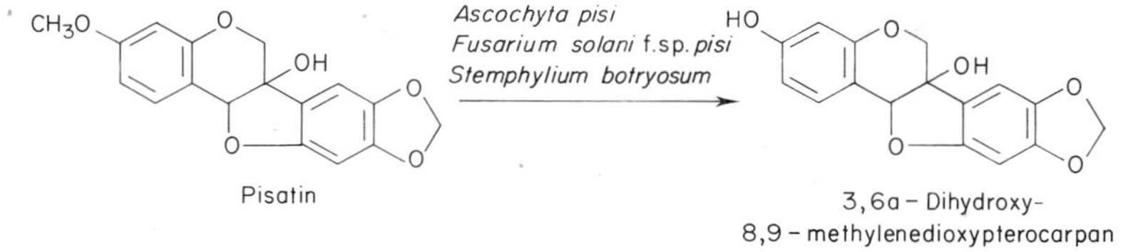

FIG. 14. Initial metabolite of pisatin produced by *A. pisi*, *F. solani* f. sp. *pisi*, and *S. botryosum*.

O-Demethylation of pisatin may exemplify one process of generalized microbial metabolism of isoflavonoids. Metabolism of 7-methoxyisoflavone by *Penicillium cyclopium* Westling (Bátkai *et al.*, 1973), and 7-hydroxy-4'-methoxyisoflavone (formononetin) and 5,7-dihydroxy-4'-methoxyisoflavone (biochanin A) by rumen fluid (Nilsson *et al.*, 1967) proceed similarly.

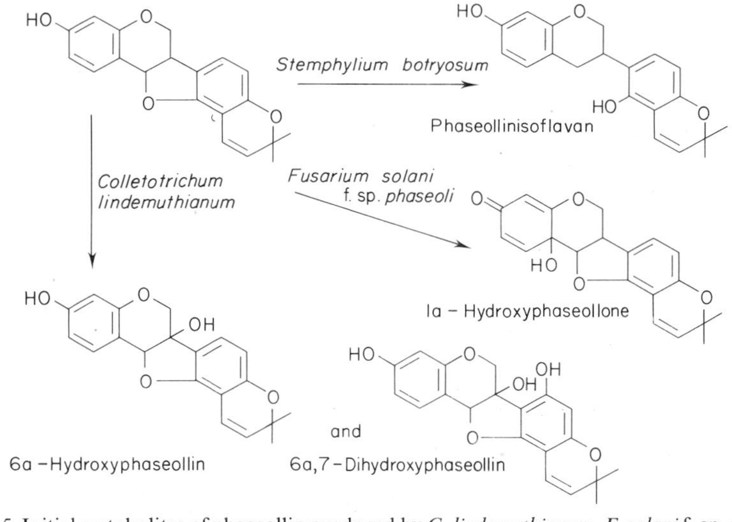

FIG. 15. Initial metabolites of phaseollin produced by *C. lindemuthianum*, *F. solani* f. sp. *phaseoli*, and *S. botryosum*.

In contrast to pisatin, phaseollin is metabolized by three distinct methods (Fig. 15). Hydroxylation at position 1a with concomitant dienone formation in ring A is the first step in the metabolism of phaseollin by *F. solani* f. sp. *phaseoli* (Heuvel *et al.*, 1974) and *Cladosporium herbarum* Pers. ex Fr. (Heuvel and Glazener, 1975). Metabolism by *C. lindemuthianum* results in hydroxylation at C-6a and C-7 (Burden *et al.*, 1974a), while metabolism by *S. botryosum* proceeds by cleavage of the benzylphenyl ether linkage in ring C to yield the analogous isoflavan, phaseollinisoflavan (Higgins *et al.*, 1974). Biosynthesis of isoflavans in higher plants is thought to be effected by the same reaction, which is possibly a widespread biological phenomenon (section III).

S. botryosum also performs the pterocarpan to isoflavan transformation on two other pterocarpan phytoalexins: 3-hydroxy-8,9-methylenedioxyptero-carpan and 3-hydroxy-9-methoxypterocarpan (Fig. 16) (Duczek, 1974; Duczek and Higgins, 1973; Higgins, 1975; Steiner and Millar, 1974). Likewise, we have recently determined that *F. solani* f. sp. *pisi* converts 3-hydroxy-2,9-dimethoxypterocarpan to its analogous isoflavan (Pueppke and VanEtten, 1976). These reactions are worthy of note, because they

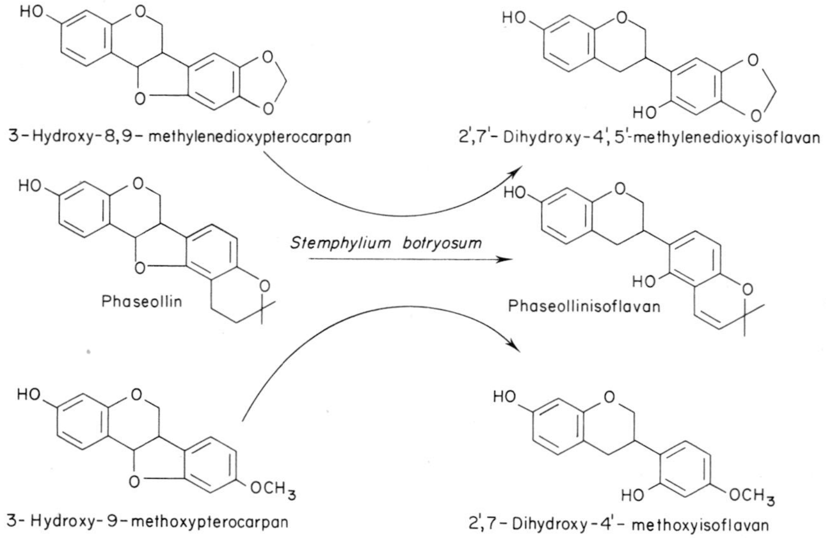

3-Hydroxy-8,9-methylenedioxypterocarpan   2',7'-Dihydroxy-4',5'-methylenedioxyisoflavan

Phaseollin   *Stemphylium botryosum*   Phaseollinisoflavan

3-Hydroxy-9-methoxypterocarpan   2',7-Dihydroxy-4'-methoxyisoflavan

FIG. 16. Initial metabolites of 3-hydroxy-8,9-methylenedioxy-pterocarpan, phaseollin, and 3-hydroxy-9-methoxy-pterocarpan produced by *S. botryosum*.

establish that a given microorganism (*S. botryosum* or *F. solani* f. sp. *pisi*) may initiate metabolism of pterocarpan phytoalexins by different methods, i.e. *O*-demethylation or benzylphenyl ether cleavage.

The route of metabolism of pterocarpans to simple molecules is unknown. On the basis of the available fragmentary data, (i) a given fungus may initiate metabolism of antifungal pterocarpans by different or identical routes, and (ii) several fungi may convert a given pterocarpan to the same metabolite. Obviously, much is to be learned before generalized pathways of metabolism (if they exist) are understood.

The fungal metabolites of other isoflavonoid phytoalexins have not been identified. Phaseollidin and kievitone are reportedly metabolized by *C. lindemuthianum* (Bailey, 1974), and phaseollinisoflavan is apparently metabolized by *C. lindemuthianum, F. solani* f. sp. *phaseoli*, and *S. botryosum* (Bailey, 1974; Heath and Higgins, 1973; Higgins *et al.*, 1974; VanEtten and Smith, 1975). Both 2',7-dihydroxy-4'-methoxyisoflavan and 7-hydroxy-

2′,4′-dimethoxyisoflavan disappear when added to cultures of *Stemphylium loti* Graham and *S. botryosum* (Bonde, 1975; Steiner and Millar, 1974).

B. ENZYMOLOGY

No fungal enzyme that transforms pterocarpans has been isolated. Even cell-free preparations which convert isoflavonoid phytoalexins to the known metabolites are difficult to obtain. Culture filtrates of *F. solani* clones allegedly transform radiolabelled pisatin (Christenson and Hadwiger, 1973), and culture filtrates of *Leptosphaerulina briosiana* (Poll.) Graham and Luttrell rapidly alter 3-hydroxy-9-methoxypterocarpan (Higgins, 1972). The products were not identified in either case. Attempts to obtain preparations from *F. solani* f. sp. *phaseoli* that convert phaseollin to a 1a-hydroxyphaseollone were unsuccessful (Heuvel and VanEtten, 1973), as were attempts to obtain a preparation from *S. botryosum* that transforms 3-hydroxy-8,9-methylenedioxypterocarpan to the corresponding isoflavan (Higgins, 1975). The difficulty in obtaining active cell-free preparations may indicate that the enzymes are particulate or labile.

The capacity of culture filtrates of *L. briosiana* to alter 3-hydroxy-9-methoxypterocarpan was due presumably to a constitutive enzyme, as activity was not dependent on prior exposure of the organism to the pterocarpan. Yet both *F. solani* f. sp. *phaseoli* and *S. botryosum* require a period of adaptation before pterocarpan metabolism commences. If cultures of *F. solani* f. sp. *phaseoli* are exposed to low concentrations of phaseollin, they subsequently rapidly metabolize concentrations of the compound which otherwise are inhibitory (Heuvel and VanEtten, 1973). Likewise, metabolic transformation of 3-hydroxy-8,9-methylenedioxypterocarpan by *S. botryosum* is facilitated by pre-exposure of the fungus to the compound (Higgins, 1975). If *S. botryosum* is exposed to inhibitory concentrations of cycloheximide concurrent with 3-hydroxy-8,9-methylenedioxypterocarpan, transformation of the pterocarpan is prevented or delayed. If cycloheximide is added after initiation of metabolism of the phytoalexin, conversion of 3-hydroxy-8,9-methylenedioxypterocarpan proceeds. These results are most easily explained if an inducible enzyme system is postulated. However, as pointed out by Higgins, the results do not permit a distinction between the induction of phytoalexin-metabolizing enzymes and other energy dependent systems which might be involved in pterocarpan metabolism, e.g. induced carrier protein transport systems.

As illustrated in Fig. 16, *S. botryosum* can metabolize at least three pterocarpan phytoalexins to their analogous isoflavans. Higgins (1975) obtained preliminary data that the same enzyme system is involved in all three transformations. The evidence is that the pterocarpans were interchangeable, both as inducers of the metabolic system, and as substrates. The transformation of all three pterocarpans was affected similarly by exposure to cycloheximide. As with isoflavonoid biosynthesis in higher plants, progress in the field of isoflavonoid phytoalexin metabolism will be enhanced greatly once the enzymes involved are purified and characterized.

C. METABOLISM AND TOLERANCE

Several fungal pathogens under certain bioassay conditions are tolerant of their host's isoflavonoid phytoalexins (section V). Whether this tolerance is due to metabolic conversion of the phytoalexin to non-inhibitory products is of paramount interest. Some experimental evidence supports this contention. If tolerant fungi detoxify phytoalexins, the metabolites should be less inhibitory than the parent phytoalexin. Mycelial growth of *A. pisi* and *F. solani* f. sp. *pisi* is tolerant of pisatin (Table V) (Christenson and Hadwiger, 1973; Cruickshank, 1962; Pueppke and VanEtten, 1974; VanEtten, 1973a), and these fungi can metabolically 3-*O*-demethylate pisatin (Fig. 14). When the antifungal activities of equimolar concentrations of pisatin and the *O*-demethyl metabolite are compared (Table V), the metabolite is less inhibitory (VanEtten and Pueppke, unpublished). Mycelial growth of *F. solani* f. sp. *phaseoli* is tolerant of phaseollin (Table VI) (VanEtten, 1973a), and this organism is able to metabolize phaseollin to 1a-hydroxyphaseollone (Fig. 15). When the antifungal activities of the phaseollin metabolite and phaseollin are compared, the metabolite is less inhibitory (Table VI) (VanEtten and Smith, 1975). The mycelial growth of *C. lindemuthianum*

TABLE V

Inhibition of radial mycelial growth of fungi on solid medium containing 3×10^{-4} M pisatin or 3,6a-dihydroxy-8,9-methylenedioxypterocarpan

	% Inhibition[a]	
Fungus	3×10^{-4} M pisatin	3×10^{-4} M 3,6a-dihydroxy-8,9-methylenedioxypterocarpan
Aphanomyces euteiches[b]	96	98
Ascochyta pinodella[b]	16	12
F. solani f. sp. *pisi*[b]	7	8
Mycosphaerella pinodes[b]	8	6
F. solani f. sp. *phaseoli*	67	6
Helminthosporium turcicum	74	35
Neurospora crassa	78	28
Penicillium expansum	36	6
Rhizopus stolonifer	74	6
S. botryosum	66	3

[a] Experimental conditions were the same as described by VanEtten, 1973a. Net growth of controls was about 24 mm, when growth of treatments were recorded (VanEtten and Pueppke, unpublished data).
[b] Pathogens of *Pisum sativum* L., a species which produces pisatin as a phytoalexin.

TABLE VI

Inhibition of radial mycelial growth of fungi on solid medium containing 10^{-4} M
phaseollin or la-hydroxyphaseollone

Fungus	% Inhibition	
	10^{-4} M phaseollin	10^{-4} M la-hydroxyphaseollone
Aphanomyces euteiches[b]	92	29
F. solani f. sp. *phaseoli*[b]	2	4[a]
Rhizoctonia solani[b]	64	8
Ascochyta pinodella	77	7
F. roseum	84	−4
Helminthosporium carbonum	45	0
Neurospora crassa	90	11
Rhizopus stolonifer	94	−2
Trichoderma viride	99	28

[a]This value from Heuvel and VanEtten, 1973; all others from VanEtten and Smith, 1975.
Net growth of controls was about 24 mm when growth of treatments was recorded.
[b]Pathogens of *Phaseolus vulgaris* L., a species which produces phaseollin as a phytoalexin.

is also tolerant of phaseollin (Bailey, 1974; Cruickshank and Perrin, 1971). Although the initial phaseollin metabolites, 6a-hydroxy- and 6a,7-dihydroxyphaseollin (Fig. 5), retain some antifungal activity, they are reportedly further metabolized to non-inhibitory metabolites (Bailey, 1974; Heuvel and Glazener, 1975). Thus it is clear that these tolerant fungi are capable of converting isoflavonoid phytoalexins to non-inhibitory products.

But is this metabolic conversion the basis for the tolerance of these organisms to the phytoalexins? It is true that a few sensitive fungi, such as *Helminthosporium carbonum* Ullstrup [originally incorrectly identified as *H. turcicum* Pass (Higgins, 1975)], apparently cannot metabolize phytoalexins. *H. carbonum* is sensitive to 3-hydroxy-9-methoxypterocarpan, 3-hydroxy-8,9-methylenedioxypterocarpan, 2′,7-dihydroxy-4′-methoxyisoflavan, and 7-hydroxy-2′,4′-dimethoxyisoflavan and cannot metabolize any of these phytoalexins (Bonde, 1975; Duczek, 1974; Higgins and Millar, 1969a). However, other fungi appear to be sensitive to certain isoflavonoid phytoalexins which they can metabolize. For instance, *S. botryosum* can metabolize pisatin to the same non-toxic metabolite produced by *A. pisi* and *F. solani* f. sp. *pisi* (Fig. 14). Yet in radial mycelial growth bioassays, *S. botryosum* is sensitive to pisatin (Table V) (Heath and Higgins, 1973). The radial mycelial growth of this fungus is also sensitive to phaseollin (Table VII) (Heath and Higgins, 1973), but tolerant of 3-hydroxy-9-methoxypterocarpan and 3-hydroxy-8,9-methylenedioxypterocarpan (Duczek, 1974; Higgins, 1972). Yet *S. botryosum* can metabolize all three pterocarpans to the analogous isoflavans (Fig. 16). The available limited data suggest that isoflavans are as

TABLE VII

Inhibition of radial growth of fungi on solid medium containing 10^{-4} M phaseollin or phaseollinisoflavan

Fungus	% Inhibition[a]	
	10^{-4} M phaseollin	10^{-4} M phaseollinisoflavan
Aphanomyces euteiches[b]	86	97
F. solani f. sp. *cucurbitae*[b]	86	87
F. solani f. sp. *phaseoli*[c]	2	12
Rhizoctonia solani[d]	63	55
S. botryosum[e]	*ca* 50	*ca* 50

[a] Bioassay conditions were similar, but not identical. In most cases net growth of controls was 20–30 mm when growth of treatments was recorded.

[b] Values from VanEtten, 1975.

[c] Values from VanEtten and Smith, 1975.

[d] Values extrapolated from dosage–response curves in Smith *et al.*, 1975.

[e] Values from Heath and Higgins, 1973.

inhibitory as their analogous pterocarpans (Table VII) (Bailey and Burden, 1973; Bonde *et al.*, 1973; Higgins, 1975; Ingham and Millar, 1973). Since both pterocarpans and isoflavans are phytoalexins (Table I), it is not surprising that they have similar antifungal activity. Thus, whereas *S. botryosum* is able to convert a phytoalexin to which it is apparently sensitive (pisatin) to a non-toxic metabolite, it is able to convert phytoalexins to which it is apparently tolerant (3-hydroxy-9-methoxypterocarpan and 3-hydroxy-8,9-methylenedioxypterocarpan) to equally inhibitory metabolites. However, in the latter cases, the inhibitory isoflavans are reportedly further metabolized (Bonde, 1975; Duczek, 1974; Higgins, 1975; Steiner and Millar, 1974).

Some of the data which might be taken as evidence for equating tolerance with the metabolic alteration of phytoalexins to non-toxic metabolites is ambiguous. This may be due to variations in experimental techniques. For instance, experimental conditions for the study of metabolic activity were not always the same as those used to determine the sensitivity of fungi to the parent phytoalexins and the fungal metabolite(s). Generally, liquid cultures of the fungi have been used to produce the fungal metabolites of phytoalexins, but the antifungal activity of the phytoalexins and the metabolites has been determined by radial mycelial growth assays. As already pointed out, the apparent sensitivity of a fungus to an isoflavonoid phytoalexin can be markedly affected by the type of bioassay conditions employed (section V).

An alternative approach which may provide direct information on the question of metabolism and tolerance is analysis of the kinetics of fungal growth versus the rate of metabolism of the phytoalexins to non-inhibitory

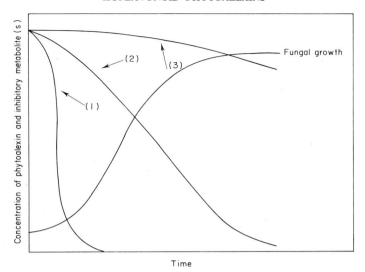

F<small>IG</small>. 17. Hypothetical relationship between growth of a fungus in liquid culture, and the concentrations of a phytoalexin and its inhibitory metabolites in the culture.

products. Figure 17 illustrates three hypothetical relationships between the growth curve of a fungus in liquid culture and the concentration of a phytoalexin and its inhibitory metabolites in the fungal culture. All else being equal, only in the first case (1) is metabolism of the compound(s) to non-inhibitory products clearly compatible with the contention that this metabolism is necessary before growth can occur. In (1) substantial metabolism of the compound(s) occurs before significant growth of the fungus. In (3) metabolism of the phytoalexin lags behind growth and indicates that other tolerance mechanisms are operative. The situation in (2) is less clearcut.

Unfortunately, no detailed analyses similar to Fig. 17 have been reported for fungi. On the basis of available data, the metabolism of phaseollin by *C. lindemuthianum* (Bailey, 1974; Bailey and Deverall, 1971) and *F. solani* f. sp. *phaseoli* (Heuvel and VanEtten, 1973), and the metabolism of 3-hydroxy-8,9-methylenedioxypterocarpan by *S. botryosum* (Higgins, 1975) are like (1) or (2) in Fig. 17. The metabolism of pisatin in liquid cultures of *F. solani* f. sp. *pisi* (Wit-Elshove and Fuchs, 1971) appears to fit pattern (3). The data of Wit-Elshove and Fuchs (1971) are particularly significant. They observed that glucose concentration markedly affected the ability of *F. solani* f. sp. *pisi* to metabolize pisatin. If the growth medium was supplemented with 3% glucose, pisatin metabolism was completely arrested, although the conditions were evidently still favourable for growth. Thus with at least one organism (*F. solani* f. sp. *pisi*), tolerance of a phytoalexin (pisatin) may not be dependent on the degradation of the phytoalexin by the fungus.

Tolerance could be dependent on such phenomena as differential uptake or binding of the phytoalexin, presence or absence of sensitive receptor sites, or other factors which may or may not be dependent on or linked to metabolic processes of the cell. Thus the metabolism of isoflavonoid phytoalexins by certain fungi may be only indirectly related to growth; the organisms may simply be attempting to use these compounds as carbon sources.

D. FUNGAL METABOLISM OF THE ISOFLAVONOID PHYTOALEXINS *IN SITU*

In vitro studies on the metabolism of isoflavonoid phytoalexins lends support to the hypothesis that in some cases the invading pathogen interacts with phytoalexins *in situ*. The likelihood of such an interaction has been questioned, primarily because the compounds are sparingly soluble in aqueous media or may be compartmentalized in tissue (Daly, 1972). Thus the invading organism may never contact the phytoalexin, although they coexist in plant tissue. Several researchers (Duczek, 1974; Heath and Higgins, 1973; Higgins, 1972; Higgins and Millar, 1970; Land *et al.*, 1975a; Steiner and Millar, 1974; VanEtten and Smith, 1975) have demonstrated that the *in vitro* fungal metabolites of phytoalexins are also present in infected tissue. However, these observations do not necessarily mean that an interaction between the invading fungus and the host's phytoalexins has taken place. The apparent fungal metabolite from infected tissue may actually be entirely of host origin (Higgins *et al.*, 1974). For example, phaseollinisoflavan was detected after spores of *S. botryosum* were incubated on *Phaseolus vulgaris* leaf tissue (Heath and Higgins, 1973; Higgins *et al.*, 1974). *In vitro*, *S. botryosum* metabolizes phaseollin to phaseollinisoflavan (Higgins, 1975, Higgins *et al.*, 1974, Fig. 15), but this compound also can be produced as a phytoalexin by the host (Bailey and Burden, 1973). In addition, in many studies the drop-diffusate technique has been utilized to collect suspected metabolites from infected tissue. This technique, while ideal for isolation studies, is artificial and may not be indicative of the natural interaction (Cruickshank *et al.*, 1974a; Keen and Horsch, 1972; Kuć, 1972).

The isolation of 1a-hydroxyphaseollone, an *in vitro* fungal metabolite of phaseollin, from *F. solani* f. sp. *phaseoli*-infected hypocotyl tissue of *P. vulgaris* seems to preclude these criticisms (VanEtten and Smith, 1975). Hypocotyls are natural infection sites, and to date it has been impossible to recover 1a-hydroxyphaseollone as a host product. Furthermore, the phaseollin-metabolizing system of this fungus is induced, and an extracellular enzyme(s) which converts phaseollin to 1a-hydroxyphaseollone was not detected (Heuvel and VanEtten, 1973). These observations strengthen the contention that *F. solani* f. sp. *phaseoli* and phaseollin come into direct contact during infection of bean hypocotyl tissue. Nevertheless, since a substantial amount of phaseollin remains in this infected tissue (VanEtten and Smith, 1975), some of the phytoalexin appears to be inaccessible to the metabolic system of the invading fungus.

It is often difficult or impossible to detect the *in vitro* fungal metabolites in infected tissue (Cruickshank *et al.*, 1974a; Higgins, 1975). Perhaps some fungal metabolites of phytoalexins have a transitory existence *in situ*, are further metabolized, and thus escape detection. Alternately, the phytoalexins may not be metabolized in these particular host–pathogen interactions, or they may not come into contact with the pathogen. Although metabolites of phytoalexins are not found in every host–pathogen interaction, where they are detected, the evidence for pathogen–phytoalexin contact is reinforced substantially.

IX. ISOFLAVONOID PHYTOALEXINS AND PATHOGENESIS

Several recent reviews pertinent to the involvement of phytoalexins in plant diseases have appeared (Cruickshank *et al.*, 1971; Deverall, 1972a,b; Ingham, 1972, 1973; Kuć, 1972; Müller, 1969; Stoessl, 1970). The reader is referred to these reviews and those by Deverall and Kuć in this volume for references and additional discussion concerning the possible involvement of phytoalexins in disease resistance. Our purpose is not to treat the subject exhaustively. Rather we shall attempt to summarize several hypotheses which attempt to explain the role of isoflavonoid phytoalexins in plant disease. Two hypotheses have dominated the phytoalexin literature. We choose to term these the differential synthesis hypothesis and the differential sensitivity hypothesis. The hypotheses are not mutually exclusive; a simplified version of each is given below.

A. DIFFERENTIAL SYNTHESIS

According to this hypothesis a susceptible reaction is characterized by a low rate of phytoalexin synthesis. Resistance, on the other hand, results from a large and rapid accumulation of phytoalexins. In any interaction resistance or susceptibility is determined by pathogen-mediated controls on phyto-alexin synthesis by the plant. Müller and Börger's (1940) original presentation of the phytoalexin concept can be interpreted in this manner, and the basic idea has been given a slightly different and more mechanistic interpretation by Hadwiger and Schwochau (1969). The differential synthesis hypothesis has been limited primarily to race-specific or varietal resistance, i.e. host–parasite interactions involving various pathogen races, each capable of attacking only certain varieties of the host. Examples of race-specific combinations are the interaction of races of the fungal pathogen *Phytophthora megasperma* var. *sojae* with varieties of soybean (*Glycine max*) and the interaction of races of *Colletotrichum lindemuthianum* with varieties of *Phaseolus vulgaris*. The data are in agreement that different races of the pathogens are more or less equally sensitive to their host phytoalexins

(Bailey, 1974; Bailey and Deverall, 1971; Keen *et al.*, 1971). Furthermore, different host varieties apparently produce the same phytoalexins (Bailey, 1974; Keen, 1971). In resistant interactions high levels of the phytoalexins accumulate quickly whereas in the susceptible interactions the phytoalexins build up much less rapidly (Bailey, 1974; Bailey and Deverall, 1971; Keen, 1971; Rahe, 1973). Control of phytoalexin synthesis by the pathogenic races has been explained in two different ways.

Keen and associates (Keen *et al.*, 1972; Keen, 1975a) made the intriguing suggestion that high levels of phytoalexins accumulate in the resistant combinations because of race-unique fungal components called specific elicitors. These hypothetical compounds are presumed to stimulate high production of phytoalexins only in host varieties which respond to the race in a resistant manner. Thus major importance is placed on the resistant reactions, where phytoalexin, levels are controlled by the pathogen. Keen (1975a) presented preliminary experimental evidence (Table III) that such specific elicitors of phytoalexins are elaborated by *P. megasperma* var. *sojae*. Yet the degree of differential stimulation of phytoalexin production in resistant versus susceptible hosts by these specific elicitors is not as substantial as in the *in situ* situation (Keen, 1971). Furthermore, cultures of each race of *P. megasperma* var. *sojae* contained many non-specific elicitors which induced more phytoalexin than did specific elicitors (Keen, 1975a).

An alternate interpretation of race-specific interactions has been advanced by Rahe and Arnold (1975). Since numerous non-specific stimuli induce phytoalexins (see section IV), they suggested that stimulation of phytoalexin accumulation in a host plant is not under specific control by the fungus. Rather, phytoalexin synthesis is simply a manifestation of non-specific disturbance of plant cells by the pathogen. Thus they imply that the control of phytoalexin synthesis may occur in susceptible reactions where only low levels of the compounds accumulate. Implicit in their theory is the potential existence of race-specific repressors of phytoalexin synthesis, which act only on susceptible varieties. Experimental evidence for such repressors, however, is yet to be established.

Two additional considerations complicate our understanding of the differential synthesis hypothesis. First, fungi can metabolize isoflavonoid phytoalexins. Second, the level of a phytoalexin in a plant may reflect an equilibrium between synthesis and plant-mediated turnover (section III). Thus, researchers who test the differential synthesis hypothesis should consider *in situ* catabolic activity of the pathogen and turnover by the plant, in addition to synthetic activity of the plant (Higgins and Millar, 1970).

B. DIFFERENTIAL SENSITIVITY

A basic premise of this hypothesis is that a susceptible reaction occurs when the microorganism is tolerant of the plant's phytoalexins, and that resistance results when the microorganism is sensitive to the plant's phyto-alexins. Thus for any plant–microbe interaction, the crucial event distinguishing susceptibility from resistance is the response of the microbe to the plant's phytoalexins. Minor emphasis is placed on how or if the microbe controls phytoalexin levels in the plant. Generally, the differential sensitivity hypothesis is not applied to race-specific or varietal resistance.

The experimental basis for this hypothesis originated with Cruickshank's 1962 study of the response of 50 fungal isolates to pisatin (Cruickshank, 1962, 1965a). Results of this and other screening studies (discussed in section V) indicate that pathogens are commonly tolerant of their host's isoflavonoid phytoalexins, whereas non-pathogens are sensitive. The correlation is not complete, however, and thus the differential sensitivity hypothesis cannot be invoked to explain all host–pathogen interactions (Cruickshank, 1965a). Nevertheless, tolerance of a phytoalexin by a pathogen may be important for pathogenicity by certain organisms. One important, and as of yet unanswered question, is whether a tolerant organism would retain its pathogenicity unaltered if it lacked this attribute.

C. ADDITIONAL CONSIDERATIONS

The hypothetical mechanisms of resistance described above are over-simplifications. When evaluating plant resistance mechanisms, most researchers have considered neither hypothesis to the exclusion of the other. Yet when individual pathogen–plant interactions are analysed, one or the other hypothesis is given greater emphasis. Hypotheses other than those outlined above are tenable. A variant of the differential sensitivity hypothesis was pointed out in studies of Higgins and Millar (1969a, 1969b, 1970) and Wit-Elshove (1968, 1969). They suggested that pathogenicity of a fungus to a particular plant depends on its ability to metabolize that plant's phytoalexins, possibly in a particular manner. Whether the differential sensitivity of fungi to isoflavonoid phytoalexins is dependent on metabolism was discussed in section VIII. Many pathogens are able to metabolize their host's phyto-alexin(s), but, as is the case of tolerance and pathogenicity, the correlation between metabolism and pathogenicity is not absolute. Fungi which can metabolize non-host phytoalexins (Bonde, 1975; Duczek, 1974; Heuvel and Glazener, 1975) and pathogens which apparently cannot metabolize their host's phytoalexins (Heuvel and Glazener, 1975; Pueppke and VanEtten, unpublished) are known. Again, as with tolerance, an important unanswered question is whether a pathogen that can metabolize its host's phytoalexin would retain its pathogenicity if it lacked this ability.

The recent discovery that plants can produce more than one phytoalexin leads to speculation that the differential synthesis hypothesis may involve qualitative, as well as quantitative aspects. Thus a plant may synthesize one set of phytoalexins in response to one potential parasite and a different group of phytoalexins in response to another parasite. An additional possibility is that resistant plants may be able to inactivate tolerance mechanisms in microorganisms. Since no single hypothesis currently explains all host–parasite interactions, it is likely that even more hypotheses will be suggested and receive experimental support in the future.

X. Conclusions

The literature on the isoflavonoid phytoalexins is impressive both in the diversity of the experimental results and the absence of universally accepted conclusions about their function in nature. One statement is safe: isoflavonoid phytoalexins are antimicrobial and are produced in infected plant tissue. In addition, *prima facie* evidence that the production of isoflavonoid phytoalexins is part of a plant's defense mechanism against microorganisms is supplied by two observations. (1) The restriction of growth of fungal pathogens in plant tissue is often temporally correlated with the accumulation of high concentrations of isoflavonoid phytoalexins (Bailey, 1974; Bailey and Deverall, 1971; Keen and Kennedy, 1974; Keen *et al.*, 1971; Pierre and Bateman, 1967; Pueppke and VanEtten, 1974; Smith *et al.*, 1975). (2) Pre-inoculation treatments which *inter alia* repress the rate of phytoalexin synthesis often can convert a normally resistant reaction to a susceptible reaction (Cruickshank and Perrin, 1967; Duczek, 1974; Keen, 1971). Conversely, pre-inoculation treatments which *inter alia* stimulate phytoalexin synthesis can sometimes render a plant less susceptible (Bridge and Klarman, 1973).

However, several workers have shown that the growth of some pathogens in phytoalexin-producing plants is restricted in the absence of significant accumulation of isoflavonoid phytoalexins (Bonde, 1975; Duczek, 1974; Higgins, 1972). In other interactions accumulation of isoflavonoid phytoalexins is subsequent to the cessation of pathogen growth (Teasdale *et al.*, 1974). In these host–pathogen interactions, other types of mechanisms apparently restrict pathogen growth. But what about those interactions where resistance is expressed concurrent with phytoalexin accumulation? Here also, perhaps other types of resistance mechanisms are operative, i.e. the accumulation of phytoalexins may be secondary or incidental to the actual resistance mechanism. As most phytoalexin researchers well know, it is not yet possible to separate the cause and effect phenomena in experiments correlating expression of resistance with phytoalexin accumulation. Yet this type of experiment constitutes the major evidence in support of the involvement of isoflavonoid phytoalexins in disease resistance.

Some research appears to provide direct experimental evidence against the involvement of phytoalexins in disease resistance. The interaction between *Aphanomyces euteiches* and *Pisum sativum* is a case in point. This pathogen, which is sensitive *in vitro* to the phytoalexin of its host, is not restricted *in situ*, although theoretically inhibitory concentrations of the phytoalexin are present (Pueppke and VanEtten, 1974). Further research is required to determine whether results such as ours on the *A. euteiches–P. sativum* interaction are indeed compatible with the phytoalexin concept of disease resistance.

Regardless of the role of the isoflavonoid phytoalexins in disease resistance, these compounds offer fertile fields for investigation in other areas of biology. The possibility that fungal components exert specific control over both the rate and direction of isoflavonoid phytoalexin biosynthesis may provide a powerful tool for investigation of isoflavonoid biosynthesis in general. Studies on the biological bases for the sensitivity or tolerance of microorganisms to isoflavonoid phytoalexins could improve our understanding of microbial physiology. In addition, the elucidation of the chemical bases for the biological activity of these compounds may suggest synthetic compounds for future agricultural or medical use. It is likely that increased knowledge in these areas will help to answer the question of ultimate concern to phytopathologists: the role of these compounds in disease resistance.

ACKNOWLEDGEMENTS

We are indebted to the Department of Genetics and Plant Breeding, Royal Agricultural College of Sweden, Uppsala, Sweden, for providing facilities for the writing of a portion of this review. We would like to thank the following authors who supplied us with unpublished information or preprints of manuscripts in press: J. A. Bailey, R. S. Burden, V. J. Higgins, J. L. Ingham, N. T. Keen, J. E. Rahe and G. Stenlid.

REFERENCES

Abawi, G. S., VanEtten, H. D. and Mai, W. F. (1971). *J. Nematol.* **3**, 301 (Abstr.).
Anderson, A. and Albersheim, P. (1974). *Proc. Am. Phytopathol. Soc.* **1**, 52 (Abstr.).
Ayers, A., Ebel, J. and Albersheim, P. (1974). *Proc. Am. Phytopathol. Soc.* **1**, 23 (Abstr.).
Bailey, J. A. (1969a). *Phytochemistry* **8**, 1393–1395.
Bailey, J. A. (1969b). *Ann. appl. Biol.* **64**, 315–324.
Bailey, J. A. (1970). *J. gen. Microbiol.* **61**, 409–415.
Bailey, J. A. (1973). *J. gen. Microbiol.* **75**, 119–123.
Bailey, J. A. (1974). *Physiol. Pl. Path.* **4**, 477–488.
Bailey, J. A. and Burden, R. S. (1973). *Physiol. Pl. Path.* **3**, 171–177.
Bailey, J. A. and Deverall, B. J. (1971). *Physiol. Pl. Path.* **1**, 435–449.
Barz, W. (1969). *Z. Naturf. Teil B.* **24**, 234–239.
Barz, W. and Adamek, C. (1970). *Planta* **90**, 191–202.
Barz, W. and Grisebach, H. (1967). *Z. Naturf. Teil B.* **22**, 627–633.
Barz, W. and Hösel, W. (1971). *Phytochemistry* **10**, 335–341.
Barz, W. and Roth-Lauterbach, B. (1969). *Z. Naturf. Teil B.* **24**, 638–647.
Barz, W., Adamek, C. and Berlin, J. (1970). *Phytochemistry* **9**, 1735–1744.

Bátkai, L., Nográdi, M., Farkas, L., Feuer, L. and Horváth, I. (1973). *Arch. Mikrobiol.* **90**, 165–166.

Ben-Aziz, A. (1967). *Science, N.Y.* **155**, 1026–1027.

Berlin, J. and Barz, W. (1971). *Planta* **98**, 300–314.

Berlin, J., Dewick, P. M., Barz, W. and Grisebach, H. (1972). *Phytochemistry* **11**, 1689–1693.

Berlin, J., Kiss, P., Müller-Enoch, D., Gierse, D. and Barz, W. (1974). *Z. Naturf. Teil C.* **29**, 374–383.

Bickoff, E. M., Loper, G. M., Hanson, C. H., Graham, J. H., Witt, S. C. and Spencer, R. R. (1967). *Crop Sci.* **7**, 259–261.

Bickoff, E. M., Spencer, R. R., Witt, S. C. and Knuckles, B. E. (1969). *U.S. Dept. Agr. Tech. Bull.* **1408**, 95 p.

Biggs, D. R. (1972). *Pl. Physiol.* **50**, 660–666.

Bonde, M. R. (1975). Ph.D. Thesis, Cornell University, Ithaca, New York.

Bonde, M., Millar, R. L. and Ingham, J. L. (1973). *Phytochemistry* **12**, 2957–2959.

Bredenberg, J. B. and Hietala, P. K. (1961a). *Acta chem. scand.* **15**, 696–699.

Bredenberg, J. B. and Hietala, P. K. (1961b). *Acta chem. scand.* **15**, 936–937.

Bridge, M. A. and Klarman, W. L. (1973). *Phytopathology* **63**, 606–609.

Burden, R. S. and Bailey, J. A. (1975). *Phytochemistry* **14**, (1389–1390).

Burden, R. S., Bailey, J. A. and Dawson, G. W. (1972). *Tetrahedron Lett.* 4175–4178.

Burden, R. S., Bailey, J. A. and Vincent, G. G. (1974a). *Phytochemistry* **13**, 1789–1791.

Burden, R. S., Rogers, P. M. and Wain, R. L. (1974b). *Ann. appl. Biol.* **78**, 59–63.

Camm, E. L. and Towers, G. H. N. (1973). *Phytochemistry* **12**, 961–973.

Chalutz, E. and Stahmann, M. A. (1969). *Phytopathology* **59**, 1972–1973.

Chang, C., Suzuki, A., Kumai, S. and Tamura, S. (1969). *Agric. Biol. Chem.* **33**, 398–408.

Chowdhury, A., Mukherjee, N. and Adityachaudhury, N. (1974). *Experientia* **30**, 1022–1024.

Christenson, J. A. (1969). *Phytopathology* **59**, 10 (Abstr.).

Christenson, J. A. and Hadwiger, L. A. (1973). *Phytopathology* **63**, 784–790.

Cocker, W., Dahl, T., Dempsey, C. and McMurray, T. B. H. (1962). *J. chem. Soc.* 4906–4909.

Creasy, L. L. and Zucker, M. (1974). *Rec. Adv. Phytochem.* **8**, 1–19.

Crombie, L., Dewick, P. M. and Whiting, D. A. (1973). *J. chem. Soc., Perkin Trans.* **1**, 1285–1294.

Cruickshank, I. A. M. (1962). *Aust. J. biol. Sci.* **15**, 147–159.

Cruickshank, I. A. M. (1963). *A. Rev. phytopath.* **1**, 351–374.

Cruickshank, I. A. M. (1965a). *Tagungsber. Dtsch. Akad. Landwirtschaftswiss.* Berlin **74**, 313–332.

Cruickshank, I. A. M. (1965b). *In* "Ecology of Soil-borne Plant Pathogens. Prelude to Biological Control" (K. F. Baker and W. C. Snyder, eds), pp. 325–334. University California Press, Berkeley.

Cruickshank, I. A. M. and Perrin, D. R. (1960). *Nature, Lond.* **187**, 799–800.

Cruickshank, I. A. M. and Perrin, D. R. (1961). *Aust. J. biol. Sci.* **14**, 336–348.

Cruickshank, I. A. M. and Perrin, D. R. (1963a). *Life Sci.* **2**, 680–682.

Cruickshank, I. A. M. and Perrin, D. R. (1963b). *Aust. J. biol. Sci.* **16**, 111–128.

Cruickshank, I. A. M. and Perrin, D. R. (1965). *Aust. J. biol. Sci.* **18**, 817–828.

Cruickshank, I. A. M. and Perrin, D. R. (1967). *Phytopath. Z.* **60**, 335–342.

Cruickshank, I. A. M. and Perrin, D. R. (1968). *Life Sci.* **7**, 449–458.

Cruickshank, I. A. M. and Perrin, D. R. (1971). *Phytopath. Z.* **70**, 209–229.

Cruickshank, I. A. M., Biggs, D. R. and Perrin, D. R. (1971). *J. Indian bot. Soc.* **50A**, 1–11.

Cruickshank, I. A. M., Biggs, D. R., Perrin, D. R. and Whittle, C. P. (1974a). *Physiol. Pl. Path.* **4**, 261–276.
Cruickshank, I. A. M., Veeraraghavan, J. and Perrin, D. R. (1974b). *Aust. J. Plant Physiol.* **1**, 149–156.
Daly, J. M. (1972). *Phytopathology* **62**, 392–400.
Day, P. R. (1974). "Genetics of the Host–parasite Interaction". Freeman, San Francisco.
Deverall, B. J. (1972a). *In* "Phytochemical Ecology" (J. B. Harborne, ed.), pp. 217–233. Academic Press, London and New York.
Deverall, B. J. (1972b). *Proc. R. Soc. London, Ser. B.* **181**, 233–246.
Dewick, P. M., Barz, W. and Grisebach, H. (1970). *Phytochemistry* **9**, 775–783.
Duczek, L. J. (1974). Ph.D. Thesis, University of Toronto.
Duczek, L. J. and Higgins, V. J. (1973). *2nd Int. Congr. Plant Pathol. Abstr.* 946.
Ebel, J., Achenbach, H., Barz, W. and Grisebach, H. (1970). *Biochim. biophys. Acta* **215**, 203–205.
Ebel, J., Hahlbrock, K. and Grisebach, H. (1972). *Biochim. biophys. Acta* **268**, 313–326.
Ellis, B. E. and Brown, S. A. (1974). *Can J. Biochem.* **52**, 734–738.
Ende, G. vanden. (1965). *TagBer. dt. Akad. LandwWiss. Berl.* **74**, 283–312.
Fawcett, C. H. and Spencer, D. M. (1969). *In* "Fungicides, An Advanced Treatise" (D. C. Torgeson, ed.), Vol. II, pp. 637–669. Academic Press, New York and London.
Ferreira, D., Brink, C. v. d. M. and Roux, D. G. (1971). *Phytochemistry* **10**, 1141–1144.
Fuchs, A. and Andel, O. M. van. (1968). *Neth. J. Pl. Path.* **74**, (*Suppl.* 1) 177–179.
Gasha, M., Tsuji, A., Sakurai, Y., Kurumi, M., Endo, T., Sato, S. and Yamaguchi, K. (1972). *J. pharm. Soc. Japan* **92**, 719–723.
Geigert, J., Stermitz, F. R., Johnson, G., Maag, D. D. and Johnson, D. K. (1973). *Tetrahedron* **29**, 2703–2706.
Grisebach, H. (1965). *In* "Chemistry and Biochemistry of Plant Pigments" (T. W. Goodwin, ed.), **79**, 279–308. Academic Press, London and New York.
Grisebach, H. (1967). "Biosynthetic Patterns in Microorganisms and Higher Plants." John Wiley and Sons, New York and Chichester.
Grisebach, H. and Barz, W. (1969). *Naturwissenchaften* **56**, 538–544.
Grisebach, H. and Hahlbrock, K. (1974). *Rec. Adv. Phytochem.* **8**, 21–52.
Grisebach, H. and Zilg, H. (1968). *Z. Naturf. Teil B.* **23**, 494–504.
Hadwiger, L. A. (1966). *Phytochemistry* **5**, 523–525.
Hadwiger, L. A. (1967). *Phytopathology* **57**, 1258–1259.
Hadwiger, L. A. (1968). *Neth. J. Pl. Path.* **74**, 163–169.
Hadwiger, L. A. (1972a). *Biochem. biophys. Res. Commun.* **46**, 71–79.
Hadwiger, L. A. (1972b). *Pl. Physiol.* **49**, 779–782.
Hadwiger, L. A. and Martin, A. R. (1971). *Biochem. Pharmac.* **20**, 3255–3261.
Hadwiger, L. A. and Schwochau, M. E. (1969). *Phytopathology* **59**, 223–227.
Hadwiger, L. A. and Schwochau, M. E. (1970). *Biochem. biophys. Res. Commun.* **38**, 683–691.
Hadwiger, L. A. and Schwochau, M. E. (1971a). *Pl. Physiol.* **47**, 346–351.
Hadwiger, L. A. and Schwochau, M. E. (1971b). *Pl. Physiol.* **47**, 588–590.
Hadwiger, L. A., Hess, S. L. and Broembsen, S. von. (1970). *Phytopathology* **60**, 332–336.
Hadwiger, L. A., Broembsen, S. von and Eddy, R. Jr. (1973). *Biochem. biophys. Res. Commun.* **50**, 1120–1128.
Hadwiger, L. A., Jafri, A., Broembsen, S. von and Eddy, R. Jr. (1974). *Pl. Physiol.* **53**, 52–63.

Hahlbrock, K., Wong, E., Schill, L. and Grisebach, H. (1970). *Phytochemistry* **9**, 949–958.

Hahlbrock, K., Ebel, J., Ortmann, R., Sutter, A., Wellman, E. and Grisebach, H. (1971). *Biochim. biophys. Acta* **244**, 7–15.

Harborne, J. B. (1973). *In* "Phytochemistry" (L. P. Miller, ed.), Vol. II, pp. 344–380. Van Nostrand Reinhold, New York.

Harborne, J. B., Boulter, D. and Turner, B. L. (eds) (1971). "Chemotaxonomy of the Leguminosae". Academic Press, London and New York.

Harper, S. H., Kemp, A. D., Underwood, W. G. E. and Campbell, R. V. M. (1969). *J. chem. Soc.* C 1109–1116.

Heath, M. C. and Higgins, V. J. (1973). *Physiol. Pl. Path.* **3**, 107–120.

Heath, M. C. and Wood, R. K. S. (1971). *Ann. Bot.* **35**, 475–491.

Hergert, H. L. (1962). *In* "The Chemistry of Flavonoid Compounds". (T. A. Geissman, ed.), pp. 553–592. The MacMillan Co., New York.

Hess, S. L. and Hadwiger, L. A. (1971). *Pl. Physiol.* **48**, 197–202.

Hess, S. L., Hadwiger, L. A. and Schwochau, M. E. (1971). *Phytopathology* **61**, 79–82.

Heuvel, J. vanden and Glazener, J. (1975). *Neth. J. Pl. Path.* **81**, 125–137.

Heuvel, J. vanden and VanEtten, H. D. (1973). *Physiol. Pl. Path.* **3**, 327–339.

Heuvel, J. vanden, VanEtten, H. D., Serum, J. W., Coffen, D. L. and Williams, T. H. (1974). *Phytochemistry* **13**, 1129–1131.

Higgins, V. J. (1972). *Physiol. Pl. Path.* **2**, 289–300.

Higgins, V. J. (1975). *Physiol. Pl. Path.* **6**, 5–18.

Higgins, V. J. and Millar, R. L. (1968). *Phytopathology* **58**, 1377–1383.

Higgins, V. J. and Millar, R. L. (1969a). *Phytopathology* **59**, 1493–1499.

Higgins, V. J. and Millar, R. L. (1969b). *Phytopathology* **59**, 1500–1506.

Higgins, V. J. and Millar, R. L. (1970). *Phytopathology* **60**, 269–271.

Higgins, V. J. and Smith, D. G. (1972). *Phytopathology* **62**, 235–238.

Higgins, V. J., Stoessl, A. and Heath, M. C. (1974). *Phytopathology* **64**, 105–107.

Hijwegen, T. (1973). *Phytochemistry* **12**, 375–380.

Hunter, L. D., Kirkham, D. S. and Hignett, R. C. (1968). *J. gen. Mictobiol.* **53**, 61–67.

Ingham, J. L. (1972). *Bot. Rev.* **38**, 343–424.

Ingham, J. L. (1973). *Phytopath. Z.* **78**, 314–335.

Ingham, J. L. and Millar, R. L. (1973). *Nature, Lond.* **242**, 125–126.

Joshi, B. S. and Kamat, V. N. (1973). *J. chem. Soc.* C. 907–911.

Keen, N. T. (1971). *Physiol. Pl. Path.* **1**, 265–275.

Keen, N. T. (1972). *Phytopathology* **62**, 1365–1366.

Keen, N. T. (1975a). *Science, N.Y.* **187**, 74–75.

Keen, N. T. (1975b). *Phytopathology* **65**, 91–92.

Keen, N. T. and Horsch, R. (1972). *Phytopathology* **62**, 439–442.

Keen, N. T. and Kennedy, B. W. (1974). *Physiol. Pl. Path.* **4**, 173–185.

Keen, N. T. and Sims, J. J. (1973). *2nd Int. Congr. Plant Pathol. Abstr.* 768.

Keen, N. T., Sims, J. J., Erwin, D. C., Rice, E. and Partridge, J. E. (1971). *Phytopathology* **61**, 1084–1089.

Keen, N. T., Partridge, J. E. and Zaki, A. I. (1972a). *Phytopathology* **62**, 768 (Abstr.).

Keen, N. T., Zaki, A. I. and Sims, J. J. (1972b). *Phytochemistry* **11**, 1031–1039.

Klarman, W. L. and Hammerschlag, F. (1972). *Phytopathology* **62**, 719–721.

Klarman, W. L. and Sanford, J. B. (1968). *Life Sci.* **7**, 1095–1103.

Koukol, J. and Conn, E. E. (1962). *J. biol. chem.* **236**, 2692–2698.

Kreuzaler, F. and Hahlbrock, K. (1972). *FEBS Letters* **28**, 69–72.

Kuć, J. (1972). *A. Rev. Phytopath.* **10**, 207–232.

Kurosawa, K., Ollis, W. D., Redman, B. T., Sutherland, I. O., Braga de Oliveira, A., Gottlieb, O. R. and Magalhães Alves, M. (1968). *J. chem. Soc., Chem. Commun.* 1263–1264.
Lampard, J. F. (1974). *Phytochemistry* 13, 291–292.
Land, B. G. van't, Wiersma-VanDuin, E. D. and Fuchs, A. (1975a). *Acta bot. neerl.* 24, 251. (Abstr.)
Land, B. G. van't, Wiersma-VanDuin, E. D. and Fuchs, A. (1975b). *Acta bot. neerl.* 24, 252. (Abstr.)
Letcher, R. M., Widdowson, D. A., Deverall, B. J. and Mansfield, J. W. (1970). *Phytochemistry* 9, 249–252.
Loper, G. M. and Hanson, C. H. (1964). *Crop Sci.* 4, 480–482.
MacDonald, R. E. and Bishop, C. J. (1952). *Can J. Bot.* 30, 486–489.
Moustafa, E. and Wong, E. (1967). *Phytochemistry* 6, 625–632.
Müller, K. O. (1956). *Phytopath. Z.* 27, 237–254.
Müller, K. O. (1958). *Aust. J. biol. Sci.* 11, 275–300.
Müller, K. O. (1969). *Zentbl. Bakt. ParasitKde Infectionskr. Hyg. Abt. 2.* 123, 259–265.
Müller, K. O. and Börger, H. (1940). *Arb. biol. Reichsanst. Land- u. Forstw. Berlin-Dahlem.* 23, 189–231.
Naim, M., Gestetner, B., Zilkah, S., Birk, Y. and Bondi, A. (1974). *J. agric. Fd Chem.* 22, 806–810.
Nilsson, A., Hill, J. L. and Davies, H. L. (1967). *Biochim. biophys. Acta* 148, 92–98.
Nonaka, F. (1967). *Agric. Bull. Saga Univ.* 24, 109–121.
Oku, H., Nakanishi, T., Shiraishi, T. and Ouchi, S. (1973). *Sci. Rep. Fac. Agric. Okayama Univ.* 42, 17–20.
Olah, A. F. and Sherwood, R. T. (1971). *Phytopathology* 61, 65–69.
Olah, A. F. and Sherwood, R. T. (1973). *Phytopathology* 63, 739–742.
Oung-Boran, P., Lebreton, P. and Netien, G. (1969). *Planta med.* 17, 301–318.
Pachler, K. G. R. and Underwood, W. G. E. (1967). *Tetrahedron* 23, 1817–1826.
Patschke, L., Barz, W. and Grisebach, H. (1966). *Z. Natursf. Teil B.* 21, 201–205.
Patterson, A. M., Capell, L. T. and Walker, D. F. (1960). "The Ring Index". *Am. Chem. Soc.,* Washington, D.C.
Paxton, J., Goodchild, D. J. and Cruickshank, I. A. M. (1974). *Physiol. Pl. Path.* 4, 167–171.
Pelter, A. (1968). *Tetrahedron Lett.* 897–903.
Pelter, A., Bradshaw, J. and Warren, R. F. (1971). *Phytochemistry* 10, 835–850.
Perrin, D. D. and Perrin, D. R. (1962). *J. Am. Chem. Soc.* 84, 1922–1925.
Perrin, D. R. (1964). *Tetrahedron Lett.* 29–35.
Perrin, D. R. and Bottomley, W. (1962). *J. Am. Chem. Soc.* 84, 1919–1922.
Perrin, D. R. and Cruickshank, I. A. M. (1965). *Aust. J. biol. Sci.* 18, 803–816.
Perrin, D. R. and Cruickshank, I. A. M. (1969). *Phytochemistry* 8, 971–978.
Perrin, D. R., Whittle, C. P. and Batterham, T. J. (1972). *Tetrahedron Lett.* 1673–1676.
Perrin, D. R., Biggs, D. R. and Cruickshank, I. A. M. (1974). *Aust. J. Chem.* 27, 1607–1611.
Pierre, R. E. and Bateman, D. F. (1967). *Phytopathology* 57, 1154–1160.
Pinkas, J., Lavie, D. and Chorin, M. (1968). *Phytochemistry* 7, 169–174.
Pueppke, S. G. and VanEtten, H. D. (1974). *Phytopathology* 64, 1433–1440.
Pueppke, S. G. and VanEtten, H. D. (1975). *J. chem. Soc., Perkin Trans.* 1, 946–948.
Pueppke, S. G. and VanEtten, H. D. (1976). *Physiol. Pl. Path.* 8, 51–66.
Rahe, J. E. (1973). *Can. J. Bot.* 51, 2423–2430.
Rahe, J. E. and Arnold, R. M. (1975). *Can. J. Bot.* 53, 921–928.
Rall, G. J. H., Englebrecht, J. P. and Brink, A. J. (1970). *Tetrahedron* 26, 5007–5012.
Rathmell, W. G. (1973). *Physiol. Pl. Path.* 3, 259–267.

Rathmell, W. G. and Bendall, D. S. (1971). *Physiol. Pl. Path.* **1**, 351–362.

Reilly, J. J. and Klarman, W. L. (1972). *Phytopathology* **62**, 1113–1115.

Ribéreau-Gayon, P. (1971). "Plant Phenolics". Oliver and Boyd, Edinburgh.

Russell, D. W. and Conn, E. E. (1967). *Archs Biochem. Biophys.* **122**, 256–258.

Sakuma, T. and Millar, R. L. (1972). *Phytopathology* **62**, 499 (Abstr.).

Schwochau, M. E. and Hadwiger, L. A. (1968). *Archs Biochem. Biophys.* **126**, 731–733.

Schwochau, M. E. and Hadwiger, L. A. (1969). *Archs Biochem. Biophys.* **134**, 34–41.

Sherwood, R. T., Olah, A. F., Oleson, W. H. and Jones, E. E. (1970). *Phytopathology* **60**, 684–688.

Shibata, S. and Nishikawa, Y. (1963). *Chem. Pharm. Bull.* **11**, 167–177.

Shutt, D. A. and Braden, A. W. H. (1968). *Aust. J. Agric. Res.* **19**, 545–553.

Sims, J. J., Keen, N. T. and Honwad, V. K. (1972). *Phytochemistry* **11**, 827–828.

Slayman, C. L. and VanEtten, H. D. (1974). *Pl. Physiol. Suppl. 134.* (Abstr.).

Smith, D. A., VanEtten, H. D. and Bateman, D. F. (1973a). *Physiol. Pl. Path.* **3**, 179–186.

Smith, D. A., VanEtten, H. D., Serum, J. W., Jones, T. M., Bateman, D. F., Williams, T. H. and Coffen, D. L. (1973b). *Physiol. Pl. Path.* **3**, 293–297.

Smith, D. A., VanEtten, H. D. and Bateman, D. F. (1975). *Physiol. Pl. Path.* **5**, 51–64.

Smith, D. G., McInnes, A. G., Higgins, V. J. and Millar, R. L. (1971). *Physiol. Pl. Path.* **1**, 41–44.

Steiner, P. W. and Millar, R. L. (1974). *Phytopathology* **64**, 586 (Abstr.).

Stenlid, G. (1970). *Phytochemistry* **9**, 2251–2256.

Stholasuta, P., Bailey, J. A., Severin, V. and Deverall, B. J. (1971). *Physiol. Pl. Path.* **1**, 177–183.

Stob, M. (1973). *In* "Toxicants Occurring Naturally in Foods", pp. 550–557. National Academy of Sciences, Washington, D.C.

Stoessl, A. (1970). *Rec. Adv. Phytochem.* **3**, 143–180.

Stoessl, A. (1972). *Can. J. Biochem.* **50**, 107–108.

Suginome, H. (1962). *Experientia* **18**, 161–163.

Suginome, H. (1966). *Bull. Chem. Soc. Jap.* **39**, 1529–1534.

Suginome, H. and Iwadare, T. (1962). *Experientia* **18**, 163–164.

Sutter, A. and Grisebach, H. (1969). *Phytochemistry* **8**, 101–106.

Swain, T. (1962). *In* "Wood Extractives" (W. E. Hillis, ed.), Academic Press, London and New York.

Teasdale, J., Daniels, D., Davis, W. C., Eddy, R. Jr. and Hadwiger, L. A. (1974). *Pl. Physiol.* **54**, 690–695.

Towers, G. H. N. (1974). *In* "MTP International Review of Science, Biochemistry Series 1". (D. H. Northcote, ed.), Vol. XI, pp. 247–276. Butterworths, London.

Uehara, K. (1960). *Bull. Hiroshima agric. Coll.* **1**, 7–10.

Uehara, K. (1963). *Bull. Hiroshima agric. Coll.* **2**, 41–44.

Uehara, K. (1964). *Ann. phytopath. Soc. Japan* **29**, 103–110.

VanEtten, H. D. (1972). *Phytopathology* **62**, 795 (Abstr.).

VanEtten, H. D. (1973a). *Phytopathology* **63**, 1477–1482.

VanEtten, H. D. (1973b). *Phytochemistry* **12**, 1791–1792.

VanEtten, H. D. (1976). *Phytochemistry* **15**, 655–659.

VanEtten, H. D. and Bateman, D. F. (1971). *Phytopathology* **61**, 1363–1372.

VanEtten, H. D. and Smith, D. A. (1975). *Physiol. Pl. Path.* **5**, 225–237.

VanEtten, H. D., Pueppke, S. G. and Kelsey, T. C. (1975). *Phytochemistry* **14**, 1103–1105.

Virtanen, A. I. and Hietala, P. K. (1958). *Acta chem. scand.* **12**, 579–581.

Wit-Elshove, A. de. (1968). *Neth. J. Pl. Path.* **74**, 44–47.

Wit-Elshove, A. de. (1969). *Neth J. Pl. Path.* **75**, 164–168.

Wit-Elshove, A. de. and Fuchs, A. (1971). *Physiol. Pl. Path.* **1**, 17–24.

Wong, E. (1968). *Phytochemistry* **7**, 1751–1758.

Wong, E. (1970). *Progr. Chem. Org. Natur. Prod.* **28**, 1–73.

Wong, E. and Grisebach, H. (1969). *Phytochemistry* **8**, 1419–1426.

Wong, E. and Latch, G. C. M. (1971). *Phytochemistry* **10**, 466–468.

Wood, R. K. S. (1967). "Physiological Plant Pathology". Blackwell Scientific Publications, Oxford and Edinburgh.

CHAPTER 14

Lignification in Infected Tissue

J. FRIEND

Department of Plant Biology, The University, Hull, England

I. INTRODUCTION

It has been generally assumed by plant pathologists that because most plant pathogenic microorganisms cannot degrade lignin, lignified cell walls will offer a barrier to pathogenesis by plant parasites. It is assumed, for example, that the hyphae of fungal parasites which attack woody cells of plants will only pass from cell to cell through pits in cell walls because the remainder of the wall, being lignified, offers an impermeable barrier. In such situations, lignin would act as a resistance factor present in the plant before infection (Butler and Jones, 1955; Tarr, 1972).

However, there are examples of lignification occurring after infection, and apparently in response to infection.

Several investigations of the host–parasite relationship between cucumber and cucumber scab caused by *Cladosporium cucumerinum* have led to the general conclusion that lignification may be an active response of host plants to pathogens and in addition that such responses may be involved in resistance reactions. It was reported by Behr (1949) that if fruits or petioles of a susceptible variety of cucumber were wounded with a scalpel before inoculation, infection was not established although there had been some hyphal growth. It was found, on anatomical investigation, that the cells surrounding the wounded tissue were lignified. If inoculation were not preceded by wounding, infection could be established. However, in later experiments, El-Din-Fouad (1956) found that there was lignification of dead cells in resistant cucumbers which could also be caused by wounding, whereas when susceptible plants were wounded there was no lignification.

Pierson and Walker (1954) showed that when resistant cucumber plants were infected there was a thickening of the cell wall around the infection site, although no examination was made for lignin. In the course of an investigation to determine the resistance-promoting effect of phenylserine against *C. cucumerinum*, Hijwegen (1963) found that infection was not established in susceptible plants when vigorous brushing of the hypocotyls was used as part of the inoculation process. When the plants were examined histologically, many patches of lignified tissue, i.e. cells with lignified walls, were found. It was established that lignification was a reaction to mechanical damage caused by brushing. Both resistant and susceptible plants were therefore inoculated by spraying with a spore suspension of the fungus and infected plants compared with uninfected controls. Histological examination of all four groups of plants showed that the only ones in which any lignification occurred were those of the infected resistant variety; this was not found in the epidermal cells but in the underlying parenchyma. Hijwegen concluded that lignification could be part of an active resistance mechanism which inhibits the parasite in its progress.

Although it is not explicitly stated in the paper, it seems to be implicit that the formation of a layer of lignified cells would act as a physical barrier between the parasitic fungus and the uninvaded cells.

Lignin was demonstrated to occur, using both biochemical and histological techniques, in cell walls of radish infected by *Peronospora parasitica* by Asada and Matsumoto (1969). In a later paper (Asada and Matsumoto, 1971) lignification was examined by u.v. microspectrophotometry and it was concluded that lignin accumulated in the middle lamellae of the parenchyma cell walls. Spectral studies, confirmed by later chemical analysis (Asada and Matsumoto, 1972) showed that there were differences between the lignin formed after infection and that present in vascular bundles and periderm of non-infected roots. The lignin in diseased parenchyma cell walls was composed mainly of guaiacylpropane units whereas that in healthy vessel walls was composed mainly of syringylpropane units. It is interesting to note that, unlike the cucumber–*C. cucumerinum* relationship studied by Hijwegen (1963), in the case of the Japanese radish–*P. parasitica* interaction there seems to be no host resistance. Infection proceeds until every parenchyma cell is invaded, although the hyphae cannot invade those cells whose walls are lignified before inoculation, i.e. the vascular bundles. It was suggested that in the case of the Japanese radish lignification of the parenchyma walls was a type of delayed resistance reaction to infection.

It has now been shown by Ride (1975) that wounding wheat leaves before inoculating them with spores of either *Botrytis cinerea* or *Mycosphaerella pinoides*, which are non-pathogens of wheat, led to the formation of lignin around the wounds. The growth of the fungi was limited to the wounds, and since wounding alone did not stimulate lignification, it was concluded that lignin was being actively synthesized in response to fungal infection. The newly synthesized lignin differed in its absorption spectrum, staining reaction and products formed by alkaline nitrobenzene oxidation from the lignin present in healthy unwounded leaves. However, when wounded wheat leaves were inoculated with the two wheat pathogens *Septoria nodorum* and *S. tritici*, it was found that lignification was much slower and the fungus was not confined to the wounded, lignified area of the leaf. It was suggested by Ride that the rapidity of lignification in wheat in response to non-pathogens and especially the close association of lignification with the approach of the fungal hyphae meant that lignin was involved in the restriction of the non-pathogenic fungi in wheat leaves. The ability of the two pathogens *S. nodorum* and *S. tritici* to spread over the wounded area of the leaf could then be related to the delay in lignification in response to these two fungi.

It appears that phenolic compounds are involved in resistance of potato tuber slices to rotting by *Pseudomonas fluorescens* (Zucker and Hankin, 1970). Tuber slices aged for 24 h in the light in the presence of cycloheximide are easily rotted either by the bacterium or by the pectic lyases purified from a bacterial culture filtrate, whereas if they are aged in the absence of cycloheximide they are resistant. Cycloheximide inhibits the increased synthesis of phenylalanine ammonia lyase (PAL) normally found in the light. A correlation was found between conditions under which cycloheximide inhibited PAL and made potato slices susceptible to maceration by *P. fluorescens*. Furthermore it was found that slices from potato cultivars which produced high levels of PAL were not macerated by the bacterium; the only two varieties which were markedly susceptible to attack by *P. fluorescens* produced far less PAL than the more resistant varieties. In addition the cycloheximide treated discs were white, unlike the untreated slices which had a tan coloured suberized layer over their cut surface. Mechanical removal of the suberized layer decreased the resistance of aged discs to maceration by pectic lyases.

It was therefore concluded that cycloheximide blocks the development of resistance by inhibiting the production of phenolic precursors of suberization, through the inhibition of PAL synthesis.

II. Lignification in the Solanaceae in Response to Fungal Infection

A. THE RESPONSE OF POTATO TUBERS TO *PHYTOPHTHORA INFESTANS*

In an analysis of the carbohydrate changes in potato tuber cell walls during the course of infection by *P. infestans* it was found by Friend and Knee (1969)

that in tuber slices of the susceptible potato cultivar King Edward a material
having the spectral properties of lignin was present in the cell walls of
infected slices and that it increased in amount as the infection increased;
some of this material was also formed in uninfected tissue. The spectroscopic
data were obtained using the difference spectrum method of Stafford (1960)
based on earlier work by Goldschmid (1954) which gives a difference
spectrum with peaks in the regions of 242, 292 and 355 nm. The peaks at
242 and 292 nm arise from non-conjugated phenols while phenols with
conjugated side chains such as the hydroxy-cinnamic acid derivatives will
account for the peak at 355 nm. It was later found (Reynolds, 1971) that cell
wall preparations from infected tuber slices could be oxidized by alkaline
nitrobenzene to yield the three characteristic aldehydes normally obtained
from lignin, namely p-hydroxybenzaldehyde, vanillin and syringaldehyde
although the actual yields of syringaldehyde were very small. The oxidation
studies supported the hypothesis that the material in infected cell walls was
lignin-like.

A series of experiments was then carried out in which biochemical changes
were measured in potatoes of both a susceptible (Majestic) and a resistant
(Orion) variety were inoculated with P. infestans.

When the deposition of the lignin-like material in Majestic and Orion
was compared (Friend, 1973; Friend et al., 1973), it was seen that the rate
of increase was very much faster in Orion than in Majestic. During the same
experiment, measurements were also made of the activity of PAL and the
results were similar, namely a more rapid increase in the enzyme level in the
resistant variety than in the susceptible and in both cases the levels in the
inoculated were higher than the corresponding uninoculated controls. It
was therefore concluded that there could be a relationship between the
increase in the level of activity of PAL and the deposition of the lignin-like
material in the cell walls of the two varieties. This conclusion was reinforced
by the results of an experiment in which tuber slices were inoculated with
mycelial discs of P. infestans. It was found that in the resistant variety, where
the fungus does not grow out much from the infection site, there were large
and relatively rapid increases in both PAL and in the lignin-like material
in that region of the tuber slices to which the fungus was confined. However,
in the susceptible variety there was a much wider spread of fungal growth
associated with a wider area of both PAL activity and the lignin-like polymer,
both of which increased relatively slowly. Thus a rapid, confined lignification
and a correlated rise in PAL activity seem to be associated with the reaction
of resistant rather than susceptible tuber slices.

PAL probably controls lignification since it is concentrated in the lignifying
xylem tissues of plants such as sycamore trees, celery petioles and pea roots
(Rubery and Northcote, 1968). Measurements of the increase of caffeic
acid-O-methyl transferase (COMT) show that in the resistant variety this
is also clearly confined to the area of the potato tuber where the fungus has

made its restricted growth (Friend and Thornton, 1974); since there is little difference between COMT levels in the infected and uninfected slices of the susceptible variety, it is possible that COMT rather than PAL may control the formation of the lignin-like material in Orion.

B. SPECIFICITY OF THE TUBER RESPONSE TO *P. INFESTANS*

In order to determine whether the formation of the lignin-like compound and the associated increase in PAL activity were really part of the resistance reaction, the effect of another race of *P. infestans*, namely 1,3,5 which will give a susceptible reaction with Orion, was compared with that of the usual race, race 4. In addition the effects of these two races of the fungus were compared on another variety of potato, Pentland Beauty (R_3) which also gives a resistant reaction with race 4 and a susceptible one with race 1,3,5 (Fig. 1). PAL activity was also determined on a susceptible variety (King Edward) inoculated with race 4 (Henderson, 1975) (Fig. 2).

Under the conditions of the experiment, the response of Orion to the two races of the fungus in terms of laying down of lignin-like material shows no difference for the first 2 days but by day 3 the resistant slices show a relatively large difference between infected and uninfected; the difference in the susceptible slices is showing a smaller difference than on day 2, which indicates that the rate of increase of deposition in the uninfected slices is faster than in the infected ones. In the case of Pentland Beauty, although there is relatively more lignin-like material in the susceptible reaction one

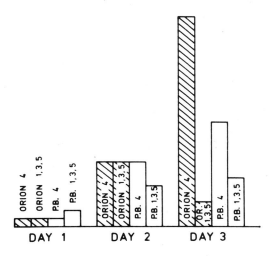

FIG. 1. Lignin levels (measured as A_{360}) for inoculated tuber discs *minus* the corresponding value for uninoculated tissue of tuber varieties Orion and Pentland Beauty inoculated with race 4 of *P. infestans* which gives a resistant reaction or with race 1,3,5 which gives a susceptible reaction.

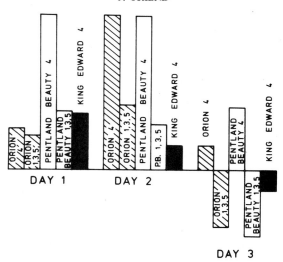

FIG. 2. Phenylalanine ammonia lyase (PAL) activity in inoculated tuber discs *minus* the values for uninoculated tissue using the potato varieties Orion, Pentland Beauty and King Edward. On both Orion and Pentland Beauty race 4 of *P. infestans* gives a resistant reaction and race 1,3,5 a susceptible reaction. King Edward is susceptible to race 4.

day after inoculation, at 2 and 3 days after inoculation there is more deposition in the resistant reaction.

In the case of PAL, the differences between the resistant and susceptible reactions are more marked in Pentland Beauty than in Orion one day after inoculation. The two varieties are showing similar reactions on the second day, and on the third day whereas in both resistant reactions there is more PAL than in the control, in the susceptible reactions and in the susceptible King Edward there is a lower level of PAL in the infected than in the un-infected tuber slices.

C. LIGNIFICATION IN LEAVES

The deposition of lignin-like material which has been described so far has been confined to tuber discs and slices, where part of the reaction is associated with wound healing. Recently, in collaboration with Professor T. Swain and Mrs M. Ward, reactions of a variety of leaves of Solanaceous plants including potato have been examined and the results are displayed in Fig. 3a–d. The technique was slightly different in these experiments in that cell walls were first isolated from the leaf using the neutral detergent method of van Soest and Wine (1967) before making the alkaline extract for difference spectrum measurements.

Two of the experiments were on the potato–*P. infestans* interaction; in one the reaction of two varieties to the same race of *P. infestans* was measured

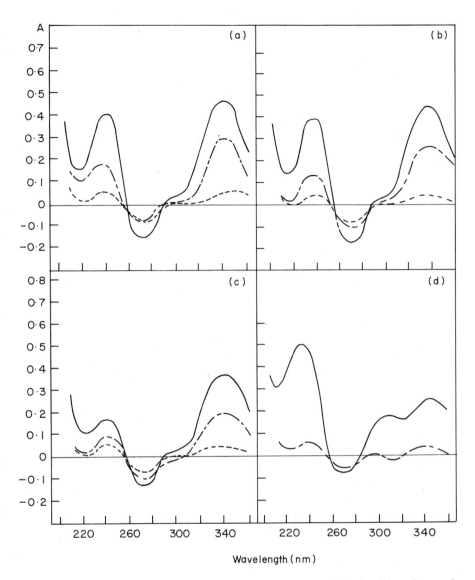

FIG. 3. Difference spectra of sodium hydroxide extracts of cell walls isolated from leaves of Solanaceous plants. (a) Orion (———); Majestic (— – —) inoculated with race 4 of *P. infestans*; Majestic control uninoculated (. . .). (b) Orion leaves inoculated with race 4 (———); race 1,3,5 (— – —); control uninoculated (. . .). (c) Majestic leaves inoculated with *Botrytis cinerea*; inoculated (———) wounded control (— – —); unwounded control (. . .). (d) Tomato leaves inoculated with *B. cinerea*; inoculated (———); wounded control (— – —).

and in the other the reaction of two races of the fungus was determined on leaves of the same variety of potato. It will be seen that in each experiment there is more lignin-like material in the resistant than in the susceptible combination.

In the other two experiments, the effect of the non-pathogen *Botrytis cinerea* on Majestic potato and tomato was investigated. It can be seen that the greater deposition of lignin-like material occurs in tomato, with Majestic potato leaf second. Here again there is evidence for deposition of lignin-like material associated with a resistance of a host to a fungus.

III. THE NATURE OF LIGNIN-LIKE MATERIAL

The spectroscopic method used for determination of the lignin-like material merely indicates that insoluble phenolic material is being dissolved by hot alkali. It was found that if the saponification were carried out for a few hours at room temperature under N_2, instead of overnight at 70°, an appreciable quantity of material was dissolved in alkali. Furthermore, after acidification of the cold alkaline extract and ether extraction it was possible to demonstrate by two-dimensional thin-layer chromatography that both *p*-coumaric and ferulic acids were present. It seems therefore that these acids are esterified to cell wall material.

Phenolic acids exist as esters of several different types in plants. Higuchi *et al.* (1967) found esterified phenolic acids as part of the structure of lignin in grasses. Hartley (1973) has found ferulic acid esters of cell wall carbohydrates in grasses; they might be esterified to either hemicellulose or cellulose. Van Sumere *et al.* (1973) have found ferulic acid esterified to protein in barley globulin. Ferulic acid is esterified to a glycoprotein in wheat flour, linked to the arabinoxylan moiety (Fausch *et al.*, 1963; Nuekom *et al.*, 1967). El-Basyouni *et al.* (1964) found that ferulic and *p*-coumaric acids were esterified in acetone powder preparations from grasses, although at that time it was presumed that the phenolic acids were esterified to enzymes and that these enzyme-bound complexes were intermediate in lignin biosynthesis.

It is not possible at present to determine to which sort of compound the phenolic acids are esterified in Solanaceous plants; speculations are perhaps best made on the basis of assumptions concerning the mechanism of the resistance reaction which are considered below. It is likely that there may be some complex with proteins since it is clear that part of the difference spectrum which we have observed is due to proteinaceous material. Some free amino acids can be shown by TLC to be present in the alkaline extract. Treatment of the cell wall preparations with pronase before saponification alters the shape of the difference spectrum; there is a considerable loss of absorbance at 242 nm which is due to non-conjugated phenols, and the spectrum more closely resembles that of a mixture of ferulic and *p*-coumaric acids. However, the presence of protein indicated by these results does not necessarily mean that it had been covalently linked to the phenolic acids.

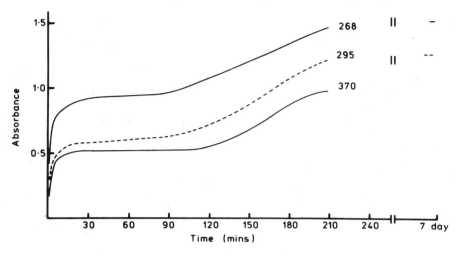

FIG. 4. The increase of absorbance with time after treating cell walls from infected Orion leaves in 0·1 NaOH at room temperature.

In more recent studies we have obtained an indication that there may be two types of esterification of the phenolic acids of cell wall material. A cell wall preparation from infected leaves of Orion potatoes was suspended in 0·1N NaOH at room temperature and the spectral changes measured at different time intervals. It will be seen (Fig. 4) that there was a rapid increase in the phenolic peaks at 268, 295 and 370 nm in the first few minutes but that there is then little change for more than 60 min after which a second and slower rate of increase occurs during the following 90 min; values then remain stationary for about 7 days. The long-term measurements are made only on the 268 and 295 nm peaks because the 370 nm peak is unstable after more than 3 h.

Although it is possible that the phenolic acids may be esterified to two different components in the cell wall, these studies give no indication of the types of component with which they are combined.

IV. INSOLUBLE ESTERS OF PHENOLIC ACIDS

A. MECHANISM OF ANTIFUNGAL ACTION

Since cell wall polysaccharide degradation catalyzed by pectic enzymes, hemicellulases and cellulases is an important factor in the invasion of host cells by pathogens (see Chapter 5), inhibition of such degradation could be an important facet of a resistance mechanism.

The two mechanisms by which alterations to host cell walls such as lignification could conceivably inhibit cell wall degradation are the formation of either a physical or a chemical barrier (Friend, 1973). The physical barrier mechanism implies that a layer of lignin or lignin-like substance is interposed between the cell wall carbohydrates and the pathogen. This barrier would be impervious so that the enzymes secreted by the pathogen are unable to come into contact with their substrates.

The chemical barrier hypothesis requires the alteration of the cell wall polysaccharide substrate such that although the enzyme and substrate could come into physical contact the chemically modified polysaccharide substrate would be unable to form the enzyme-substrate complex necessary for the hydrolytic or lytic reaction to proceed. One implication of the chemical barrier hypothesis is a smaller total chemical change than is necessary for the formation of a physical barrier.

It is known that modification of the structure of polygalacturonic acid can alter its ease of hydrolysis by fungal polygalacturonases; for example, the polygalacturonase of *Aspergillus niger* hydrolyses the partial acetyl ester of polygalacturonic acid slowly and incompletely (Deuel and Stutz, 1958). It is therefore possible to propose that hydroxylated cinnamic acids found in resistant potato cell walls after fungal attack are esterified to at least one of the carbohydrate polymers of the cell wall. Potato pectin contains galactose as its major component (Friend and Knee, 1969; Hoff and Castro, 1969). It is assumed that the galactose forms a galactan containing 1,4 linked β-D-galactopyranose units because galactose is released when either the potato wall or the pectic fraction is treated with an extracellular enzyme from *P. infestans* which has β-1,4-galactosidase activity (Knee and Friend, 1968, 1970; Knee, 1970; Cole, 1970) and in addition Wood and Siddiqui (1972) have isolated a galactan containing a predominance of β-1,4-linked galactopyranose units from potato tubers. Although it is not yet clear whether the galactan is covalently linked to the rhamnogalacturonan moiety of the pectin or is a separable polymer, hydrolysis of the galactan appears to be a major factor in tuber cell wall degradation by *P. infestans* (Friend and Knee, 1969; Knee, 1970). If the galactan were modified by esterification of C_2 or C_3 hydroxyls of the galactosyl residues by hydroxy cinnamic acids, it might then no longer be able to act as a substrate for the fungal galactanase. Such a mechanism could well explain the resistance of tuber cell walls to degradation by *P. infestans*. Zucker and Hankin (1970) postulated that the resistance of tuber slices, which have been aged in the light, to maceration by pectate lyases is due partly to the formation of a suberized layer on the surface and also to the presence of an intrinsic resistance factor. The use of the term suberization raises the question of the role of phenolics in the structure of suberin. Kolattukudy (1975) has stated that one third of potato suberin consists of aliphatic compounds while the rest appears to be phenolic; it is not clear whether the phenolic and aliphatic moieties are covalently bonded or are separate entities. Lignin has been shown by staining methods to be present in the suberized layers present in wound barriers in cut potato tubers (Craft and Audia, 1962). Since there is little deposition of the characteristic aliphatic components of suberin in wound periderm for several days after the excision of potato tuber slices (Kolattukudy and Dean, 1974), it is likely that the so called suberized layer found by Zucker and Hankin 24 h after excision was a lignin-like polymer. In this case one has also to consider that insoluble esters of hydroxylated cinnamic acids are present and that these could be involved in the resistance of aged tuber slices to pectate lyase in a chemical barrier mechanism.

If this is so it is necessary to suggest that the hydroxy cinnamic acids are esterified to the galacturonan (rhamnogalacturonan) component of the pectin. The C_2 or C_3 hydroxyls of the galacturonosyl residues would be a likely site for esterification and would also correspond with the acetyl groups found in sugar beet pectin. Since the outer layers were the most resistant part of the discs to pectate lyase, it is necessary to make an additional postulate, namely that there is a greater esterification of cell wall pectin at the outer layers of tuber discs.

To date there is no direct evidence for the esterification of either the galactan or the galacturonan components of cell walls by phenolic acids. The carbohydrate ester of ferulic acid isolated from grass cell walls (Hartley, 1973) contains xylose, arabinose and glucose units and it can be released from the wall by treatment with cellulase preparation which also had hemicellulase activity. However, it is not clear whether the ferulic acid ester is attached to the cellulose or to the hemicellulose components of the cell wall (Hartley et al., 1973). So far we have been unable to demonstrate the presence of similar phenolic-carbohydrate esters in potato cell walls by cellulase treatment. However if the phenolic acids are esterified to either the galactan or galacturonan component of the potato pectin one would hardly expect cellulase to degrade such complexes. Nor, if the function of phenolic acid esterification is to inhibit degradation by polygalacturonases, pectate lyases and galactanases, would it be expected that these pectic enzymes would catalyze the release of ferulic acid-carbohydrate esters.

B. OCCURRENCE IN THE PLANT KINGDOM

Ferulic and p-coumaric acids are esterified to cell walls in all members of the Graminae so far examined (Hartley and Harris, 1975) and it is noteworthy that they are present in grass plants which give a negative phloroglucinol/HCl test for lignin. In a range of lower plants Mrs M. Ward and Professor Swain have found that both hydroxybenzoic and hydroxycinnamic acids are esterified in the insoluble fraction after methanol extraction

302   J. FRIEND

TABLE I

Methanol-insoluble esterified phenolic acids in lower plants

	p-Hydroxy-benzoate	Proto-catechuate	Vanil-late	p-Coum-arate	Caf-feate	Ferul-ate
Angiopteris tyrofeuca	+	−	−	+	+	+
Adiantum tenerum	+	−	−	+	+	+
Cycas revoluta	+	−	+	+	−	+
Zamia spp	+	?	−	+	−	+
Juniperus procumbens	+	−	−	+	+	+
Tsuga canadiensis	+	−	+	+	−	+
Pinus strobus (plus unidentified pale blue fluorescing compound)	+	−	−	+	−	+

(Table I). Although there is no evidence that insoluble cinnamic acid esters have a protective function against fungal attack in the grasses, it is still possible that in the lower plants they could act as a primitive defence mechanism. This action would depend upon the nature of the compounds to which they are esterified.

In view of the widespread distribution of insoluble cinnamyl and other phenolic esters in the plant kingdom, many reports of lignin occurrence may well be suspect. It would be advisable not to use methods of analysis such as the difference spectrum method and alkaline nitrobenzene oxidation which will also estimate esterified cinnamic acids (cf. Higuchi *et al.*, 1967) but rather to use the more recently developed methods of Miksche and his colleagues (e.g. Erikson *et al.*, 1973).

ACKNOWLEDGEMENTS

I wish to thank Professor T. Swain and Mrs M. Ward for their collaboration in some of the experiments reported in the paper. I also wish to thank the Agricultural Research Council and the Science Research Council for financial support of much of this work.

REFERENCES

Asada, Y. and Matsumoto, I. (1969). *Ann. Phytopath. Soc., Japan* **35**, 160–167.
Asada, Y. and Matsumoto, I. (1971). *Physiol. Pl. Path.* **1**, 377–383.
Asada, Y. and Matsumoto, I. (1972). *Phytopath. Z.* **73**, 208–214.
Behr, L. (1949). *Phytopath. Z.* **15**, 92–123.
Butler, E. J. and Jones, S. E. (1955). "Plant Pathology". Macmillan, London.
Cole, A. L. J. (1970). *Phytochemistry* **9**, 337–340.
Craft, C. C. and Audia, W. V. (1962). *Bot. Gaz.* **123**, 211–219.
Deuel, H. and Stutz, E. (1958). *Adv. Enzymol.* **20**, 341–382.
El-Basyouni, S. Z., Neish, A. C. and Towers, G. H. N. (1964). *Phytochemistry* **3**, 627–639.

El-Din-Fouad, M. K. (1956). *Meded. Landb Hoogesch. Wageningen* **56**, 1–55.

Erickson, M., Larson, S. and Miksche, G. E. (1973). *Acta chem. scand.* **27**, 127–140.

Fausch, H. Kundig, W. and Neukom, H. (1963). *Nature, Lond.* **199**, 287.

Friend, J. (1973). *In* "Fungal Pathogenicity and the Plant's Response" (R. J. W. Byrde and C. V. Cutting, eds), pp. 383–396. Academic Press, London and New York.

Friend, J. and Knee, M. (1969). *J. exp. Bot.* **20**, 763–775.

Friend, J. and Thornton, J. D. (1974). *Phytopath. Z.* **81**, 56–64.

Friend, J., Reynolds, S. B. and Aveyard, M. A. (1973). *Physiol. Pl. Path.* **3**, 495–507.

Goldschmid, O. (1954). *Analyt. Chem.* **26**, 1421–1423.

Hartley, R. D. (1973). *Phytochemistry* **12**, 661–665.

Hartley, R. D. and Harris, P. J. (1975). Paper presented at SCI Agriculture Group Meeting—Nutritive Value of Plant Fibre.

Hartley, R. D., Jones, E. C. and Wood, T. M. (1973). *Phytochemistry* **12**, 763–766.

Henderson, S. J. (1975). Ph.D Thesis, University of Hull.

Higuchi, I., Ito, Y. and Kawomura, I. (1967). *Phytochemistry* **6**, 875–881.

Hijwegen, T. (1963). *Tijdschr. PlZiekt.* **69**, 314–317.

Hoff, J. E. and Castro, M. D. (1969). *J. Agric. Fd Chem.* **17**, 1328–1331.

Knee, M. (1970). *Phytochemistry* **9**, 2075–2083.

Knee, M. and Friend, J. (1968). *Phytochemistry* **7**, 1289–1291.

Knee, M. and Friend, J. (1970). *J. gen. Microbiol.* **60**, 23–30.

Kolattukudy, P. E. (1975). *In* "Recent Advances in the Chemistry and Biochemistry of Plant Lipids" (T. Galliard and E. I. Mercer, eds), pp. 203–246. Academic Press, London and New York.

Kolattukudy, P. E. and Dean, B. B. (1974). *Pl. Physiol.* **54**, 116–121.

Nuekom, H., Providoli, L., Gremli, H. and Hui, P. A. (1967). *Cereal Chem.* **44**, 238–251.

Pierson, C. F. and Walker, J. C. (1954). *Phytopathology* **44**, 459–465.

Reynolds, S. B. (1971). Ph.D Thesis, University of Hull.

Ride, J. P. (1975). *Physiol. Pl. Path.* **5**, 125–134.

Rubery, P. H. and Northcote, D. H. (1968). *Nature, Lond.* **219**, 1230–1234.

Stafford, H. A. (1960). *Pl. Physiol.* **35**, 108–114.

Tarr, S. A. J. (1972). "The Principles of Plant Pathology", Macmillan, London.

Van Soest, P. J. and Wine, R. H. (1967). *J. Ass. off. agric. Chem.* **50**, 50–55.

Van Sumere, C. F., De Pooter, H. Haider Ali and Degraw-van-Bussel, M. (1973). *Phytochemistry* **12**, 407–411.

Wood, P. J. and Siddiqui, I. R. (1972). *Carbohydrate Res.* **22**, 212–220.

Zucker, M. and Hankin, L. (1970). *Ann. Bot.* **34**, 1047–1062.

CHAPTER 15

Nucleic Acid Metabolism in Biotrophic Infections

J. A. CALLOW

Department of Plant Sciences, University of Leeds, Leeds, England

I. INTRODUCTION

There are two basic objectives in the study of nucleic acid metabolism in plants infected by pathogenic fungi. The first is a characterization of the plant's response to infection at the level of macromolecular synthesis. This "biochemical symptomatology" may yield useful information on the underlying basis of the metabolic changes that occur in diseased tissue, and on the developmental aspects of disease progression, but tells us little of the mechanisms which control infection in the first place. Indeed, measurable changes in nucleic acid synthesis often do not take place until several days after initiation of infection. The second objective is an examination of the fundamental macromolecular events that must be involved in the determination of host–pathogen specificity through gene-for-gene type interactions. Day (1974) has recently developed a simple model to describe the genetic and regulatory basis of such interactions. To illustrate this model, consider as an example the interaction between an avirulent parasite and a host possessing the corresponding resistance genes. The model requires that the products of avirulence genes (possibly nucleic acid) cross the interface between host and parasite and reach the host nucleus. These products may then interact with the resistance genes (sensor genes) resulting in the "switching-on" of an activator RNA molecule. The binding of such activator RNAs to structural gene cistrons will then possibly lead to their de-re-pression, resulting in *de novo* protein synthesis and the development of new metabolic pathways leading to the expression of resistance through phyto-alexin accumulation for example (Hadwiger and Schwochau, 1969). Alternatively, the same mechanisms could lead to induced susceptibility in the case of a virulent pathogen. Whilst such models are not easy to test, a combination of genetically suitable material and the appropriate techniques has recently yielded exciting results (see section V).

In this review it is the intention to examine some of the more recent work on nucleic acid metabolism in biotrophic infections with these two basic objectives in mind. Amongst the questions to be considered are:

(1) What are the qualitative and quantitative changes in nucleic acid synthesis and degradation in the host–parasite complex, where are such changes located within the cell, and what is their chronology in relation to the infection process and establishment of the host–parasite interaction?

(2) Are the changes that occur of host, or parasite origin?

(3) How are such changes controlled at the transcriptional level?

(4) What are the parallel changes in the enzymes of nucleic acid synthesis and degradation?

(5) Is the transfer of "information" across host–parasite interfaces, as predicted by current models of the gene-for-gene hypothesis (Day, 1974), in the form of nucleic acid?

II. Nucleic Acid Synthesis in Non-neoplastic, Rust and Powdery Mildew Diseases

A. EARLY STUDIES ON TOTAL RNA AND DNA SYNTHESIS

Much of the early work on nucleic acid metabolism in infected plants has been reviewed elsewhere (Heitefuss, 1966a, 1968). There is, however, a danger of over generalizing from these early observations and, since they illustrate several crucial points in this area of research, it is proposed to examine briefly some of these earlier reports.

Three main types of technique were used in the initial studies on total RNA and DNA synthesis. Total RNA and DNA were estimated by u.v. absorption, phosphate analysis, or nucleotide analysis, after chemical extraction from the tissue (Millerd and Scott, 1963; Quick and Shaw, 1964;

Malca *et al.*, 1964). Synthesis of RNA and DNA were measured by determining the specific activity of extracted nucleic acids following ^{32}Pi administration (Rohringer and Heitefuss, 1961; Heitefuss, 1965, 1966b). Finally, cytophotometric techniques were used to estimate DNA, RNA and nuclear proteins in individual nuclei (Whitney *et al.*, 1962; Bhattacharya *et al.*, 1965).

It was first demonstrated by Allen (1923) that host nuclei and nucleoli in mesophyll cells of rust-infected wheat leaves became enlarged in the early stages of infection before collapsing later on. Working with susceptible Little Club and resistant Khapli wheat, workers in Shaw's group showed that 6 days after infection the loss of RNA in control leaves as they senesced was arrested in infected tissues. Subsequently there followed a real increase in RNA content of infected leaves, paralleled by an increased respiration rate (Quick and Shaw, 1964). There was little change in the DNA content following infection, and an increased respiration rate in resistant wheat was not accompanied by an increased RNA content. These results confirmed the conclusions of earlier investigations on the same host–parasite combination, where ^{32}Pi was fed to the tissue and the specific activity of the RNA estimated following chemical extraction (Rohringer and Heitefuss, 1961). Large increases in specific activity of RNA were obtained $4\frac{1}{2}$ days after infection at the time of first symptom production but after 7 days, RNA specific activities were little different from controls. Again, using the temperature-sensitive Sr6 wheat system, Heitefuss (1965, 1966b) was able to demonstrate increased amounts and specific activities of RNA from compatible associations only. In the case of powdery mildew infections Millerd and Scott (1963) claimed significant increases in RNA levels in mildewed barley leaves 2 days after inoculation. The maximum increase above control leaf levels coincided with the maximum development of the fungus at sporulation. Thereafter the RNA content decreased more rapidly in mildewed leaves. Similar observations with mildewed leaves have been made by Malca *et al.* and Plumb *et al.* (1968). These results would appear to indicate, therefore, that in both rust and powdery mildew infections there is a substantial synthesis of RNA in the host–parasite complex more or less coincident with symptom production. However, interpretation of these results is complicated by several important considerations. Firstly, it is known that fungi accumulate phosphate in the form of high molecular weight polyphosphates (Bennet and Scott, 1971b). Such molecules tend to associate, or bind to nucleic acids (Wang and Mancini, 1966). Hence estimation of nucleic acids as DNA-P or RNA-P could lead to an over-estimation of nucleic acid content in infected tissues. The same problem applies in the case of ^{32}Pi feeding and it has been shown that more ^{32}Pi is incorporated into polyphosphates in infected leaves than in controls (Wolf, 1968). Secondly, while increased rates of isotopic precursor incorporation into nucleic acids could genuinely reflect increased rates of synthesis, the same effect could also be obtained if precursor pool sizes were reduced. Reduction in host intracellular phosphate pools due to sequestration of Pi by the fungus has been implicated in the control of starch biosynthesis in infected tissue (MacDonald and Strobel, 1970).

Another more serious problem concerns the origins of the reported increases in RNA synthesis. Do the increases represent host or fungal nucleic acid, or both? This question was examined by Millerd and Scott (1963), who claimed that the increases in RNA synthesis, observed in their experiments with mildewed barley leaves, largely represented synthesis of host RNA, since brushing the leaves to remove the majority of the mycelium resulted in only a small diminution in the amount of RNA extracted. As the authors themselves admit, and as pointed out by Quick and Shaw (1964), brushing the leaf does not remove haustoria from the epidermal cells. Haustoria appear to be rich in ribosomes and Whitney et al. (1962) and Plumb et al. (1968) have calculated that the nucleic acid content of haustoria in a heavily mildewed leaf can account for a substantial portion of the infected leaf total RNA content. The origins of the increased RNA content in rusted wheat leaves have been examined by Quick (1960). Comparison of the nucleotide base ratios of RNA extracts from infected and control leaves, with those of fungal uredospores, showed that the infected RNA was intermediate in base ratio between fungal and control leaf RNA, the implication being that a substantial portion of the RNA in the infected leaf is fungal RNA. The contribution of host and parasite to changes in nucleic acid synthesis has also been examined in bean leaves infected by Uromyces phaseoli (Heitefuss, 1968). Tissue discs consisting of a central rust pustule and a ring of uninfected tissue (which later developed into green islands) contained approximately 40% more RNA, 9 days after infection, than control tissue. Comparable discs from which the fungal pustule had been removed continued only 15% more RNA than controls. It seems clear then that much of the increased RNA content of an infected leaf can be accounted for by the fungal contribution, but that there is a smaller proportion which may be genuinely due to increased host synthesis. The interpretation of these results from Uromyces infected bean leaves may, however, be complicated by the fact that there is a small degree of host cell hypertrophy around infection sites (Heitefuss, 1968).

Less equivocal results of a somewhat different nature have been obtained using cytophotometric techniques. In the wheat rust system, Whitney et al. (1962) demonstrated that host DNA content remained unchanged until 15 days after infection when there was a 60% decrease. However, host nuclear RNA content was doubled after 6 days compared with controls and then declined as the pustule increased in size. Further studies (Bhattachaya et al., 1965) confirmed these observations and showed that there was significant increases in host nuclear RNA content after only 48 h, reaching a peak at 6 days, then declining. In contrast to the early chemical determinations, increases in nuclear RNA content were also observed in mesophyll cells of resistant varieties. In addition, significant losses in nuclear histones were observed in infected wheat cells. This work, subsequently substantiated by autoradiographic techniques (Bhattacharya and Shaw, 1967), indicates then that there is an increase in host nuclear RNA within 48 h of infection of susceptible wheat, accompanied by changes in histones which may reflect changes in gene activity. More recent work by Chakravorty and Shaw (1971) has shown that there is also an apparent increase in the rate of RNA synthesis within 48 h of infection of flax cotyledons by *Melampsora lini*, in the absence of any detectable net change in RNA content. This increase in the rate of RNA synthesis would appear to be host in origin since at 48 h the growth of the rust fungus is limited and the base composition of the newly synthesized RNA reflects that of the host rather than the parasite.

To summarize briefly then, it would appear that the total RNA content of rust and mildew infected plants is significantly higher than controls. This difference to some extent is exaggerated since, in nearly all cases, workers have used mature tissues which were starting to senesce. Hence an unknown portion of the so-called increase in RNA content could, in reality, reflect a delay in the normal senescent decline in RNA content (Heitefuss, 1965). The major quantitative differences are obtained several days after inoculation and are coincident with the expression of visual symptoms and changes in other aspects of host metabolism. It must be concluded then that a substantial portion of the increased RNA levels several days after infection represents fungal RNA. However, a small amount of biochemical and cytological evidence has shown that there are early changes in host RNA synthesis, both in compatible and incompatible associations, which are unlikely to be detected by simple extraction techniques followed by chemical estimation. What then is the significance of these changes? One might speculate here, that the early increase in host RNA synthesis results from changes in gene expression, and that these are then responsible for the altered patterns of metabolism found in infected host tissues or the inception of new metabolic pathways leading to resistance. However, any rational approach to this problem requires a thorough investigation of the types of nucleic acid being synthesized.

B. SYNTHESIS OF SPECIFIC RNA MOLECULES

It is known that all living cells contain three major classes of ribonucleic acid, ribosomal RNA (rRNA), transfer RNA (tRNA) and "messenger" RNA (mRNA). Of these three classes, rRNA is by far the most abundant, accounting for some 80% of the total (Davidson, 1972), and in green plant cells there are several distinct types of rRNA associated with 3 distinct types of ribosome. Depending on leaf age (Callow *et al.*, 1972), some two-thirds to one-half of the total rRNA is found in 80S cytoplasmic ribosomes in the form of 1.3×10^6 and 0.7×10^6 daltons molecular weight RNA, whilst the majority of the remainder is found in the 70S chloroplast ribosomes in the form of 1.1×10^6 and 0.56×10^6 daltons molecular weight RNA (Ingle *et al.*, 1970). In addition, mitochondrial ribosomal RNA is present, but accounts for a relatively small proportion of the total.

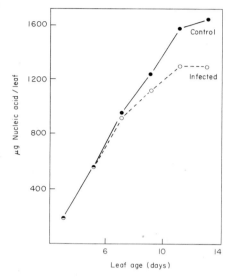

FIG. 1. Total phenol-extractable RNA in uninfected leaves, and leaves infected with *Erysiphe cichoracearum* when 3 days old: (●———●), uninfected, control, (○———○) infected. (After Callow, 1973.)

Any adequate description of the effects of fungal infection on host nucleic acid metabolism requires that these different nucleic acids be separated and a number of studies have demonstrated striking effects of fungal infection on ribosomal RNA metabolism. Figure 1 illustrates the effect of powdery mildew (*E. cichoracearum*) on the net accumulation of total RNA in cucumber leaves infected at a young, 3-day-old stage (Callow, 1973). During the course of leaf expansion there was a net accumulation of RNA, which was retarded in infected leaves. When the total RNA was fractionated by polyacrylamide gel electrophoresis, control leaves showed a net accumulation of both cytoplasmic and chloroplastic rRNAs as the leaves expanded (Fig. 2). Pulse-labelling with ^{32}Pi (Fig. 3) showed that by the time the leaves had completed their expansion (13 days) there was a marked reduction in the incorporation of label into the chloroplastic (1.1 M $+ 0.56$ M) rRNAs although incorporation into the cytoplasmic (1.3 M $+ 0.7$ M) rRNAs in the absence of net synthesis (i.e. turnover) was still proceeding. Following powdery mildew infection of 3-day-old leaves, the fungus retarded slightly the net accumulation of cytoplasmic 1.3 M $+

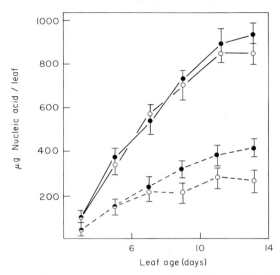

FIG. 2. Cytoplasmic and chloroplastic rRNA per leaf, i.e. μg 1·3M + 0·7M rRNA, and 1·1M + 0·56M rRNA, in control leaves, and leaves infected with *E. cichoracearum* when 3 days old (calculated from gel scans shown in Fig. 3). (●———●) cytoplasmic rRNA, control, (○———○) cytoplasmic rRNA, infected, (●– – –●) chloroplastic rRNA, control, (○– – –○) chloroplastic rRNA, infected. 95% confidence limits are shown. (After Callow, 1973.)

0·7 M rRNAs (Fig. 2) but there was an earlier, and more marked effect on chloroplastic (1·1 M + 0·56 M) rRNA synthesis (reduced by 30% 6 days after infection). There was also a much reduced incorporation of ^{32}Pi into the chloroplastic rRNA at this time compared with controls (Fig. 3). Table I shows that the ratio of cytoplasmic (1·3 M + 0·7 M):chloroplastic (1·1 M + 0·56 M) rRNA is virtually constant over the phase of leaf expansion in control leaves, whereas infection resulted in increased ratios, reflecting the primary effect of the fungus in decreasing the synthesis or net accumulation of chloroplastic rRNA. The effect of powdery mildew infection on the decline in rRNA levels in senescing leaves was also examined by infecting 10-day-old cucumber leaves approaching full expansion. Again, although the levels of both chloroplastic and cytoplasmic rRNA were reduced, infection had a more drastic effect on chloroplastic rRNA levels than on cytoplasmic rRNA levels (Table I). Thus the major effect of cucumber powdery mildew infection on rRNA synthesis would appear to be a retardation of chloroplastic rRNA synthesis, and an acceleration of the rate of senescence. It should be noted that in these experiments the fungal contribution was minimized by brushing the surface mycelium and spores from the infected leaves.

These results are in partial agreement with those of other workers. Rubin *et al.* (1971) found only reduced total (i.e. 70S + 80S) ribosome contents in

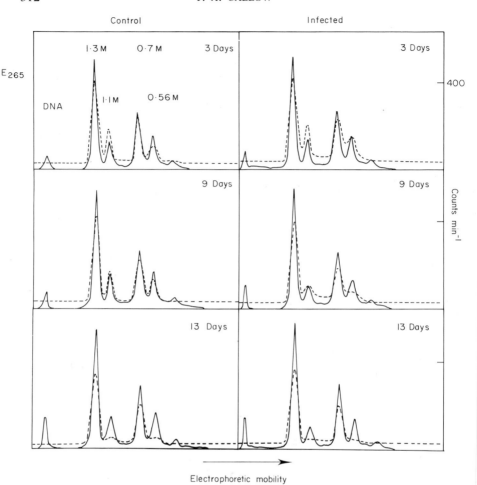

FIG. 3. Representative polyacrylamide gel electrophoresis scans at 265 nm, of RNA from uninfected cucumber leaves, and leaves infected with *E. cichoracearum* when 3 days old (————), and incorporation of ^{32}Pi (– – – – – –). 25 µg nucleic acid were loaded onto each gel (2·4%) and separation was for 2·5 h at 4mA/tube. Other details as in Callow (1973).

mildewed leaves. Bennet and Scott (1971a) clearly show that one distinct effect of powdery mildew infection of barley leaves is a rapid loss in chloroplast ribosomes. However, conflicting results were obtained for the cytoplasmic ribosomes, an approximately two-fold increase in the level of 80S cyto-plasmic ribosomes being obtained 9 days after infection. It is noteworthy that the barley leaves used in the experiments of Bennet and Scott (1971a) were probably more heavily infected than the infected cucumber leaves (J. Bennet, personal communication), also indicated by the more dramatic

TABLE I

Cytoplasmic: chloroplastic rRNA ratios in control, uninfected cucumber leaves, and in leaves infected with *E. cichoracearum* after 3 days (Expt 1) and 10 days (Expt 2). Each value is the average of 3 determinations made on 3 separate leaves (Callow, 1973)

Leaf age	Experiment 1		Leaf age	Experiment 2	
(days)	Control	Infected	(days)	Control	Infected
3	2·1	2·1	10	2·6	2·6
5	2·8	2·6	15	2·1	2·6
7	2·2	2·6	22	2·3	4·4
9	2·3	3·4	28	3·9	33·4
11	2·3	3·1	34	3·4	50·5
13	2·3	3·2			

effects on chloroplast ribosomes obtained in their experiments. Hence, brushing the leaf surface to remove the fungus was possibly less effective in the experiments with barley mildew since haustorial development would be more intense. It is suggested therefore that the enhanced 80S ribosome levels found in the experiments of Bennet and Scott (1971a) probably represent, in large part, ribosomes of the parasite, which would be typical, eukaryotic, 80S ribosomes (Leary and Ellingboe, 1971). Unfortunately no cytophotometric data comparable to that for rust infections appears to exist for powdery mildew infections. This would establish whether host cells were unequivocally engaged in more rapid ribosome synthesis. One other report on the effects of powdery mildew infection on RNA synthesis (Oku *et al.*, 1972), is quite anomalous in that it is claimed that within 24 h of inoculation there is a rapid enhancement in chloroplastic ribosomal RNA synthesis. However, since these workers present no convincing evidence that they are in fact dealing with chloroplast nucleic acids, the significance of this observation is not clear.

A further demonstration of the effects of powdery mildew infection on chloroplastic RNA metabolism was obtained by Dyer and Scott (1972). Chloroplast polyribosome levels were reduced in infected, susceptible lines of barley within 24 h of inoculation. Whilst such changes could be obtained by interference in the metabolism of the ribosomes, or from changes in the rates of messenger RNA synthesis or degradation, the authors themselves conclude that decreased availability of amino acids or other substrates such as ATP could produce the same effects.

Changes in rRNA synthesis in relation to rust infection have also been examined. Wolf (1967, 1968) and Chakravorty and Shaw (1971) have shown increased incorporation of ^{32}Pi into rRNA in rust-infected leaves, but the value of the work is somewhat reduced since no attempts were made to fractionate the rRNA into its cytoplasmic and chloroplastic components.

Tani *et al.* (1973) showed that whilst there were no net increases in either cytoplasmic or chloroplastic nucleic acids in rust-infected oat leaves, there was an increased amount of cytoplasmic rRNA in infected leaves, relative to that of senescing controls, and after the third day there was relatively a more rapid loss of chloroplastic rRNA. The authors interpret these results in terms of a retardation of senescence effect during the first few days, followed by a more rapid rate of senescence whilst fungal rRNA (indistinguishable from host cytoplasmic rRNA) increases. Results from infection of a resistant oat variety, using pulses of ^{32}Pi, appeared to indicate that the early stages of infection resulting in the resistant, hypersensitive response, were associated with an approximately 70% stimulation of host cytoplasmic and chloroplastic rRNA synthesis, although this was not detectable in terms of net synthesis.

The balance of evidence would appear to indicate, then, that infection of susceptible tissues by rust fungi in particular results in a small, early increase in rRNA synthesis in host cells, which is not necessarily revealed in terms of net synthesis of RNA, but which may be demonstrated by pulse-labelling or cytophotometric techniques. Similar evidence is not available for powdery mildew infections.

In later stages of both rust and powdery mildew infections the major effect appears to be an acceleration of leaf senescence resulting in a reduction in chloroplast nucleic acid content, although in terms of total rRNA there may be an apparent increase due largely to the synthesis of fungal nucleic acids, but also possibly due to some extra host rRNA synthesis (Quick and Shaw, 1964; Hamilton, 1969).

The relevance of the early increase in host RNA synthesis to the establishment of disease is a matter for some speculation. Much of the increase appears to be in the form of new ribosomes and at the present time no convincing demonstration of enhanced rates of transcription of other nucleic acids, especially mRNA, is available. If the early increase in host RNA synthesis is required for the synthesis of only a few new proteins concerned with the establishment of a functional host–parasite interaction, then such increased transcription rates would probably be undetectable. It may of course be that the increased host RNA synthesis is not primarily concerned with inception of disease but that it merely represents a secondary consequence of infection, resulting perhaps from changes in the levels of hormones in the diseased tissue.

III. Nucleic Acid Synthesis in Neoplastic Tissues, Induced by Smut Fungi

A. RNA SYNTHESIS

The majority of diseases in which nucleic acid metabolism has been investigated are caused by rust or powdery mildew fungi. Very little attention has been given to diseases caused by smut fungi, despite the fact that they exhibit certain advantages. Thus, smuts can be grown axenically with ease so that it is possible to examine the nucleic acids of the fungus in isolation, and therefore potentially account for the origins of changes in the host–parasite complex. In addition, certain smut diseases produce interesting changes in host development such as in the formation of galls or neoplasms, which imply profound changes in nucleic acid metabolism.

When meristematic tissues of maize (*Zea mays* L.) seedlings are inoculated with sporidia of the maize smut fungus, *Ustilago maydis* (DC.) Corda, two types of symptom are produced. Leaf development is retarded, whilst there are localized increases in growth of the leaf sheaths which swell up to form fleshy, gall-like structures (Callow and Ling, 1973). Microscopical examination shows that leaf mesophyll cells are disorientated and contain few plastids, and that the fleshy leaf sheaths have undergone substantial hyperplasia (Callow and Ling, 1973). Electron micrographs of the gall tissues shows that the hyphae are predominantly intracellular in the early stages of infection, and are surrounded by an intact host plasmalemma and a cell wall apposition (Bracker and Littlefield, 1973), continuous with the host cell wall, and that many of the host cells are filled with dense cytoplasmic contents (Fig. 4, Callow, unpublished observations).

In spite of the observed stunting and retardation of leaf development, infected seedlings have the same protein content as control seedlings, which is presumably a reflection of the enhanced growth of the leaf sheaths and an increase in fungal protoplasm. Neoplastic leaf sheaths in fact contain three times the amount of protein per unit dry weight compared with controls (Callow and Ling, unpublished).

Investigations of the rRNA metabolism of infected maize seedlings with time (Callow and Ling, in preparation), show that there is a progressive disappearance of chloroplastic ribosomes from infected plants, relative to cytoplasmic ribosomes (Fig. 5), which is also reflected in terms of absolute amounts of nucleic acid (Fig. 6). In examinations of individual types of tissue 7 days after inoculation (Fig. 7), there is a large (lamina 2) or total (lamina 3, leaf sheath 2) inhibition of chloroplastic rRNA synthesis in leaf and sheath tissues. The amount of cytoplasmic rRNA in the neoplastic leaf sheaths is increased three-fold, expressed either on an organ, or per unit fresh weight basis (Table II). In neoplastic leaf sheaths then, there is a total inhibition of chloroplastic ribosome synthesis, but a stimulation in cytoplasmic ribosome synthesis.

Since the cytoplasmic rRNAs of *U. maydis* are typically eukaryotic (i.e. are identical in molecular weight to plant cytoplasmic rRNAs), gel electrophoresis cannot separate the fungal rRNAs from those of the plant cytoplasmic ribosomes. Three lines of evidence show, however, that the majority of the increased cytoplasmic rRNA content of neoplastic leaf sheaths is genuinely of host origin.

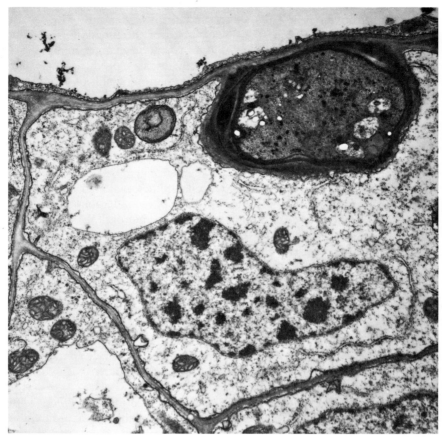

FIG. 4. Electron micrograph of neoplastic gall tissue from maize seedlings infected with *U. maydis*. The micrograph shows the fungal hypha surrounded by a host cell wall apposition, and intact host plasmalemma, and dense, ribosome-containing cytoplasm in the host cell.
× 11 250

(1) Electron micrographs (Fig. 4) show that host cells contain more dense ribosome populations than controls.

(2) The base composition of the cytoplasmic rRNA extracted from infected leaf sheaths is quite different from that of fungal cytoplasmic rRNA, and is more like that of controls (Table III). The small difference in rRNA base composition between control and infected leaf sheaths, may be due to the fact that rRNA extracted from control leaf sheaths also contains a small proportion (10–15%) of chloroplastic rRNA.

(3) Host nucleoli in infected sheath tissues are larger (Callow and Ling, 1973), and have a greater dry mass (Callow, 1975, Fig. 8) than control nucleoli. Since the size of the nucleolus is, in part, a measure of the

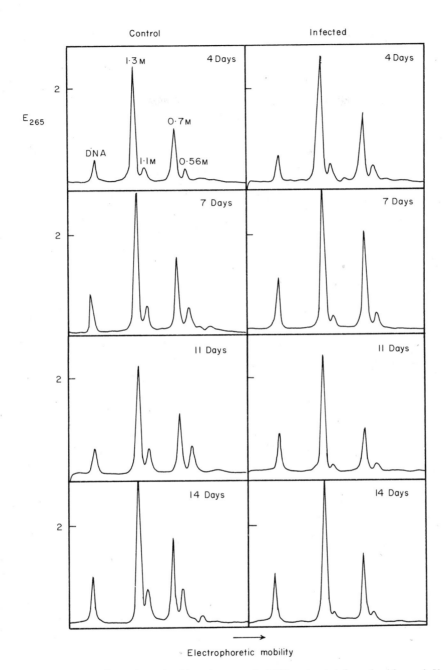

FIG. 5. Representative polyacrylamide gel scans of rRNA extracted from healthy and *U. maydis*—infected maize seedling at different times after inoculation. 25 μg of RNA were loaded onto each (2·4%) gel and electrophoresis was carried out for 2·5 h at 4mA/tube. (Callow and Ling, unpublished observation.)

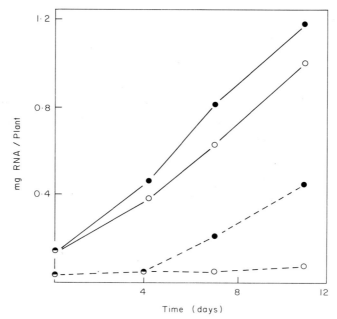

FIG. 6. Absolute content per plant of cytoplasmic rRNA (●———●, control; ○———○ infected) and chloroplastic rRNA (●– – –● control; ○– – –○ infected) following the inoculation of 7-day-old *Zea mays* seedlings with sporidia of *U. maydis*. Results derived from gel traces exemplified in Fig. 5. Each point is the average of 3 determinations on RNA extracts from 3 different plants. (Callow and Ling, unpublished observations.)

rate of ribosome synthesis, this indicates that host cells are engaged in a more active synthesis of ribosomes.

B. DNA SYNTHESIS AND ENDOPOLYPLOIDY

Several investigations have demonstrated profound changes in host nuclei in tissues infected by fungi and bacteria. In rust-infected wheat leaves, and *Brassica* root hairs infected by *Plasmodiophora*, increases in size of both nuclei and nucleoli are associated with increases in RNA synthesis, but not in DNA synthesis (Allen, 1923; Bhattacharya *et al.*, 1965; Bhattacharya and Williams, 1971; Whitney *et al.*, 1962). However, in those cases where there are substantial changes in cell differentiation leading to neoplasia, as in *Agrobacterium*—induced crown gall (Kupila and Therman, 1968; Broekhaert and van Parijs, 1973), *Phytomonas*—induced galls of tomato (Friedman and Francis, 1942), *Exobasidium*—induced galls of *Rhododendron* (Catarino, 1971), *Rhizobium*—induced root nodules (Mitchell, 1965), and *Plasmodiophora*—induced club-root galls (Williams, 1966), large increases in DNA synthesis have been reported.

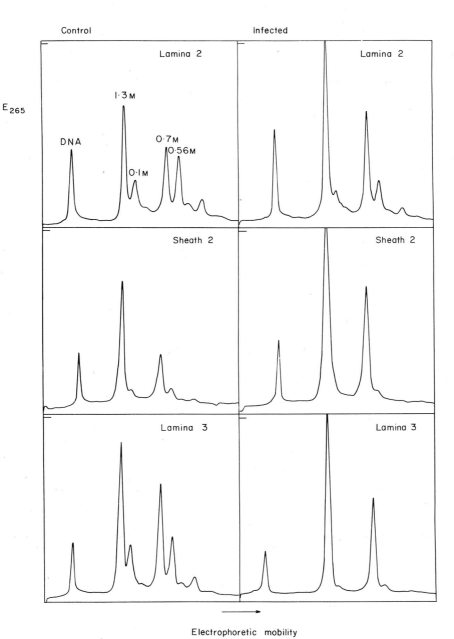

FIG. 7. Representative polyacrylamide gel scans of rRNA extracted from various tissues of 14-day-old healthy (control) maize seedlings, or 14-day-old seedlings infected with sporidia of *U. maydis* after 7 days. 25 μg of rRNA were loaded onto each (2·4%) gel and electrophoresis was carried out at 4mA/tube for 2·5 h (Callow and Ling, unpublished results).

TABLE II

Ribosomal RNA contents of the neoplastic second leaf sheath of 14-day-old maize seedlings infected after 7 days with *U. maydis*

	Control	Infected
1·3 M + 0·7 M (μg/organ)	25	74
1·3 M + 0·7 M (μg/g fresh wt.)	121	434
1·1 M + 0·56 M (μg/organ)	4	0
1·1 M + 0·56 M (μg/g fresh wt.)	18	0
total rRNA (μg/g fresh wt.)	139	434

TABLE III

Nucleotide composition of total ribosomal RNA isolated from the second leaf sheaths of 14-day-old maize seedlings infected with *U. maydis* after 7 days, and total ribosomal RNA isolated from cultured *Ustilago* sporidia. RNA samples were hydrolysed in 1N HCl at 100 °C, and the mixture of purine bases and pyrimidine nucleotides was separated by paper chromatography in a solvent containing methanol, ethanol, hydrochloric acid and water (50:25:6:19, v/v). Spots were localized under u.v. light, eluted and measured spectrophotometrically, as described by Smith and Markham (1950) (Callow and Ling, in preparation).

	moles/100 moles nucleotide				
	C	A	G	U	%GC
Control	25·8 ±0·5[a]	25·2 ±0·4	29·9 ±0·1	19·1 ±0·6	55·7 ±0·3
Infected	26·5 ±0·4	24·0 ±0·8	30·6 ±0·7	18·8 ±1·0	57·1 ±0·6
Fungus	24·0 ±1·3	27·0 ±0·7	28·1 ±1·1	21·0 ±1·1	52·0 ±0·5

[a] ± standard deviation.

DNA synthesis in *Ustilago*—induced maize neoplasms has been investigated by biochemical extraction and Feulgen microspectrophotometry (Callow, 1975; Callow and Ling, 1975, in preparation). In terms of extractable DNA, there is a three-fold increase in DNA content per unit dry weight in infected tissue. Whilst some of this increase is undoubtedly due to fungal

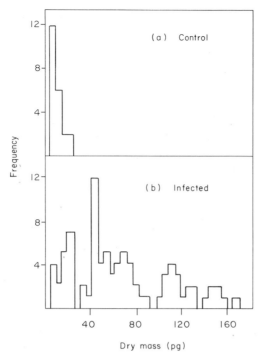

Fig. 8. Nucleolar dry masses of cells from the 2nd leaf sheath of 14-day-old maize seedlings, infected with sporidia of *U. maydis* after 7 days. (a) control cells, (b) cells from neoplastic, infected tissue. Dry masses were measured by interferometry on a Vickers M86 Scanning Microdensitometer/Interferometer equipped with laser illumination. Nucleolar dry masses were not corrected for underlying or overlying nucleoplasm (Callow, 1975).

DNA, Feulgen microspectrophotometry clearly shows that host nuclear DNA content has increased (Callow, 1975). The majority of nuclei in control leaf sheath tissues have undergone somatic polyploidy (D'Amato, 1964) to the 4C DNA level (Fig. 9a) i.e. nuclei in non-meristematic cells double their DNA content without the ensuing mitotic division.

In infected sheath tissue a large heterogeneity in nuclear size and DNA content was observed. Small cells, close to the vascular stele had small nuclei with 2C or 4C DNA contents (Fig. 9b), suggesting a meristematic function, as observed in crown gall tissues (Broekhaert and van Parijs, 1973). However, the large, vacuolate cells that comprise the major portion of infected tissue (Callow and Ling, 1973), contained large nuclei, often distinctly lobed, with large, prominent nucleoli. Three new size classes of DNA content were found associated with these nuclei, at the 16C, 32C and 64C levels, with a small number of 8C nuclei (Fig. 9c). Intermediate values were also found, which may indicate that some nuclei were still undergoing

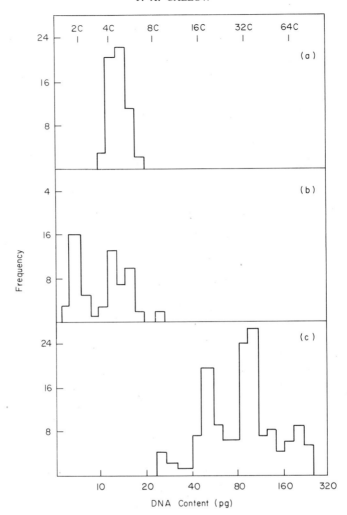

FIG. 9. DNA contents of nuclei from the 2nd leaf sheath of 14-day-old maize seedlings infected with sporidia of *U. maydis* after 7 days; (a) control, uninfected tissue, (b) infected tissue, vascular region, (c) infected tissue, cortical region. (Callow, 1975.)

DNA synthesis, or that there was some degree of under replication of the DNA.

The development of endopolyploidy thus appears to be a characteristic feature of those infections that result in neoplasia. A common feature of such infections is that neoplastic growth is associated with perturbations in the hormonal status of the tissue (Braun, 1957; Butcher *et al.*, 1974; Catarino, 1971; Turian, 1961). In both *Ustilago*—and *Exobasidium*—

induced galls, secretion of growth hormones by the fungus itself may be partly responsible (Wolf, 1952; Turian, 1961; Norberg, 1968). The role of hormones in neoplastic disease is further emphasized by the similarity between these neoplasms and hormonally induced callus formation by differentiated tissues. In tobacco pith callus, endopolyploidy is induced by IAA and kinetin (Patau et al., 1957), and small islands of meristematic cells (2C–4C DNA contents) are found, similar to those developed in maize smut neoplasms and crown gall tumours (Broekhaert and van Parijs, 1973). It has been speculated elsewhere (Callow, 1975), that this distribution of cells of varying ploidy might reflect a heterogeneous distribution of hormones within the infected tissue, due to uneven distribution of fungal mycelium.

Somatic polyploidy may result from endomitosis (endomitotic—chromosomal duplication), or polyteny (supernumary chromonemal replication) (D'Amato, 1964). It has not been possible to distinguish between these two processes in maize smut infections since no polyploid nuclei have been observed in mitosis.

C. GENE AMPLIFICATION OR GENE UTILIZATION?

The results of the biochemical investigations on rRNA synthesis show that neoplastic maize leaf sheath tissue accumulates cytoplasmic ribosomes 3–4 times faster than control tissues. This conclusion is confirmed by the interferometric studies on nucleolar mass (Fig. 8). It is now well established that all the genes for cytoplasmic rRNA synthesis are clustered within the nucleolar organizer region (NOR) of the chromosomes (Birnstiel et al., 1971). Hence nucleolar mass is a direct measure of the rate of ribosome accumulation. Although relatively little is known of the mechanisms by which ribosome synthesis is controlled, one can envisage two ways in which this enhanced synthesis of ribosomes could occur. Firstly, limited evidence would tend to suggest that plants contain more rRNA genes than required for normal development (Ingle, 1973), i.e. there appears to be spare capacity for rRNA synthesis within the plant nucleus. Hence the three- to four-fold increase in host ribosome synthesis in neoplastic tissues might simply result from a more extensive use of the available genes. The second possibility is that the number of rRNA genes per nucleus or cell might be increased in one of two ways. As a result of endopolyploidy the number of NORs, and hence the number of rRNA genes, will increase in approximate proportion to the rest of the genome. A more intriguing possibility is that the rRNA genes are specifically increased in number relative to the rest of the genome, via the process of gene amplification. The latter suggestion stems from the work on the oocyte of Xenopus where the requirements for a rapid rate of rRNA synthesis are satisfied by an approximately 3 000-fold specific amplification of the rRNA genes (Perkowska et al., 1968). There is, however, little evidence for the occurrence of this phenomenon in animal somatic tissues, and whilst there would appear to be some degree of flexibility in the numbers of copies of rRNA genes possessed by plants (Timmis and Ingle, 1974), there is no convincing demonstration of rRNA gene amplification in plants (Ingle, 1973).

Initial attempts have been made to examine the control of rRNA synthesis in maize smut-infected corn seedlings. The possibility that the enhanced rates of rRNA synthesis and increased nucleolar sizes in infected tissues are associated with an increase in the number of rRNA genes (over and above the increase due to polyploidy) has been examined by the technique of RNA/DNA hybridization (Callow and Ling, 1975, in preparation). Radioactive 1·3 M + 0·7 M maize rRNA, purified by polyacrylamide gel electrophoresis was incubated with single-stranded DNA isolated from control, or infected tissues, under conditions permitting hybridization between the rRNAs and the complementary base sequences present in the DNA. By using different concentrations of rRNA it was possible to determine the saturating level of rRNA, at which all complementary sites in the DNA were filled, and hence it is possible to determine the proportion of each genome coding for rRNA. Typical saturation curves for control and infected DNA are shown in Fig. 10. It is apparent that at saturating rRNA levels, there is an approximately 33% increase in the proportion of DNA from neoplastic tissue containing sequences complementary to rRNA. This would appear to indicate a small increase or amplification in the number of rRNA gene cistrons per genome, but in the face of increases of the order of 3 000 in other systems (Ingle, 1973), it can hardly be termed gene amplification and it is quite possible that there are alternative explanations for this effect.

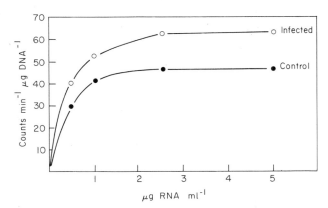

FIG. 10. DNA/RNA hybridization saturation curves. A 2:1 (w/w) mixture of [32]Pi-labelled 1·3 M rRNA and 0·7 M rRNA (specific activity 13 000 cpm/μg) isolated from maize root tips, was hybridized with 20 μg of extensively purified DNA isolated from control, or infected second leaf sheaths of maize, as described by Scott and Ingle (1973). (Callow and Ling, unpublished results.)

An apparent increase in the proportion of DNA containing complementary sequences to RNA could result from the under replication of non-ribosomal portions of the genome during the course of endopolyploidy. Alternatively, it should not be ignored that DNA from infected plants will contain a small proportion of fungal DNA. Should a greater proportion of this DNA hybridize with plant rRNA than the authentic plant DNA, then this would also result in an increased hybridization per unit weight of DNA. In fact, where studied, it would appear that the percentage of the total genome containing rRNA genes is greater in fungi than in higher plants (Birnstiel *et al.*, 1971). Any explanation of the 33% increase obtained in Fig. 10 based on contaminating fungal DNA would of course require that maize cytoplasmic rRNA hybridized with the same efficiency to fungal DNA as maize DNA.

The results of the hybridization experiment indicate that there is no substantial change in the proportion of the host genome devoted to rRNA synthesis, but one of the consequences of endopolyploidy will be a proportional increase in the number of NORs per nucleus. The question may be asked therefore, is the rate of rRNA synthesis directly related to the number of NORs, i.e. the number of rRNA genes?

Assuming that the number of NORs per nucleus increases in direct proportion to the degree of ploidy, as in euploidic hyacinths (Timmis *et al.*, 1972) and that nuclear mass is in direct proportion to DNA content (i.e. degree of ploidy), we have attempted to answer this question by estimating the relationship between nuclear and nucleolar masses as determined by quantitative interferometry (Callow, unpublished observations). A high degree of correlation ($r = 0.891$) was observed in such comparisons, indicating a close relationship in this tissue between the rate of ribosome production and the number of genes available.

Finally, preliminary attempts have been made to examine the RNA polymerase activity associated with neoplastic growth. An approximately 10-fold increase in RNA polymerase activity per unit of DNA was found in chromatin preparations from infected plants compared with controls (Callow and Neve, unpublished results). These results indicate that the rate of transcription or utilization of genes is greater in infected tissues although it has yet to be established what proportion of this extra RNA polymerase activity is associated with rRNA synthesis.

IV. RIBONUCLEASE ENZYMES AND DISEASE

A frequently observed response of plant tissues to senescence, mechanical injury and stress in the form of osmotic shock, heat, or water shortage, is that the level of RNase activity increases (see Wyen *et al.*, 1971). Such increases are generally associated with degradation of RNA, but in *Melampsora*—infection of flax cotyledons, and stem rust infections of wheat, a quite different relationship has been observed (Chakravorty *et al.*, 1974a,b; Scrubb *et al.*, 1972).

Following infection of flax or wheat tissues, there is a characteristic, bi-modal increase in RNase activity. An "early" increase in RNase is found between 2 to 4 days after inoculation, in both compatible and incompatible associations. The subsequent accumulation of "late" RNase, beginning 5 days after inoculation, but detectable as a change in enzyme catalytic properties after only 3 days, is specifically elicited in the compatible interactions. Using the temperature-sensitive Sr6 system, a positive correlation between increased "late" RNase activity and the ability of the host to support the pathogen was obtained. The increase in "late" RNase was not simply due to a non-specific wounding response since in wounded flax cotyledons, a variety of hydrolases, RNase, DNase, phosphodiesterase, and acid and alkaline phosphatases were increased in activity, whereas in infected tissue, the increase was confined to the "late" RNase. Furthermore in both flax and wheat infections there were important qualitative differences in the catalytic properties (DEP and temperature sensitivity, nucleotide preferences) between the "early" and "late" RNases, the RNase that accumulates through wounding, the RNase present in uninfected tissue and the RNase extracted from the fungus.

The significance of these changes in RNase activity, accompanying the growth of the pathogen, must be considered in the light of a parallel increase in total RNA content found in infected flax and wheat tissues, a proportion of which is considered to be due to an increase on host nucleic acids (Chakravorty and Shaw, 1971; Rohringer and Heitefuss, 1961). There are three possible alternatives. Firstly, the late RNase might not be functional in the infected host tissue due to the presence of inhibitors (Scrubb *et al.*, 1972). Secondly, the balance between synthesis of RNA and its degradation by the RNase could lie in favour of net synthesis. Thirdly, it has been suggested that both the "early" and "late" RNase (quite different in properties to the control enzyme) could be involved in the post-transcriptional processing of RNA in either susceptible, or resistant combinations. (Chakravorty *et al.*, 1974). Clarification of these alternatives awaits more definitive evidence on the origins of these different enzymes.

V. Nucleic Acid Transfer and the Possible Role of RNA in Host–Parasite Specificity

Whilst much of the work presented in this review is of value in elucidating the molecular aspects of host–parasite interaction, there is little doubt that many of the changes in nucleic acid metabolism observed are the consequences of infection, rather than being centrally involved in the inception of the functional host–parasite interaction. Recently, however, it has been claimed that RNA produced by a specific gene for disease resistance is directly responsible for initiating a resistant response in a susceptible host (Rohringer *et al.*, 1974). In these experiments, wheat leaves containing the Sr6 allele for stem rust resistance were inoculated with spores of the avirulent C17 (56) wheat rust race at a temperature above 25 °C. When the leaves were densely colonized, the temperature was reduced to 20 °C, resulting in an extensive resistant response whereby host cells undergo necrosis. Crude nucleic acid preparations from such leaves were then injected into genotypically resistant leaves (Sr6) or genotypically susceptible leaves (sr6) both inoculated with the virulent C45 (65A) rust race. The leaf cells were assayed after 6 h for the presence of necrotic cells, indicating resistant interactions. An active extract was defined as one that induces significantly more necrotic cells in the genotypically resistant Sr6 leaves than in the genotypically susceptible leaves.

Results with crude nucleic acid preparations showed that significantly more necrotic sites were produced when the preparation was injected into genotypically resistant plants than when it was injected into susceptible plants. The crude extract was further purified, and RNA was shown to be the active principle. Furthermore the activity was restricted to the Sr6/P6 gene-for-gene system. Sr6/P6 extract was ineffective in the Sr11 bioassay system, conversely Sr11/P11 extract was inactive in the Sr6 bioassay system.

Thus, the results presented by these workers demonstrate that some form of RNA is directly involved in the resistant reaction to wheat stem rust, with a specificity comparable to the specificity of the host/parasite system. In current models of the gene-for-gene interaction, such as developed by Day (1974), it is proposed that in an interaction between, for example, an avirulent parasite and a host possessing the corresponding gene for resistance there is a transfer of the products of avirulence genes across the host–parasite interface. Such products could then interact with the host genetic system resulting in the "switching-on" of an "activator" RNA. The binding of such "activator" RNA to structural gene operons would then lead to their de-repression resulting in *de novo* protein synthesis and the expression of metabolic pathways leading to a resistant response. The exciting results of Rohringer *et al.* (1974) can be considered in the light of this scheme. The active, resistance-inducing RNA extract obtained from the resistant inter-action could represent the RNA product of the parasite avirulence gene passing between parasite and host. Alternatively the RNA could represent the "activator" RNA involved in "switching-on" of host structural gene operons, or indeed, the mRNA products of the gene operons themselves. Deciding between these alternatives would require a more complete charac-terization of the active RNA involved and represents an exciting prospect for future work.

Experiments in which disease reaction has apparently been altered through the application of DNA to host plants have been recently reported (McIntyre *et al.*, 1975; Yamamoto and Matsuo, 1976).

McIntyre *et al.* (1975) found that crude DNA-containing extracts isolated from both virulent and avirulent cells of *Erwinia amylovora*, would effectively protect "Bartlett" pear seedlings and shoots against subsequent challenge inoculation with virulent *E. amylovora* cells. It was demonstrated that DNA was the active component of such extracts, other polyanions, polycations and DNA degradation products failed to exert any protective effect. Salmon sperm and pear DNA were also ineffective indicating a substantial degree of specificity, but a reduced degree of protection was also obtained with DNAs isolated from *Erwinia herbicola*, and *Xanthomonas campestris*, but not from *Pseudomonas tabaci*. DNA from *E. amylovora* was more effective when isolated by techniques designed to reduce physical shearing.

The physiology and biochemistry of this protective effect is unknown but appears to depend on host responses since a 0·5 h time lag is required between application of the DNA and the challenge inoculation for protection to be established. *Erwinia amylovora* DNA does not appear to inhibit the growth or virulence of the bacterium *per se*. One of the major questions to be answered concerns the fate of the exogenous DNA in the plant. Does it remain intact inside the cell, and if so, is it transcribed?

A somewhat different approach was used by Yamamoto and Matsuo (1976). Attempts were made to transfer major gene resistance to *Phytophthora infestans* from a resistant potato cultivar (hybrid 96–56) to a susceptible cultivar (Norin No. 1), by painting DNA-containing fractions onto potato tissues (leaf midribs and tubers). When untreated Norin No. 1 tissues are challenged by a virulent race of *P. infestans*, only lesions result. DNA-treated Norin No. 1 tissues, however, produced both lesions and hypersensitive flecks (denoting resistance). In addition, whereas tissues of resistant hybrid 96–56 normally only produce hypersensitive flecks when challenged by a virulent race of *P. infestans*, hybrid tissue treated with DNA-containing fractions isolated from susceptible Norin No. 1 tissues produced both hypersensitive flecks and lesions. These results might be taken to indicate that genetic transformations has taken place, that genes controlling the induction of both resistant and susceptible responses can be transferred from donor to recipient and that these new genes can be expressed in the host cells. If such conclusions are correct then this demonstration could have important consequences in the breeding of resistant plants. However, on the basis of the results reported by Yamamoto and Matsuo such far reaching conclusions are entirely premature. These authors fail to produce convincing evidence to support their claim that DNA is the active component in the DNA-containing extracts. The DNA used in such experiments must be rigorously purified to remove polysaccharide, protein and other nucleic acid contaminants. Simple controls using nuclease enzymes should be employed. If DNA is shown to be the active component then the specificity of the response should be examined. Is DNA acting non-specifically as a giant polyanion or does it act in a manner related to its biological specificity? Will any type of DNA such as calf thymus exert the same effect or is the response limited to potato DNA? Indeed, a rather important control in such experiments would be to examine the effect of applying DNA isolated from one variety to tissues of the same variety. Again, what happens to the exogenous DNA inside the plant, does it remain intact, or is it degraded? If degraded, will DNA breakdown products applied to the plant also exert the same effect as the intact molecule? Until these and similar questions are answered it is difficult to assess the status of such studies either from the point of view of the plant pathologist seeking a deeper understanding of the fundamental relationship between host and parasite, or for the plant breeder seeking novel ways of incorporating disease resist-ance into crop plants.

ACKNOWLEDGEMENTS

Much of the work described in this review was undertaken with the assistance of Miss Irene Ling, and with financial support from the Agricultural Research Council.

REFERENCES

Allen, R. J. (1923). *J. agric. Res.* **26**, 571–604.
Bennet, J. and Scott, K. J. (1971a). *FEBS Letters* **16**, 93–95.
Bennet, J. and Scott, K. J. (1971b). *Physiol. Pl. Path.* **1**, 185–198.
Bhattacharya, P. K. and Shaw, M. (1967). *Can. J. Bot.* **45**, 555–563.
Bhattacharya, P. K. and Williams, P. H. (1971). *Physiol. Pl. Path.* **1**, 167–175.
Bhattacharya, P. K., Naylor, J. M. and Shaw, M. (1965). *Science* **150**, 1605–1607.
Birnstiel, M. L., Chipchase, M., and Spiers, J. (1971). *Progr. Nucleic Acid. Res. Molec. Biol.* **11**, 351–389.
Bracker, C. E. and Littlefield, L. J. (1973). *In* "Fungal Pathogenicity and the Plant's Response" (R. J. W. Byrde and C. V. Cutting, eds), pp. 159–313. Academic Press, London and New York.
Broekhaert, D. and Van Parijs, R. (1973). *J. exp. Bot.* **24**, 820–827.
Braun, A. (1957). *Soc. Exptl. Biol. Symposia* **11**, 132–142.
Butcher, D., El-Tighani, S. and Ingram, D. (1974). *Physiol. Pl. Path.* **4**, 127–140.
Callow, J. A. (1973). *Physiol. Pl. Path.* **3**, 249–257.
Callow, J. A. (1975). *New Phytol.* **75**, 253–257.
Callow, J. A. and Ling, I. T. (1973). *Physiol. Pl. Path.* **3**, 489–494.
Callow, J. A., Callow, M. E. and Woolhouse, H. W. (1972). *Cell. Differentiation* **1**, 79–90.
Catarino, F. M. (1971). *Port. Acta biol.* **12**, 53–64.
Chakravorty, A. K. and Shaw, M. (1971). *Biochem. J.* **123**, 551–558.
Chakravorty, A. K., Shaw, M. and Scrubb, L. A. (1974a). *Physiol. Pl. Path.* **4**, 335–358.
Chakravorty, A. K., Shaw, M. and Scrubb, L. A. (1974b). *Nature* **247**, 577–580.
D'Amato, F. (1964). *Caryologia* **17**, 41–42.
Davidson, J. N. (1972). "The Biochemistry of the Nucleic Acids" (7th edition). Chapman and Hall, London.

Day, P. R. (1974). "Genetics of Host–Parasite Interaction". Freeman, San Francisco.
Dyer, T. A. and Scott, K. (1972). *Nature* **236**, 237–238.
Friedman, B. A. and Francis, T. Jr. (1942). *Phytopathology* **32**, 762–772.
Hadwiger, L. A. and Schwochau, M. E. (1969). *Phytopathology* **59**, 223–227.
Hamilton, W. E. (1969). M.A. thesis, University of Saskatchewan, Canada.
Heitefuss, R. (1965). *Phytopath. Z.* **54**, 379–400.
Heitefuss, R. (1966a). *A. Rev. Pl. Pathol.* **4**, 221–224.
Heitefuss, R. (1966b). *Phytopath. Z.* **55**, 67–85.
Heitefuss, R. (1968). *Neth. J. Pl. Path.* **74**, (Suppl. 1), 9–18.
Ingle, J. (1973). *In* "Biosynthesis and its Control in Plants" (B. V. Milborrow, ed.), Ann. Proc. Phytochem. Soc. Vol. 9, pp. 69–91. Academic Press, London and New York.
Ingle, J., Possingham, J. V., Wells, R., Leaver, C. J. and Loening, U. E. (1970). *Soc. Exptl Biol. Symposia,* **24**, 303–325.
Kupila, A. S. and Therman, E. (1968). *Adv. Morphogen.* **7**, 45–78.
Leary, J. V. and Ellingboe, A. H. (1971). *Phytopathology* **61**, 1030–1031.
Macdonald, P. W. and Strobel, G. A. (1970). *Pl. Physiol.* **46**, 126–135.
Malca, I., Zscheile, F. P. and Gulli, R. (1964). *Phytopathology* **54**, 1112–1116.
McIntyre, J. L., Kuć, J. and Williams, E. B. (1975). *Physiol. Pl. Path.* **7**, 153–170.
Millerd, A. and Scott, K. J. (1963). *Aust. J. biol. Sci.* **16**, 775–783.
Mitchell, J. P. (1965). *Ann. Bot.* **29**, 371–376.
Norberg, S. O. (1968). *Symp. Bot. Upsaliensis,* **19**, 1–117.
Oku, H., Ouchi, S. and Shiraishi, T. (1972). *Symp. Biol. Hung.* **13**, 277–282.
Patau, K., Das, N. K. and Skoog, F. (1957). *Physiol. Pl.* **10**, 949–966.
Perkowska, E., Birnstiel, M. L. and Macgregor, H. C. (1968). *Nature* **217**, 649–650.
Plumb, R. T., Manners, J. G. and Myers, A. (1968). *Trans Br. mycol. Soc.* **51**, 563–573.
Quick, W. A. (1960). M.A. thesis. University of Saskatchewan, Canada.
Quick, W. A. and Shaw, M. (1964). *Can. J. Bot.* **42**, 1531–1540.
Quick, W. A. and Shaw, M. (1966). *Can. J. Bot.* **44**, 777–778.
Rohringer, R. and Heitefuss, R. (1961). *Can. J. Bot.* **39**, 263–267.
Rohringer, R., Howes, N. K., Kim, W. K. and Samborski, D. J. (1974). *Nature* **249**, 585–588.
Rubin, B. A., Axenova, V. A. and Huyen, N. D. (1971). *Acta Phytopathol. Acad. Scientarum Hung.* **6**, 61–64.
Scott, N. S. and Ingle, J. (1973). *Pl. Physiol.* **51**, 677–684.
Scrubb, L. A., Chakravorty, A. K. and Shaw, M. (1972). *Pl. Physiol.* **50**, 73–79.
Smith, J. D. and Markham, R. (1950). *Biochem. J.* **46**, 509–513.
Tani, T., Yoshikawa, M. and Naito, N. (1973). *Phytopathology* **63**, 491–494.
Timmis, J. N. and Ingle, J. (1974). *Nature New Biol.* **244**, 235–236.
Timmis, J. N., Sinclair, J. and Ingle, J. (1972). *Cell Differentiation.* **1**, 335–339.
Turian, G., (1961). *Conf. on Scientific Problems of Plant Protection,* 35–39.
Wang, D. and Mancini, D. (1966). *Biochim. biophys. Acta.* **129**, 231–239.
Williams, P. M. (1966). *Phytopathology* **56**, 521–524.
Whitney, H. S., Shaw, M. and Naylor, J. M. (1962). *Can. J. Bot.* **40**, 1533–1544.
Wolf, F. T. (1952). *Proc. natn. Acad. Sci., U.S.A.* **38**, 106–111.
Wolf, G. (1967). *Phytopath. Z.* **59**, 101–104.
Wolf, G. (1968). *Neth. J. Pl. Path.* **74**, (Suppl. I), 19–23.
Wyen, N. V., Erdei, S. and Farkas, G. L. (1971). *Biochim. biophys. Acta.* **232**, 472–483.
Yamamoto, M. and Matsuo, K. (1976). *Nature* **259**, 63–64.

Author Index

The numbers in italic are those pages where References are listed in full

A

Abawi, G. S., 254, 262, *283*
Abdul-Baki, A. A., 94, *102*
Aberhart, D. J., 21, *22*
Achenbach, H., 252, *285*
Adamek, C., 252, 253, *283*
Adams, M. J., 186, *190*
Adityachaudhury, N., 245, *284*
Agrios, G. N., 162, *190*
Ainsworth, G. C., 170, *190*
Aist, J. R., 44, 56, 71, *77*, *78*, 79, 88, 93, *99*
Akai, S., 56, *78*, 162, *190*
Akinrefon, O. A., 112, *115*
Albergbina, A., 95, *99*, 112, *115*
Albersheim, P., 58, *77*, 80, 81, 82, 83, 84, 85, 88, 89, 90, 91, 92, 98, *99*, *100*, *101*, *103*, 106, 111, 113, 114, *115*, 148, *158*, 255, 256, 257, *283*
d'Alessio, U., 151, *156*
Alexander, M., 98, *101*
Allard, R. W., 27, *40*
Allen, E., 229, *236*
Allen, E. H., 214, *222*
Allen, P., 44, *78*
Allen, P. J., 54, 56, 66, *77*, 122, 127, *131*
Allen, R. J., 307, 318, *329*
Allen, R. F., 184, *190*
Allington, W. B., 184, *190*
Allison, D. W., 86, *102*
Andel, O. M. van., 240, *285*
Anderson, A., 114, *115*, 255, 257, *283*
Anderson, A. B., 201, *205*
Anderson, A. J., 88, 89, 90, 98, *99*, *101*
Andrews, J. H., 45, 72, *77*
Antonelli, E., 124, 129, *131*
Aoki, H., *158*
Appel, O., 169, *190*
Ap Simon, J. W., 21, *22*
Arens, K., 179, *190*, *191*
Arh-hwang, Chen., 202, *206*

Arigoni, D., 144, *157*
Armstrong, D. J., 128, *133*
Arnold, R. M., 221, *223*, 280, *287*
Arntzen, C. J., 147, 155, *156*, *157*
Aronson, J., 233, *236*
Arulpragasam, P. V., 4, *10*
Asada, Y., 292, *302*
Aspinall, G. O., 81, 82, *99*
Atkins, I. M., 27, *40*
Audia, W. V., 301, *302*
Aveyard, M. A., 294, *303*
Axenova, V. A., 311, *330*
Ayers, A., 255/256, 257, *283*
Ayers, W. A., 86, *99*

B

Backman, P. A., 154, *156*, *158*
Bailey, J. A., 168, *191*, 209, 210, 211, 216, 219, 221, *222*, 243, 244, 254, 255, 257, 259, 260, 261, 262, 263, 265, 271, 272, 275, 277, 278, 280, 282, *283*, *284*, *288*
Bailiss, K. W., 128, *131*
Baker, E. A., 164, 176, *191*, *192*
Baker, K. F., 11, *22*
Balasubramani, K. A., 91, *99*
Baldwin, J. K., 146, *159*
Ballio, A., 150, 151, *156*
Barash, I., 147, *156*
Barna, B., 74, *78*, 208, 219, *223*, 226, *236*
Barnaby, V. M., 85, *101*
Barnes, G., 4, 5, 6, 8, *9*, *10*
Barrett, D. H. P., 30, 35, *40*
Bartnicki-Garcia, S., 233, *236*
Barz, W., 247, 248, 250, 251, 252, 253, *283*, *284*, *285*, *287*
Basham, H. G., 80, 88, 89, 90, 94, 95, 96, 97, 98, 99, *99*, *100*, *102*, 108, 109, 110, 111, *115*, *116*
Bateman, D. F., 79, 80, 84, 85, 86, 88, 89, 90, 91, 92, 93, 94, 95, 96, 97, 98, 99, *99*,

Hanssler, G., 98, *101*
Hapner, K., 142, *158*
Harborne, J. B., 241, 249, *286*
Hardegger, E., 215, *222*
Hargreaves, J. A., 220, 221, *223*
Harmon, D. L., 29, 30, 34, 38, *40*
Harper, S. H., 242, *286*
Harrar, J. G., 162, *192*
Harris, P. J., 301, *303*
Hart, H., 161, 184, 187, 189, *191*
Hartley, R. D., 301, *303*
Harvey, C. C., 56, *77*
Harvey, R. B., 173, 187, *191*
Hasegawa, S., 84, *101*
Hashimoto, J., 119, *132*
Hashimoto, S., 85, *101*
Hatanaka, C., 84, *102*
Haugh, M. F., 147, *156*
Hawkins, A. R., 123, 124, *133*
Hawkins, L. A., 173, 187, *191*
Heath, I. B., 56, 61, 70, *77, 78*
Heath, M. C., 56, 70, 71, 74, *78,* 91, *101,* 254, 259, 270, 271, 272, 275, 276, 278, *286*
Heather, W. A., 166, 169, *191*
Heitefuss, R., 306, 307, 308, 309, 326, *330*
Helgeson, J. P., 118, 120, *132*
Hellerquist, C. G., 150, *156*
Henderson, S. J., 295, *303*
Hergert, H. L., 245, *286*
Hess, S. L., 212, *222,* 252, 255, 258, *285, 286*
Hess, W. M., 137, 139, *158*
Heuvel, J. vanden., 261, 271, 273, 275, 277, 278, 281, *286*
Heybroek, H. M., 188, *192*
Hietala, P. K., 202, 203, *206,* 217, 218, 222, 223, 240, 243, 245, *284, 288*
Higgins, V. J., 211, *222, 223,* 240, 242, 243, 254, 255, 259, 261, 270, 271, 272, 273, 275, 276, 277, 278, 279, 280, 281, *282, 285, 286, 288*
Higinbotham, N., 144, *157*
Higinbotham, H., 121, *132*
Hignett, R. C., 240, *286*
Higuchi, I., 298, 302, *303*
Hijwegen, T., 250, *286,* 292, *303*
Hildebrand, D. C., 15, *23*
Hildebrant, A. C., 44, *78,* 174, *192*
Hill, J. L., 271, *287*
Hilton, J. L., 154, *157*
Hinman, R., 127, *132*
Hinson, W. H., 13, *22*

Hirst, E. L., 84/85, *101*
Hirst, J. M., 3, 4, *10*
Hislop, E. C., 85, *101*
Hodges, T. K., 143, *157*
Hoff, J. E., 300, *303*
Holt, R. W., 13, *22*
Honkanen, E., 203, *206*
Honward, V. K., 210, 223, 244, 251, *288*
Hooker, A. L., 25, 29, 30, 32, 35, *40, 41,* 146, *157*
Horner, C. E., 91, *102*
Horsch, R., 254, 278, *286*
Horsfall, J. G., 98, *101*
Horton, J. C., 86, 87, *101*
Horváth, I., 271, *284*
Hösel, W., 252, *283*
Howard, B., 214, *222,* 227, *236*
Howes, N. K., 30, *40,* 74, *78,* 327, *330*
Hsu, E. J., 86, *101*
Hsu, S. C., 19, *22*
Huang, J. S., 143, 148, *157*
Huang, P. Y., 143, 148, *157*
Hughes, D., 227, *236*
Hui, P. A., 298, *303*
Hunter, L. D., 240, *286*
Hursh, C. R., 168, 184, 187, *191*
Huyen, N. D., 311, *330*

I

Ichimi, T., 84, *101*
Idle, D. B., 180, *191*
Imaizumi, K., 214, *223,* 226, 227, *236*
Imamura, A., 179, *192*
Imamura, H., 201, *206*
Imaseki, H., 120, 121, *132, 133*
Ingham, J. L., 208, 210, 211, 217, 218, 221, *222, 223,* 240, 242, 243, 244, 254, 276, 279, *284, 286*
Ingle, J., 310, 323, 324, 325, *330*
Ingram, D., 322, *329*
Ingram, D. S., 44, 45, 46, 50, 51, 52, 53, 54, 55, 56, 57, 60, 74, 76, *78,* 88, *102, 116*
Inman, R. E., 123, *131*
Ishita, Y., 147, *157*
Ishizaka, N., 226, 228, 231, *237*
Israel, H. W., 82, *103*
Ito, Y., 298, 302, *303*
Iwadare, T., 267, *288*

Markham, R., 320, *330*
Marks, G. C., 177, 187, *192*
Marlow, W., 21, *22*
Marré, E., 119, *132*, 152, *157*
Martin, A. R., 255, 257, *285*
Martin, J. T., 162, 163, 164, 170, 174, 175, 176, 177, *191*, *192*
Martin, N. E., 93, *103*
Martinson, C. A., 147, *156*
Masamune, A., 227, *236*
Masamune, T., 214, *223*, 226, 229, 231, *236*, *237*
Mason, P. A., 45, *78*
Mathre, D. E., 169, *192*
Mathiaparanam, P., 154, *157*
Matsumoto, I., 292, *302*
Matsumoto, T., 147, *157*
Matsunaga, A., 227, *236*
Matsuo, K., 328, *330*
Matsuura, T., 214, *223*
Mauri, M., 150, *156*
Maw, G. A., 98, *101*
Maxie, E. C., 121, *131*
Maxwell, D. P., 91, *102*
Mayo, G. M. E., 29, 35, 36, 37, *40*, *41*
Mayama, S., 130, *132*
Mazzenchi, V., 95, *99*
McClendon, J. H., 93, 94, *102*
McColloch, R. J., 84, *102*
McCoy, M. S., 31, 32, *40*
McDonald, I. R., 118, *132*
McHardy, W. E., 168, *191*
McInnes, A. G., 211, *223*, 242, *288*
McIntyre, J. L., 328, *330*
McKay, M. B., 179, 184, *192*
McKee, R., 229, *236*
McKeen, W. E., 5, 6, *10*, 56, 70, *78*, 88, *102*, 173, *191*
McLaughlin, D. J., 61, *77*
McLean, F. T., 179, 180, *192*
McMurray, T. B. H., 243, *284*
McNabb, H. S. Jr., 188, *192*
Melander, L. W., 172, *192*
Melouk, H. A., 91, *102*
Mence, M. J., 174, *192*
Mercer, P. C., 115, *116*
Meredith, D. S., 3, *10*
Mertz, S. M., 147, *157*
Metlitskii, L. V., 214, *223*, 226, 227, *236*
Meyer, F. H., 118, *132*
Meyer, W. L., 154, *157*
Miksche, G. E., 302, *303*
Milborrow, B. V., 152, *156*

Millar, R. L., 79, 80, 84, *100*, 210, 211, *222*, *223*, 242, 243, 244, 254, 255, 261, 270, 272, 273, 275, 276, 278, 280, 281, *284*, *286*, *288*
Miller, L., 84, *102*
Miller, R. J., 146, 147, *156*, *159*
Millerd, A., 122, *132*, 306, 307, 308, *330*
Mitchell, H. K., 204, *205*, *206*, 218, *222*, *223*
Mitchell, J. P., 318, *330*
Mitchell, N., 70, *78*
Mohammed, R., 85, *101*
Mohri, R., 155, *158*
Mondal, M. H., 147, *158*
Moore, L. D., 91, 98, *102*
Moore, R. E., 157
Moore, W. C., 1, *10*
Moran, F., 86, *102*
Morré, D. J., 94, *102*
Moseman, J. G., 29, 30, 31, *40*
Mothes, K., 127, *132*
Motoyama, K., 84, *101*
Mount, M. S., 71, *78*, 94, 95, 96, 98, *102*, *103*, 109, 111, *116*
Moustafa, E., 248, *287*
Mueller, W. C., 168, *191*
Muggli, R., 82, *102*
Muhlethaler, K., 82, *102*
Mukherjee, N., 245, *284*
Mullen, J. M., 84, 85, 86, 88, 89, 90, 91, 98, 99, *102*
Muller, K., 225, *236*
Müller, K. O., 207, 217, *223*, 239, 240, 253, 256, 263, 265, 270, 279, *287*
Müller-Enoch, D., 252, *284*
Murai, A., 214, *223*, 226, *236*
Muramatsu, T., 85, *101*
Murphy, J. V., 91, *99*
Muse, B. D., 91, 98, *102*
Muse, R. R., 91, 98, *102*
Mussell, H. W., 86, 94, 95, 96, *102*, 111, *116*
Myers, A., 98, *100*, 112, *115*, 307, 308, *330*

N

Nagai, I., 179, *192*
Nagel, C. W., 84, *101*
Nagdy, G. A., 178, *192*
Naim, M., 245, *287*
Nair, V. M. G., 188, *193*
Naito, N., 314, *330*

Pueppke, S. G., 243, 251, 254, 257, 261, 262, 263, 271, 272, 274, 282, 283, *287*, *288*
Pugsley, A. T., 29, 30, *40*
Pupillo, P., 95, *99*, 112, *115*
Puranik, S. B., 169, *192*
Purnell, T. J., 6, 7, *10*

Q

Quick, W. A., 306, 307, 308, 314, *330*

R

Raa, J., 195, 199, *206*
Radulescu, E., 178, *192*
Rahe, J. E., 219, *223*, 257, 280, *287*
Rall, G. J. H., 251, *287*
Randazzo, G., 151, *156*
Rathbone, M. P., 120, *132*
Rathmell, W. G., 212, 216, 221, *223*, 247, 249, 254, 257, 258, *287*, *288*
Read, W. H., 170, *190*
Rebel, H., 150, *158*
Reddy, M. N., 80, *102*
Redman, B. T., 244, *287*
Reese, E. T., 85, *102*
Rehfeld, D. W., 130, *132*
Reilly, J. J., 255, *288*
Renz, J., 21, *23*
Reynolds, S. B., 294, *303*
Ribéreau-Gayon, P., 247, 249, 251, *288*
Ricard, J., 127, *132*
Rice, E., 262, 280, 282, *286*
Rich, P. H., 154, *157*
Rich, S., 162, 178, *192*
Ride, J. P., 292, *303*
Ridge, I., 96, *102*
Ries, S. M., 148, 149, *158*
Riker, A. J., 177, 187, *192*
Rimmer, S. R., 88, *102*
Rishbeth, J., 4, *10*
Robards, A. W., 113, *116*
Roberts, F. M., 176, *192*
Roberts, M. F., 174, 175, 176, 177, *192*
Robinson, T., 205, *206*
Rodrigo, W. R. F., 4, *10*
Roerig, S., 82, *101*
Rogers, P. M., 254, *284*
Rohringer, R., 30, *41*, 74, *78*, 125, 128/129, 129, *132*, *133*, 307, 326, 327, *330*

Rombouts, F. M., 84, *102*
Romig, R. W., 185, *192*
Rosenbaum, J., 171, *192*
Rossi, C., 151, *156*
Roth-Lauterbach, B., 251, 252, *283*
Roux, D. G., 250, *285*
Rowe, J. W., 202, *206*
Rowell, J. B., 35, 37, *41*, 72, *78*
Roy, D. N., 202, *206*
Royle, D. J., 6, *10*, 43, *78*, 179, 180, 181, 182, 183, *192*
Rubery, P. H., 294, *303*
Rubin, B. A., 311, 330
Ruesink, A. W., 95, *102*, 111, *116*
Ruppell, E. G., 179, 184, *192*
Russell, D. W., 251, *288*
Russell, G. E., 3, *10*, 169, *192*
Russell, W. A., 30, *40*
Russi, S., 151, *156*, *157*

S

Sachs, T., 118, *133*
Sackston, W. E., 98, *100*
Sadava, D., 82, *102*
Sadowski, P., 149, *158*
Sakai, S., 120, *132*
Sakuma, T., 226, 231, *237*, 270, *288*
Sakurai, A., 118, 119, *132*, *133*
Sakurai, Y., 245, *285*
Salemink, C. A., 150, *158*
Salpeter, M. M., 82, *103*
Samaddar, K. R., 136, 143, *158*
Samborski, D. J., 30, *40*, 74, *78*, 122, 124, 125, 127, 128/129, 129, 130, *132*, *133*, 327, *330*
Sando, C. E., 171, *192*
Sands, D. C., 87, *100*
Sanford, J. B., 243, 255, 262, *286*
Santurbano, B., 151, *156*
Sargent, J. A., 44, 45, 46, 47, 50, 51, 52, 53, 54, 55, 56, 57, 60, *78*, 88, *102*, 114, 115, *116*
Sargent, J. A. S., 74, 76, *78*
Sarko, A., 82, *102*
Sassa, T., 155, *158*
Sato, N., 214, *223*, 226, 229, 230, 231, *236*, *237*
Sato, S., 93, 94, *102*, 245, *285*
Sawai, K., 147, *157*
Saxena, J. M. S., 25, 29, 32, *40*
Sayre, R. M., 122, *131*

Starr, M. P., 86, *102*
Starratt, A. N., 155, *158*
Staub, T., 56, *78*, 174, 175, *192*
Steindl, D. R. L., 189, *192*
Steiner, P. W., 272, 273, 276, 278, *288*
Steiner, G. W., 136, 137, 142, *158*
Stenlid, G., 263, *288*
Stermitz, F. R., 213, *222*, 244, 246, *285*
Stevens, G. J., 88, 95, 96, 97, 99, *102*, 107, 108, 109, 110, 111, 112, 113, *116*
Steward, F. C., 82, *103*
Stewart, W. W., 153, *158*
Stholasuta, P., 254, 262, *288*
Stob, M., 263, *288*
Stoessl, A., 128/129, *133*, 203, 204, *206*, 212, 214, 218, *222*, *223*, 227, *237*, 243, 271, 272, 278, 279, *286*, *288*
Stoll, A., 21, *23*
Stone, G. M., 185, *191*
Stothers, J., 227, *237*
Stothers, J. B., 21, *22*, 214, *222*
Stowe, B. B., 118, *133*
Straley, C. S., 149, *158*
Straley, M., 149, *158*
Strange, R. N., 9, *10*
Strobel, G. A., 85, *103*, 136, 137, 138, 139, 142, 143, 146, 147, 148, 149, *156*, *157*, *158*, 308, *330*
Strouse, B., 86, *102*
Stuckey, R. E., 73, *78*
Sturdy, M. L., 95, *100*
Stuteville, D. L., 80, *102*
Stutz, E., 81, 84, *100*, 300, *302*
Subramanian, S., 228, *234*
Sugaya, T., 155, *158*
Suginome, H., 243, 262, 267, *288*
Sugiyama, N., 155, *158*
Sullivan, T. P., 72, *78*
Sutherland, I. O., 244, *287*
Sutter, A., 247, 251, *286*, *288*
Suzuki, A., 263, *284*
Svahn, C. M., 202, *206*
Svensson, S., 150, *156*
Swain, T., 252, *288*
Swinburne, T. R., 91, 98, *103*

T

Tagawa, K., 84, 85, 94, *101*, *103*
Takasugi, M., 214, *223*, 226, 231, *236*, *237*
Talboys, P. W., 98, *103*

Talmadge, K. W., 81, 82, 83, 84, 85, 88, 89, *100*, *101*, *105*, 106, 113, *115*, 148, *158*
Tam, J., 154, *157*
Tamura, S., 118, 119, *132*, *133*, 263, *284*
Tanaka, H., *158*
Tanaka, N., 201, *206*
Tanaka, S., 143, 145, *158*
Tani, T., 314, *330*
Tapke, V. F., 169, *192*
Tarr, S. A. J., 162, *192*, 291, *303*
Tatum, L. A., 146, *158*
Taylor, P. A., 153, *158*
Teakle, D. S., 189, *192*
Teasdale, J., 256, 257, 282, *288*
Templeton, G. E., 154, 155, *157*, *158*
Thaller, V., 209, *222*
Therman, E., 318, *330*
Thies, G., 118, 120, *132*
Thimann, K. V., 95, *102*, 111, *116*, 118, *133*
Thomas, C. A., 214, 222,
Thomas, G. G., 6, *10*, 43, *78*, 179, 180, 181, 182, 183, *192*
Thompson, J. F., 154, *157*
Thomson, R. H., 202, *206*
Thornton, J. D., 295, *303*
Timmis, J. N., 324, 325, *330*
Tipton, C. L., 147, *158*
Tirokata, H., 155, *158*
Toda, J. K., 202, *206*
Tomiyama, K., 214, *223*, 226, 227, 228, 229, 230, 231, *236*, *237*
Tommerup, I. C., 44, 45, 46, 50, 51, 52, 53, 54, 55, 56, 57, 60, 74, 76, *78*, 88, *102*, *116*
Tonolo, Al., 150, *156*
Toussoun, T. A., 11, *23*
Towers, G. H. N., 247, *284*, *288*, 298, *302*
Tribe, H. T., 95, 96, *103*, 108, *116*
Triebold, H. O., 171, *193*
Trocha, P., 130, *133*
Troughton, J. H., 163, *192*
Tschernoff, V., 150, *158*
Tschesche, R., 197, *206*
Tseng, T. C., 95, 96, *103*, 109, *116*
Tsuji, A., 245, *285*
Turian, G., 322, 323, *330*
Turner, B. L., 241, *286*
Turner, E. M. C., 204, *206*
Turner, M. T., 94, *103*
Turner, N. C., 119, *133*, 150, 152, *158*

Subject Index

A

Acer pseudoplatanus, 83
Aeromonas liquefaciens,
    control of polygalacturonic acid *trans*
      eliminase, 86
Agrobacterium tumefaciens, 117
Airborne diseases, 12
Alternaria, 4, 9
   kikuchiana, 143, 155
   mali, 147
   solani, 155
   tenuis, 154
   toxin, of, 145, 154
   zinniae, 155
Alternaria blotch, on apple, 147
Antifungal action,
   mechanism of, 299–301
Antifungal activity,
   of mansonones E and F, 202
Antifungal compounds,
   in gramineae, 202–205
   in wood, 200–202
Aphanomyces euteiches, 86, 262, 263, 264,
    268, 269, 274, 275, 276
Apple canker, 167
Apple scab, 9
   Ascochyta,
      pinodella, 274, 275
      pisi, 271
Aspergillus niger, 85, 300
"Attacking mechanism", 79
Avena, 111
Avenacin, 204

B

Bacterium tabacum, 184
Barrier tissue,
   lignified tissue as, 188 (*see also*
     Lignification)
   and resistance, 186–189

Bean rust, 122
Biotrophic organisms, 129
   diseases caused by, 122
Biotrophic fungi, 43–77
   definition of, 43
   gene-for-gene interaction, 74
   infection by, 43–77
   infection structures, formation of, 43,
     44, 46, 48, 49
Black rot disease of sweet potato, 121
Botrytis, 8, 9, 220, 221
   cinerea, 5, 6, 7, 9, 19, 20, 21, 92, 164,
     170, 197, 220, 221, 293
   fabae, 220
   tulipae, 19, 20, 21, 197
Bremia lactucae, 44, 45, 72, 73
   germ tube and appressorium, 46, 47
   haustoria of, 61, 66, 67, 68, 69, 70
   infection of lettuce by, 45–58
   intracellular infection, structures of,
     58–70
   penetration of host by, 54–58
   pre-penetration responses of host
     cell to, 47, 53
   penetration of walls of Brassica root
     hairs by, 56–71
   primary vesicle structure of, 58–60, 62,
     63, 75, 76
   secondary vesicle, structure of, 61
     64, 65
Brown rot, 171

C

Cane spot, 164
Capsidiol, 214
Cell wall hydrolysis, by pathogens, 79–99
Ceratocystis
   fimbriata, 120, 121
   ulmi, 88, 150, 188, 201, 202
Cerocospora beticola, 179, 184